Environmental Permits

Environmental Permits

A TIME-SAVING GUIDE

Donna C. Rona, P. E.

VNR VAN NOSTRAND REINHOLD COMPANY
New York

Library of Congress Catalog Card Number 87-27417
ISBN 0-442-27838-1

Printed in the United States of America

Designed by Stan Rice, Autospec, Inc.

Van Nostrand Reinhold Company Inc.
115 Fifth Avenue
New York, New York 10003

Van Nostrand Reinhold Company Limited
Molly Millars Lane
Wokingham, Berkshire RG11 2PY, England

Van Nostrand Reinhold
480 La Trobe Street
Melbourne, Victoria 3000, Australia

Macmillan of Canada
Division of Canada Publishing Corporation
164 Commander Boulevard
Agincourt, Ontario M1S 3C7, Canada

16 15 14 13 12 11 10 9 8 7 6 5 4 3 2 1

Library of Congress Cataloging-in-Publication Data

Rona, Donna C., 1954-
Environmental permits.

Bibliography: p.
Includes index.
1. Environmental permits—United States. I. Title.
KF3775.R66 1988 353.0077'2 87-27417
ISBN 0-442-27838-1

Contents

Preface

Environmental permits enable regulatory agencies to control the disturbance and degradation of the environment caused by man's activities. Created by governments through legislation, the permit processes are administered by elected officials. Environmental legislation is relatively new as an independent field of the law: the laws themselves are primarily a conglomeration of older legal doctrines from other fields of law, modified and adapted for particular situations (Landau and Rheingold 1971). Like other laws, environmental laws are meant to serve and protect the rights and well-being of the public. However, like other laws, they have created confusion, a proliferation of costly paperwork, and some inequities.

Numerous statutes, dealing with most aspects of pollution, exist at every level of government—so many, in fact, that they frequently conflict and overlap. These statutes establish a public policy toward polluters. They also empower the regulatory bodies that issue permits. It would be impossible to compile a list of requirements for each type of permit. Therefore, this book will acquaint the reader with the common aspects of environmental permits: their terminology, components, and application processes. This book covers the permit process from initial agencies/applicant contacts through application parts and procedures, to application approval. Special emphasis has been placed on bringing together copies of laws and lists of agencies as appendices. The appendices give the reader easy access to materials that will help clarify the permit process. If you follow the general rules and methodologies set forth, you should be able to approach the permit process with confidence and achieve results efficiently and relatively inexpensively.

1

Introduction

ENVIRONMENTAL PERSPECTIVE

To understand the serpiginous, often confusing, and sometimes contradictory nature of environmental science, permit standards, and evaluation criteria, you must first understand how the various components of our environment interact. Nature functions within cycles (we will exclude external catastrophes such as meteor impacts). These cycles vary widely in space and time, defying simple interrelations. On a global spacial scale and a million-year time scale, the structure of the earth and the composition of the atmosphere are constantly changing. The continents are changing position on the globe, as a result of which gases are given off, water is absorbed and formed, and the chemical composition of the oceans is altered. On a shorter time scale, glaciation not only affects world atmospheric conditions, sea level, and erosion patterns, but also species dominance, local food webs, and other relationships within ecosystems. In a still shorter time frame and a more limited geographic scale, climatic patterns, erosion, natural fires, and other natural phenomena change how the environment functions. On a still narrower time and space scale, individuals live, die, and contribute to the next portion of the food web as food or detritus. On a global scale, short-duration events, such as tidal waves, volcanic activity, and droughts, may affect environments dramatically.

Natural chemical, biological, hydrologic, geological, and meteorological cycles, each with their own subcycles, all have their own time and space frame. But all affect each other—none are independent (see figures 1-1, 1-2).

Now the actions of man are introduced into these cycles (see figure 1-3). It is certainly an awesome task to understand and evaluate the impact of any one activity on all these in the presence of cycles. Whereas certain impacts are physically obvious on a short time scale, other impacts, especially over the long term, are as yet unknown, or the prediction is very complex.

WHY ENVIRONMENTAL PERMITS ARE NEEDED

The term *environmentalism* is generally used to describe a movement and a philosophy to protect the quality of life through conservation and control of

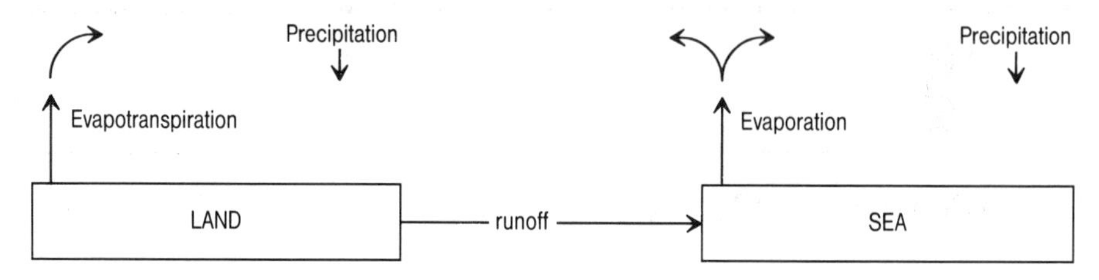

Source: After Ehrlich et al., 1977.

1-1. The hydrologic cycle.

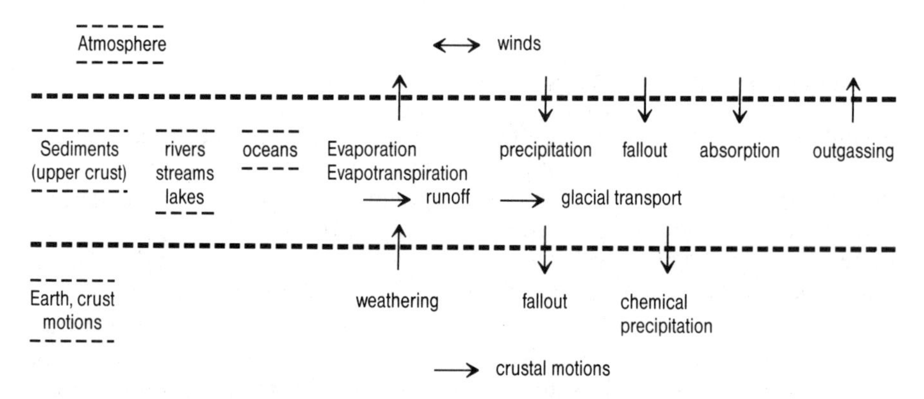

Source: After Ehrlich et al., 1977.

1-2. Transport mechanisms in the world.

land use and resources and the prevention of pollution. Environmentalists are concerned about how man's actions affect the ecosystems and the natural cycles just described. The philosophy of modern environmentalism was established by Thomas Jefferson, Ralph Waldo Emerson, Henry David Thoreau, and other early naturalists. Not until the time of Theodore Roosevelt, however, did our system of conservation through the establishment of national parks and forests, wildlife refuges, and the like begin to develop. Organizations such as the Sierra Club and the Audubon Society broadened public awareness of conservation and pollution issues, and this national concern prompted pervasive federal legislation in the early 1970s. Consequently, many activities now require environmental permits that try to ensure the protection of our natural environment.

> Even the most naive among us must recognize by now that the powerful industrialist and developer, and even the not so powerful small entrepreneur, cannot be expected to do what is right and environmentally sound simply out of decency. The free-enterprise system has shown itself capable of initiating positive and creative new alternatives in community design, but it needs a progressive legal framework of governmental support to encourage, and insist on, development in harmony with environmental needs and the public interest. Then even the best-intentioned private

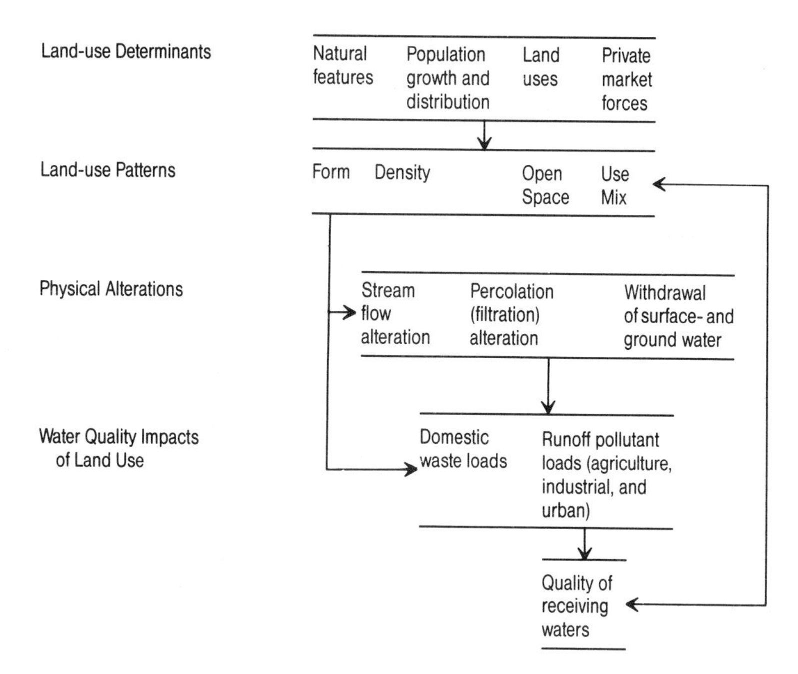

Source: After Canter, 1977.

1-3. Schematic of the relationship between land use and water quality.

> developer cannot have more than a piecemeal effect on any region. New laws and new roles for regional government are an inescapable imperative. (Tomioka and Tomioka 1984)

Society's desire for a safe and healthy environment and its perpetuation for future generations has generated the laws that now govern our actions with regard to the environment. Land-use plans and policies, preserve-area designations, and coastal construction restriction zones are only a few of the many methods used to protect the environment. While planning and zoning ordinances provide a general guide to environmental considerations within a region, laws set standards for conduct and construction, and the permit procedure assures the public that these laws are being upheld.

The process of obtaining and granting permits probably originated soon after people began to discover that all craftsmen, all building materials, and all designs were not alike. No doubt, after a chief's new shelter fell apart, primitive standards for construction were established. Without standards, and a means for ensuring that those standards were upheld, no one knew if he was safe. This is the foundation of environmental permits. Without regard for the environment, builders, miners, developers, and others have created imbalances in the natural system. Air, land, and water have been polluted and defiled. Plant and animal species have been driven to extinction.

Although the first smoke abatement law was passed in 1273, the ideal of pollution control resolutions was adopted in 1869 by the Massachusetts Board of Health: "We believe all citizens have an inherent right to the enjoyment of pure and uncontaminated air and water and soil; that this right should be regarded as belonging to the whole community; and that no one should be allowed to trespass upon it by his carelessness or his avarice or even his ignorance."

Federal legislation for the protection of the environment in this country began with the Forest Reserve Act of 1891 (figure 1-4) and has proceeded steadily up to the present.

1-4. Federal legislation regulating development.

Date Enacted	*Name of Act*	*Reference*
1891	Forest Reserve Act	26 Stat 1095, 16 USC 471
1897	Organic Act of 1897	16 USC 475
1899	Rivers and Harbors Act of 1899	30 Stat 1148, 1150–1155, 33 USC 403
1899	Refuse Act	30 Stat 1152
1912	Public Health Service Act of 1912	37 Stat 309
1916	National Park Service Act of 1916	39 Stat 535
1920	Federal Water Power Act	41 Stat 1063, 16 USC 791–823
1920	Mineral Leasing Act of 1920	41 Stat 437
1924	Oil Pollution Act of 1924	43 Stat 604, 33 USC 431–437
1932	Emergency Conservation Act	
1934	Fish & Wildlife Coordination Act	48 Stat 401, 16 USC 661–666
1940	Ohio River Valley Sanitation Compact	
1948	Water Pollution Control Act	PL 80–845, 62 Stat 1155, 33 USC 1151
1950	Fish Restoration and Management Project Acts	64 Stat 430, 16 USC 777
1952	Public Law	PL 82–579, 33 USC 1159
1952	Saline Water Conservation Act	66 Stat 328, 42 USC 1951–58
1952	Federal Water Pollution Control Act	66 Stat 755
1954	Atomic Energy Act	68 Stat 919, 42 USC 2011
1955	Air Pollution Control Act	PL 84–159, 42 USC 7401 et seq
1956	Fish and Wildlife Act of 1956	70 Stat 1119, 15 USC 713
1960	Multiple Use Sustained Yield Act of 1960	PL 86–517, 74 Stat 215, 16 USC 528–531

Date Enacted	Name of Act	Reference
1961	Federal Water Pollution Control Act Amendments of 1961	PL 87–88, 33 USC 1151 et seq
1961	Wetlands Act of 1961	PL 87–383, 75 Stat 813, 16 USC 715
1961	Oil Pollution Act of 1961	75 Stat 402, 33 USC 1001–15
1963	Clean Air Act	PL 88–206, 42 USC 7401 et seq
1964	Wilderness Act of 1964	PL 88–577, 78 Stat 890, 16 USC 1131–36
1964	Water Resources Act	PL 88–379, 42 USC 1961
1964	Water Resources Research Act of 1964	PL 88–379, 78 Stat 329, 42 USC 1961
1965	Water Quality Act of 1965	PL 89–234, 33 USC 1151
1965	Federal Water Pollution Control Act of 1965	PL 84–660, 70 Stat 498, 33 USC 1251 et seq
1965	Water Resources Planning Act	42 USC 1962 et seq
1965	Motor Vehicle Air Pollution Control Act	PL 90–148, 42 USC 7401 et seq
1965	Solid Waste Disposal Act of 1965	PL 89–272, 42 USC 7401 et seq
1965	Fish and Wildlife Coordination Act	PL 89–72, 79 Stat 216, 16 USC 662
1966	Clean Water Restoration Act of 1966	PL 89–753, 80 Stat 1256, 33 USC 1153
1966	Marine Resources and Engineering Act	PL 89–454, 80 Stat 203, 30 USC 1101–08
1966	Clean Air Amendments of 1966	PL 89–675, 80 Stat 954, 42 USC 1857
1967	Air Quality Act of 1967	PL 90–148, 81 Stat 485, 42 USC 1857
1967	Clean Air Act of 1967	42 USC 1857
1968	Air Act to Require Aircraft Noise Abatement Regulation	PL 90–411, 49 USC 1431
1969	National Environmental Policy Act	42 USC 4321 et seq
1969	Tax Reform Act of 1969	PL 91–172, 83 Stat 487, 26 USC 1,2,11
1969	Endangered Species Conservation Act of 1969	PL 91–135, 83 Stat 283, 16 USC 668
1970	Clean Air Amendments of 1970	PL 91–604, 42 USC 740
1970	Environmental Education Act of 1970	
1970	Mining & Minerals Policy Act of 1970	

Continued

1-4. Continued.

Date Enacted	Name of Act	Reference
1970	Resource Recovery Act of 1970	PL 91–512, 42 USC 3251
1970	Water Quality Improvement Act of 1970	PL 91–224, 84 Stat 91, 33 USC 1151
1970	Environmental Quality Improvement Act (Title II of PL 91–224)	
1972	Water Pollution Control Act Amendments	PL 92–500, 33 USC 1251 et seq
1972	Marine Protection Research & Sanctuaries Act of 1972	PL 92–532, 86 Stat 1052, 33 USC 1401 16 USC 1431–4
1972	Noise Control Act	PL 92–574, 86 Stat 1234, 42 USC 4901–18
1972	Federal Environmental Pesticide Control Act	PL 92–516, 7 USC 136
1972	Coastal Zone Management Act	16 USC 1451–64
1972	Ports and Waterways Safety and Health Act	33 USC 1221
1972	Marine Mammal Protection Act	PL 92–522, 86 Stat 1027, 16 USC 1361
1973	Endangered Species Act	PL 93–205, 87 Stat 884, 16 USC 1531
1973	Emergency Petroleum Allocation Act of 1973	PL 93–159, 15 USC 751
1974	Energy Supply and Environmental Coordination Act of 1974	PL 93–319, 88 Stat 246, 15 USC 791–8 42 USC 1857
1974	Safe Drinking Water Act	PL 93–523, 42 USC 300f
1974	Clean Air Act Amendments	PL 93–319, 42 USC 1857
1975	Agriculture-Environment and Consumer Protection Appropriation Act of 1975	PL 93–563, 15 USC 713
1976	Resource Conservation Recovery Act	PL 94–580, 42 USC 6921
1976	Toxic Substances Control Act	PL 94–469, 15 USC 2601
1976	Federal Land Policy and Mangement Act	PL 94–514, 26 USC 279
1976	Solid Waste Disposal	42 USC 6901
1976	National Forest Management Act	PL 94–588, 16 USC 1600
1977	Federal Water Pollution Control Act Amendments of 1977	PL 95–217, 91 Stat 1566–1609, 33 USC 1251
1977	Clean Air Act Amendments	PL 95–95, 91 Stat 685, 15 USC 792 42 USC 7401
1977	Toxic Substances Control Act	PL 94–469, 15 USC 2601

Date Enacted	Name of Act	Reference
1977	Surface Mining Control and Reclamation Act	PL 95–87, 91 Stat 445, 30 USC 1201
1978	Quiet Communities Act of 1978	PL 95–608, 25 USC 1901
1978	Fish & Wildlife Improvement Act of 1978	PL 95–616, 92 Stat 3110, 16 USC 460
1979	Safe Drinking Water Act Amendments	PL 96–63, 42 USC 300
1980	Comprehensive Environmental Response, Compensation, and Liability Act of 1980	PL 96–510, 94 Stat 2767, 42 USC 9601
1980	Acid Precipitation Act	PL 96–294, 42 USC 8901–12
1980	Safe Drinking Water Act Amendments	PL 96–502,
1983	Clean Air Act Amendments	PL 98–213, 42 USC 7625–6
1984	Solid Waste Act Amendments	PL 98–616,
1984	Water Resources Research Act	PL 98–242, 42 USC 7801
1987	Water Quality Act of 1987	PL 100–4, 33 USC 1251

PL = Public Law
Stat = Statute Law
USC = United States Code

HEALTH, SAFETY, AND CONSUMER ACTS (date and title only)

1877 Dangerous Cargo Act
1893 Safety Appliance Acts
1906 Federal Food, Drug, and Cosmetic Act
1906 Federal Meat Inspection Act
1957 Poultry Products Inspection Act
1958 Federal Aviation Act
1960 Motor Vehicle Exhaust Study Act
1960 Federal Hazardous Substances Labeling Act
1965 Service Control Act
1966 National Traffic and Motor Vehicle Safety Act
1966 Federal Metal and Nonmetalic Mine Safety Act
1968 Natural Gas Pipeline Safety Act
1968 Radiation Control for Health and Safety Act
1968 Flammable Fabric Act
1969 Federal Coal Mine Health and Safety Act
1969 Child Protection and Toy Safety Act
1969 Federal Mine Safety and Health Act
1970 Occupational Safety and Health Act
1970 National Materials Policy Act
1970 Federal Railroad Safety Act; Hazardous Materials Transportation Control Act
1970 Poison Prevention Packaging Act
1971 Federal Boat Safety Act
1971 Lead-Based Paint Poisoning Prevention Act

Continued

1-4. *Continued.*

HEALTH, SAFETY, AND CONSUMER ACTS (date and title only)

1971	National Cancer Act
1972	Ports and Waterways Safety Act
1972	Consumer Products Safety Act
1974	National Mobile Home Construction and Safety Act
1974	Motor Vehicle and School Bus Safety Act
1975	Hazardous Materials Transportation Act
1975	Rail Safety Improvement Act
1979	Hazardous Liquid Pipeline Safety Act

Sources: From: Hendricks et al., 1975; Landau and Rheingold, 1971; Advisory Commission on Intergovernmental Relations, 1981; Heer and Hagerty, 1977; CRS Report 84-44, 1984; Bogen, K. T., Coordinator of Regulatory Risk Analysis, Report 81-209, 1981; SPR, 1981; EPA, personal communication, 1985; Library of Congress, personal communication, 1985.

Environmental permits were established to protect our resources from destruction by the unknowing or uncaring. Unfortunately, there are no black-and-white rules for environmental protection, no unquestionable standards, and no references for absolute authority. The process would be considerably simpler if there *were* absolutes. In the absence of absolutes, the permit process has become an elaborate system, requiring an evaluation process that changes frequently and relies on technical expertise and judgments. The lack of absolutes stems from the many interdependent variables in the environment, as described at the beginning of this chapter, all of which are subject to natural alterations as well as those induced by man (figure 1-5). This lack of absolutes is the source of the majority of costly aspects of the environmental permit process.

PROJECTS REQUIRING ENVIRONMENTAL PERMITS

The need for environmental permits is obvious for some projects and less obvious for others. Today almost any project, from demolishing a house to building a community, requires numerous permits from state, county, or city offices. Zoning approvals of construction permits are required to ensure public safety and welfare. The larger the project, the greater the variety and range of agencies involved in granting permits. Each permit or required agency approval seeks to protect some aspect of the environment (see figure 1-6, for example).

In general, any project that physically alters the environment, such as dredging a lake or removing a forest, will require some environmental permits. A specific list of such projects would take up volumes and would vary in each state and municipality. Some states and municipalities have "one-stop" environmental agencies, which enable permit applicants to complete properly all required permits at once. Unfortunately, most areas are without such information brokerages, and the applicant is left to discover (or omit) regulatory agencies on his own.

1-5. Environmental attributes.

Air
- Diffusion factor
- Particulates
- Sulfur oxides
- Hydrocarbons
- Nitrogen oxide
- Carbon monoxide
- Photochemical oxidants
- Hazardous toxicants
- Odor

Water
- Aquifer safe yield
- Flow variations
- Oil
- Radioactivity
- Suspended solids
- Thermal pollution
- Acid and alkali
- Biochemical oxygen demand (BOD)
- Dissolved oxygen (DO)
- Dissolved solids
- Nutrients
- Toxic compounds
- Aquatic life
- Fecal coliform

Land
- Soil stability
- Natural hazard
- Land-use patterns

Ecology
- Large animals (wild and domestic)
- Predatory birds
- Small game
- Fish, shellfish, and water fowl
- Field crops
- Threatened species
- Natural land vegetation
- Aquatic plants

Sound
- Physiological effects
- Psychological effects
- Communication effects
- Performance effects
- Social behavior effects

Human Aspects
- Life styles
- Psychological needs
- Physiological systems
- Community needs

Economics
- Regional economic stability
- Public-sector review
- Per capita consumption

Resources
- Fuel resources
- Nonfuel resources
- Aesthetics

Source: Jain, Urban, and Stacey, 1977.

Figure 1-7 illustrates the range of permits and the corresponding agencies that exist in just one state. As a starting point to evaluate which, if any, environmental permits may be needed for a given project, evaluate it in terms of its effects on air, land, and water in order to anticipate the categories of environmental permits that may be required. Consider the potential consumption, alteration, and pollution of air, land, and water caused by the project. Then consider the biological ramifications of the project, then its effects on the community and the compatibility of the proposed plan with existing land-use plans and archaeological, architectural, or aesthetic resources. These considerations are factors in the granting of most environmental permits (if not the subjects of their own permits). Thus a logical sequence of evaluations should form the basis for a pursuit of environmental permits.

Example: A developer is considering 160 acres of land for a housing and shopping-center development. The land contains a lake and ten acres of virgin forest. Each of the above-mentioned elements is taken into consideration.

Air: The low-density housing planned for this area should not significantly affect air quality. No parking garages or heavy industry is planned. Noise pollution should not be a factor.

1-6. EPA activities that might affect an energy project.

(Permits, Waivers, Variances, Approvals, and Certifications)

Resource Conservation and Recovery Act
- Permit for long-term storage or disposal of hazardous waste

Safe Drinking Water Act
- Underground Injection Permit
- Sole Source Aquifer designation and federally funded project review

Clean Water Act
- NPDES permit
- 403 (c), ocean discharge review
- 301 (c), 301 (k), 301 (g), variances
- Fundamentally Different Factors variances
- 316 (a), thermal discharge variance
- 316 (b), cooling water intake demonstration
- Section 404 program

Marine Protection, Research, and Sanctuaries Act
- Ocean dumping permit

Clean Air Act
- PSD permit
- New Source Review preconstruction permit
- SIP modification review
- Preconstruction permit for NESHAP pollutants
- Compliance waiver from NESHAP regulations
- Emerging technology waiver
- Innovative technology waiver
- Delayed Compliance Order
- New fuel or fuel additive waiver
- Automobile emission-standards waivers

Consolidated Permit Issuance Procedures for NPDES, RCRA, PSD, UIC, and 404 Programs

Toxic Substances Control Act
- Premanufacture Notification

National Environmental Policy Act
- New Source EIS preparation
- Review of other agency EISs

Endangered Species Act
- Consultation with FWS/NOAA as part of consolidated process

Miscellaneous
- Bus noise emission approval

Land: The consumption of most of the ten acres of virgin forest is inevitable if this particular project is to be economically feasible. Alteration of the terrain by grading will be necessary. Present drainage from surrounding lands must be altered or modified as part of this development. Ground pollution by septic tanks will be avoided by using the sewer system of a nearby city (capacity has already been found to be adequate).

1-7. Summary of North Carolina environmental permits.

	Permits	*Responsible Agency*	*Normal Process Time in Days*
Water quality	NPDES Permit	Environmental Management	100
	Non-Discharge Permit	Environmental Management	30
	401 Certification	Environmental Management	60
	Water Use Permit	Environmental Management	30
	Well Construction Permit	Environmental Management	
	State Lakes Construction Permit	Parks & Recreation	20
Land quality	Sedimentation Control	Land Resources	30
	Dam Safety Permit	Land Resources	30
	Mining Permit	Land Resources	30
	Geophysical Exploration	Land Resources	10
	Oil/Gas Exploration	Land Resources	10
Air quality	State Air Quality Permit	Environmental Management	60
	PSD Permit	Environmental Management	120
	Open Burning Permit	Environmental Management	1
	Burning Permit	Forest Resources	1
Coastal quality	CMAM Major Development State Dredge and Fill	Coastal Management	55
		Coastal Management	55
	CAMA Minor Development	Local Government	22
	Easement to Fill	Department of Administration	55
Facilities	Oil Refinery Facility Permit	Environmental Management	160
	Hazardous Waste Facility	Department of Human Resources	90
	Solid Waste Disposal Facility	Department of Human Resources	75
	Tax Certification/Resource Recycle Facility	Department of Human Resources	60
	Tax Certification/Pollution Abatement	Local Government	NA
Health	Public Water Supplies	Department of Human Resources	30
	Impoundment Permit	Department of Human Resources	20
	Sanitary Landfill	Department of Human Resources	75
	Septic Tank (less than 3,000 gallons)	Local Government	NA

Water: The lake will be drained, contoured to fit a design scheme, and refilled. Depth and volume will be reduced. Present surface-water drainage to the lake (its only water source) will be altered. No direct pollution is anticipated. Potable water will come from wells.

Next the developer considers the ecological impact of these elements (figure 1-8):

Air: No pollution impact
Land: Habitat reduction, modification, and/or elimination
Water: Habitat reduction and modification. Use of local groundwater supply

The developer now considers it likely that environmental permits may be required for the following activities:

1. Tree removal
2. Drainage pattern alteration
3. Fish and wildlife habitat destruction (land and lake)

1-8. Physical parameters to be considered in project evaluation.

	Air	*Land*	*Water*
Consumption	Not usually considered	Physical: removal of land from biologically productive uses (i.e., covering grass land with pavement); removal of constituents (e.g., mining) Chemical: removal of nutrient content of land (e.g., removal of topsoil)	Physical: surface—removal from existing uses (e.g., reducing river flow); ground—altering recharge (e.g., covering open land with pavement) Chemical: removal of chemical components
Alteration	Physical: addition of suspended particles; alteration of air-flow patterns Chemical: addition of chemical constituents to air	Physical: change of physical character of land (e.g., desert to golf course) Chemical: change of chemical character of land (e.g., use of fertilizers, waste disposal)	Physical: changing physical character (location, temperature, flooding, erosion, drainage, etc.) Chemical: addition or removal of chemical constituents (pH, salinity, minerals, etc.)
Pollution	Physical: addition of suspended particles; alteration of air-flow patterns Chemical: addition of toxic or obnoxious chemical constituents Noise	Physical: changing physical character of land Chemical: addition of toxic or obnoxious chemical constituents	Physical: changing physical character of water Chemical: addition of toxic or obnoxious chemical constituents

Of course, there may be additional areas of concern, but the developer can now at least begin to seek local agency officials in the identified areas and initiate discussions. These officials should be knowledgeable about related permit requirements and will direct the developer to other agencies requiring permits.

OBTAINING ENVIRONMENTAL PERMITS—OVERVIEW

The process involved in obtaining an environmental permit varies, depending on the nature of the project, its size, and location. At best, obtaining an environmental permit is one more bureaucratic process requiring forms, drawings, filing fees, and time. At worst, the process is expensive, time consuming, and labor intensive, bordering on a scientific research project. Most delays or confrontations during the process arise from ill-planned, poorly located projects. Occasionally the permit process serves to point out design or location problems. The permit process, however, was not designed to be an environmental consulting service, and its use as such is costly in time and capital.

The time and effort required to obtain a permit should be justified by its purpose. The protection of the environment is certainly a goal that justifies a bureaucratic process that makes government, developers, builders, and private citizens conscious of the environmental impact of their actions. It does not, however, automatically justify a permit process that is cumbersome and confusing. In the past, it was to the advantage of the governmental agencies to remain vague, omnipotent entities capable of exercising control in many areas not specifically detailed in the law but covered by the "spirit" of the law. This behavior could delay a project indefinitely; it was sometimes used to kill a project economically, or force a developer into conceding with the agency personnel. Lawsuits and maturity in the permit process have helped stop this behavior, although to the average permit applicant, the process may seem even more unclear, as published standards and established procedures are constantly deleted, updated, and in some cases, set at random by each agency, independent of the others.

In all cases, obtaining environmental permits adds to the cost of a project. Between 50 and 60 percent of all environmental permit applications are granted with little more than paperwork processing and payment of permit fees. When obtained in conjunction with traditional building permits, the additional costs to the project are minimal. The number of relatively simple permit processes is increasing as developers, planners, engineers, and architects develop greater environmental awareness and demonstrate sensitivity in project designs and site selection.

The remaining 40 to 50 percent of the projects requiring permits experience delays because of a variety of factors. Poor planning and blatant disregard for environmental degradation account for some of the delays and difficulties. Most delays, however, stem from two general problems: the applicant's failure to prepare an unambiguous application and provide all required data, or the deterioration of the agency/applicant relationship into that of adversaries. In most cases these delays are caused or aggravated by permit applicants.

The environmental permit process can affect a project in three ways:

1. Time and expense
2. Additions or alterations to plans
3. Demise of the project

The time and cost requirements of obtaining environmental permits vary greatly. Figure 1-9 summarizes typical cost categories for each phase of the permit process. Time translates into expense, but time itself is a resource that many applicants value as much as capital. Time involved includes that for application preparation, waiting time until approval, delays in other phases of the project, and hours spent on the project by various personnel (such as draftsmen, estimators, engineers, biologists, and typists). Capital expenses include direct outlays in application fees, laboratory fees, legal fees, technical-expert fees, telephone, general overhead costs, paperwork, and travel.

The initial time and cost requirements of permit applications and the supporting documentation that is routinely required can be readily budgeted with a little

1-9. Cost categories in the permit process.

Process Categories	*Typical Costs*
Preapplication meeting	Time and transportation
Application preparation	Gathering (i.e., obtaining required aerial photos, affidavits, surveys, etc.) Calculations (i.e., cubic yards of fill, acres involved, etc.) Drawing (i.e., preparing suitable sketches) Writing (i.e., application forms)
Application submission	Application fees Postage
Application review	Time and transportation Gathering (obtaining missing information) Writing (providing information)
Requests for additional information	Gathering (may involve field studies, modeling, surveys, etc.) Calculating Drawing Technical experts Writing (general correspondence, reports, rebuttals, etc.) Time and transportation
Public hearings	Time and transportation Display materials
Application approval	Permit fees Performance bonds
Application denial (if contested)	Legal fees Court fees Technical-expert fees Gathering Calculating Drawing Writing Time and transportation

experience. The costs involved in requests for additional information or major documentation and litigation, however, can seldom be budgeted. While certain types of projects in environmentally sensitive areas will encounter every possible delay, the degree and final extent of delays in any project cannot be predetermined. (It is for this reason that technical consultants involved in permit application typically work on a time basis, not a lump-sum basis; see chapter 3.)

Additions or alterations to project plans may also result from permit reviews. Although changes are costly, they occasionally facilitate the granting of a permit and prevent permit denial or subsequent litigation (which can be even more costly). Such alterations may range from the addition of a green swale along a roadway to establishing conservation areas within project boundaries. Proposed construction techniques or materials may be altered. Project timing (coincident with breeding cycles or migration routes, for example) may be a factor in certain habitats. Typically, these changes occur as part of a compromise arrived at after agency review but before a formal permit denial is issued. The decision whether or not to make these changes is frequently a decision to continue with a project or let it die.

Project demise can and does occur because of environmental considerations and can sometimes be avoided by compromise. More typically, however, projects that meet with a total denial of permits show a lack of environmental consideration in the initial design phase and a total lack of early communications with environmental agencies. This translates into wasted time and money.

The environmental permit process is only a small portion of the overall project process. It begins with the application procedure. It does not end, however, until the project is completed. While this book will concentrate on the process of obtaining environmental permits, it should be noted that environmental agencies will be involved through completion of a project by way of inspections, monitoring (both during construction and/or for a period following completion), and/or submission of "as builts" (plans of all details of the project as constructed). Failure to comply with all details and stipulations of the permits can result in the project being stopped at any time during construction, in fines or penalties, or even the requirement to remove structures or fill that have already been placed.

THE LAWS

> Much of the law is permissive rather than imperative; it does not consist of commands but of rules for securing desired legal consequences, conferring rights, creating obligations, and attaining other legal results. In this aspect law is a positive instrument rather than a negative restraint, but an instrument of cooperation rather than of authority. (McKnight et al. 1974)

Federal laws generated by the administrative branches of the federal government are generally in the form of policy statements. An act broadly defines policy, prohibited activities, and penalties, as well as assigning the agency responsible for enacting and policing the provision of the act. Federal acts often set up study commissions as well. Laws can be designed as policy statements and directives for the total environment, a specific element of the environment (e.g., air or water), or a portion of industry (e.g., automobile air pollution or energy-related water pollution).

The responsible agency formulates regulations that set forth details of how the law will be enacted, allowed and prohibited activities, standards, review and permit procedures, and penalties and procedures for dealing with violations. Technical details, such as allowed pollutant levels, may be part of the regulations or may be issued separately as guidelines.

Federal laws are assigned a Public Law number, in the form of a Title, when passed by Congress. They are then placed into the United States Code (USC) in a system whereby acts and their revisions are grouped by subject. The following laws contain environmental legislation: Title 15—Commerce and Trade; Title 16—Conservation; Title 25—Indians; Title 26—Internal Revenue Code; Title 30—Mineral Lands and Mining; Title 33—Navigation and Navigable Waters; Title 42—The Public Health and Welfare; Title 49—Transportation. The United States Code is published by the Superintendent of Documents in volumes available in most major (and all legal) libraries. Volumes of the code are revised at least once each calendar year and issued on a staggered quarterly basis. The agency regulations are published in the *Federal Register*, a daily publication of federal agency regulations, proposed and final, and other legal documents of the executive branch. The *Federal Register* is available by subscription from the Government Printing Office or at most major libraries. Agency guidelines, application forms, and sometimes, information brochures, are available through agency offices.

1-10. State laws example: Florida.

Department of Environmental Regulations	
Ch. 17-3	Rules of D.E.R., pollution of waters
Ch. 17-4	Rules of D.E.R., permit requirements, dredge and fill, construction
Ch. 18-6	Rules of T.I.T.F. state wilderness system
Department of Natural Resources	
Ch. 16-3	Rules for establishing mean high water
Ch. 16B-24	Rules for procedures for application for coastal construction permits
Ch. 16B-25	Rules for procedures for coastal construction and excavation
Administrative Codes	
Ch. 161	Beach and Shore Preservation Act
	Part I. Regulation of Construction, Reconstruction and other Physical Activity
	Part II. Beach and Shore Preservation District
Ch. 253	Land Acquisition Trust Fund
Ch. 403	Environmental Control
	Part I. Pollution Control
	Part II. Electrical Power Plant Siting
	Part III. Interstate Environmental Control Compact
	Part IV. Resource Recovery and Management
	Part V. Environmental Regulation
Ch. 74-171	Establishment of Biscayne Bay Aquatic Preserve
Ch. 77-120	Creates Section 403.07 2 Florida Statutes to use soil and plants as water-level indicator
Ch. 120	Administrative Procedure Act, Rulemaking and Judicial Review
Ch. 258	State Parks and Preserves
Ch. 298	Drainage and Water Management
Ch. 370.16	Oysters and Shellfish: regulations
Ch. 373	Water Resources
	Part I. State Water Resource Plan
	Part II. Permitting Consumptive Uses of Water
	Part III. Regulation of Wells
	Part IV. Management and Storage of Surface Waters
	Part V. Finance and Taxation
	Miscellaneous Provisions

State, county, and city laws may be assigned by a similar procedure, or, more typically, they may be more specific, self-contained units. (States and municipalities have the option to adopt more, but not less, stringent guidelines than those required federally.) There is no uniform notation system throughout the nonfederal governmental system, so laws are often difficult to locate and coordinate with preexisting laws, and inconsistencies are not uncommon. Figure 1-10 is an example of one state's laws relating to the environment.

2

Types of Permits

As stated in chapter 1, environmental permits are required for the performance of any function that will physically alter the environment. The permit application, evaluation, and issuance process assures the public (through its agencies) that the environment will not be threatened or damaged by that function, just as building permits ensure that the public health and safety will not be threatened by a construction project. A building permit is issued after an agency engineer reviews the structural plans and determines that the structure will not collapse or that there is adequate water to meet fire requirements. Most building concerns have been categorized, standardized, and transformed into building codes. A designer, architect, or engineer, following code requirements carefully, is almost assured of receiving a building permit. This is not the case, however, with environmental permits. The field is relatively new and inexact, but perhaps more importantly, there are few, if any, absolute ways of quantifying environmental health and well-being. Steel, concrete, water, electrical, and air-conditioning specifications are far simpler to quantify. Environmental variables are numerous, and each site has its own peculiar characteristics. Our understanding of toxic and carcinogenic levels of pollutants, as well as the health and well-being of ecosystems, is still developing. Thus, when someone applies for an environmental permit in a given area, there are often only vague

2-1. Local programs example.

Coastal Facilities Act
- County Comprehensive Plan
- Municipal Master Plan
- Energy Development Plan
- Shoreline Master Plan
- Port District Plan
- District Floor-Control Plan
- Protective Controls for Port Development
- Waterfront Development Plan
- Socioeconomic Growth Plans

Source: U.S. Department of the Interior, Geological Survey, Report 79-1481, 1979.

and sometimes contradictory guidelines as to acceptable and unacceptable projects, or acceptable and unacceptable pollution levels for a given toxin. Certainly there are no universally acceptable environmental "codes." This is one reason environmental permits may appear to be "sleeping monsters."

Environmental permits are not always permits, per se. They may be termed "approvals" or "letters of permission." However, while these distinctions might manifest themselves in the fees and paperwork involved, the effects of all three are the same, and the permit, approval, or letter of permission must be granted prior to project construction. For the purposes of this book, therefore, approvals and letters of permission will be lumped under the general category of "permits."

CLASSIFICATION

Since permits are issued to perform very specific functions, a plethora of permits might be required for a single project. Often, multiple agencies (state, federal, and local) will require permits that appear to overlap (see figures 2-1 through 2-4). In some areas multi-agency agreements have created "clearinghouse" agencies to assist the applicant in determining all permits required. In other cases, one agency literally acts as an "agent" by accepting, processing, and reviewing applications for completeness, and holding hearings for the other agencies being given "approval/denial" review authority. Bear in mind, however, that it is the *applicant's* responsibility to ensure that all necessary permits are secured.

Permits can be categorized generally by the portion of the environment they seek to protect: air, land, water, biological, or regional.

Air pollution permits usually deal with the discharge of toxic or noxious fumes or particulates into the air. At one time only factories were required to obtain such permits. Now, however, large parking garages, cement plants (even temporary ones), and many other facilities must do so as well.

Air pollution can be defined as the presence of airborne substances (gases and particulates) at levels that cause injury to property or life. The emissions most commonly controlled by air-pollution regulations include carbon monoxide (CO), hydrocarbons (HC), oxides of sulfur (SO_x), oxides of nitrogen (NO_x), total suspended particles (TSP), ozone (O_3), lead (Pb), and asbestos. The National Ambient Air Quality Standards (NAAQS) describe allowable concentrations of various pollutants.

Areas of a state with air quality below the NAAQS for any contaminant on the list are referred to as *nonattainment areas.* These areas are subject to a permit program under the State Implementation Plan (SIP) of each state under the federal Clean Air Act (CAA). The permits seek to control major stationary sources in nonattainment areas. The principal conditions for the issuance of a permit include:

1. A net reduction in total emissions by regulation on a source-specific basis or by area-wide regulation

2-2. Environmental permit responsibilities within the Department of Natural Resources.

BUREAU OF GEOLOGY
- Environmental Geology Division
  - Determines geologic suitability of sanitary landfill sites and hazardous waste disposal facilities; reviews studies on nuclear waste repositories
- Mining and Reclamation Division
  - Regulates surface mining and reclamation for sand, gravel, clay, limestone, fill material, coal, and lignite
- Mineral Lease Division
  - Issues permits for geophysical, seismic, or other exploration/operations on state-owned land and water bottoms
  - Leases state-owned land for exploration and production of oil and gas and other minerals

BUREAU OF LAND AND WATER RESOURCES
- Land and Water Resources Division
  - Licenses water-well drillers
  - Delineates capacity use areas for management of ground water
  - Authorizes construction/alteration of dams
- Regional Water Resources Division
  - Allocates available surface water through prior appropriation doctrine

BUREAU OF POLLUTION CONTROL
- Water Quality Division
  - Regulates discharge of wastewater to waters of the state (national pollution discharge elimination system, NPDES)
  - Regulates discharge of wastewater into underground waters (underground injection control, UIC)
  - Issues operating permit for systems with no discharge
  - Issues pretreatment permit for discharging wastes into publicly owned treatment works
  - Certifies that proposed projects will not violate water quality standards of state streams (water quality certification)
- Air Quality Division
  - Regulates construction and operation of air emission equipment
- Solid/Hazardous Waste Division
  - Regulates construction and operation of treatment, storage, and disposal facilities for solid and hazardous waste (equivalent to Resource Conservation and Recovery Act (RCRA) requirements)

Source: Mississippi Research and Development Center, 1983.

2. Compliance of all sources owned or operated by a permit applicant or an approved SIP order
3. Application of technology to achieve the lowest available emission rate (LAER)

Nonattainment area permits are designed to offset the additional volume of pollutant generated by new or modified stationary sources through legally enforceable reduction in the quantity of that same pollutant emitted by existing sources elsewhere (Harrison 1984).

Noise pollution has been well defined in the Occupational Safety and Health Act (OSHA) and the Noise Control Act of 1972. These acts established guidelines for evaluating exposure to noise of a given intensity (measured in decibels—dbs) for a given duration, in terms of the physical and psychological effects on humans (figure 2-5). Of course, the effects on flora and fauna have yet to be established.

2-3. Local environmental agencies example.

Regional
- Regional Planning Council
- Water Management District
- Environmental Council

County and/or Municipal
- Environmental Management Department
- Water and Sewer Department
- Planning Department
- Building and Zoning Department

2-4. EPA permit programs.

Name	*Abbreviation*	*Coverage*	*Act*
Hazardous Waste Management Program	HWM	Generation, transportation, treatment, storage, disposal of hazardous waste (also known as TSD)	Resource Conservation Recovery Act
Underground Injection Control Program	UIC	Well injection/protection of drinking-water aquifers	Safe Drinking Water Act
National Pollutant Discharge Elimination System	NPDES	Discharge of wastewater into waters of the U.S.	Clean Water Act
Dredge or Fill Program	404	Discharge of dredge or fill material, often in wetlands	Clean Water Act
Prevention of Significant Deterioration	PSD	Emission of pollutants from sources in attainment areas	Clean Air Act
New Source Review	NSR	Emission of pollutant from new and modified sources in nonattainment areas	Clean Air Act

Source: Guide to the Proposed Consolidated Permit Regulations EPA, C-3, June 1979, and interviews with EPA.

Land-type permits include the dumping of solid or liquid wastes on or into the ground and alterations to the surface of the land. Dumping-type permits include solid-waste (refuse) or hazardous-waste (chemicals) dumpsites. Although solid-waste permits are generally handled by state and local permit programs, the numerous controls on hazardous-waste generators, transporters, and disposal sites are regulated through the Resource Conservation and Recovery Act (RCRA). The disposal of liquids by injection wells is regulated by Part C of the Safe Drinking Water Act (SDWA). Alterations of the land's surface and the resulting effects on storm-water runoff and flooding require another permit largely controlled by state and local legislation.

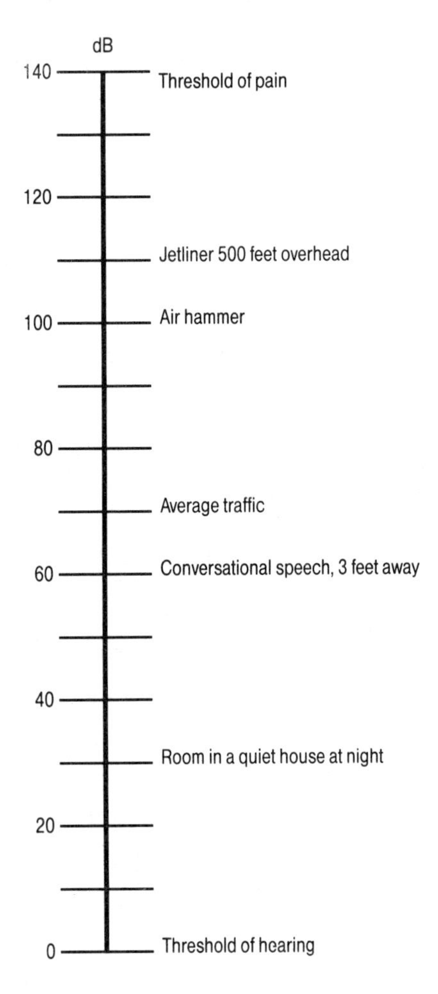

Source: After Vesilind, 1977.

2-5. Noise levels in the environment.

Water-type permits cover discharge into water, construction in or on water, watercourse creation or alteration, walls, sewage disposal, storm-water retention, and water management (including salt-water intrusion management). These permits cover water and wetlands.

The River and Harbor Act of 1899 first gave the U.S. Army Corps of Engineers authority over the "navigable waters of the United States." Later this definition of "waters of the United States" was extended to include wetlands, tributaries to navigable waters, and other waters, including isolated wetlands and lakes, intermittent streams, and prairie potholes, the degradation or destruction of which could affect interstate commerce.

The federal Clean Water Act prohibits pollutant discharge from any discernible, confined, and discrete conveyance (termed a *point source*) into the waters of the United States, unless state and/or federal permits are obtained. Permits ensure that specified standards are upheld limiting pollutant loads, and requiring reporting and monitoring to ensure accountability. Pollutants are classified as toxic, conventional (including biochemical oxygen demand (BOD), fecal coliform bacteria, pH, and oil and grease), and nonconventional (including chemical oxygen demand (COD), phosphorus, ammonia, chloride, color, and iron).

Beyond these federal permits, state and local permits are required for activities that pollute or otherwise affect waters of their jurisdiction.

Biological-type permits are required for tree removal, destruction or significant alteration of a habitat, use of hazardous chemicals, and species introduction. These permits vary widely from region to region and state to state.

Regional-type permits include a range of combination permits that seek to protect a particular type of environment. They are usually administered by a regional board consisting of representatives of various political and environmental agencies. Examples of this type of permit are coastal or wetland permits, in which multiple environmental factors are considered in the review process. Such multidisciplinary environmental reviews can cover a range of factors, from breeding population and brooding area maintenance to drinking water safety; from water exchange (flushing) to boat access; from smog to groundwater. The legislation that enables this regional type of permit and the experts that review and rule on applications consider the complex interactions that occur within that environment. They also consider the individual application (and rules applying to development within that region) in light of those interactions.

Numerous miscellaneous permits cover the manufacture or importation of certain chemical substances (Toxic Substances Control Act—TSCA) and similar environmental hazards that do not fall into the categories above. However, these permits are less frequently encountered and will not be discussed here.

Permits are issued by state, local, regional, and federal agencies. As part of the review process, documentation is sometimes required for agency evaluation prior to permit issuance. This documentation may include an Environmental Impact Statement (EIS) if federal (or certain state) permits, or permits affected by federal funding jurisdictions, or restrictions, are being issued. *The EIS is not a permit or a permit application—it is a document produced in support of an application.* Permits seldom (if ever) cover a comprehensive environmental "package" of physical and biological environment. They are issued by agencies empowered by legislation to oversee specific, and usually narrowly defined, portions of the environment, such as air quality or water quality. The all-inclusive EIS (which is used by one or more agencies in decision making) is a packaging and relative weighing of *all* environmental effects. It will be discussed in detail in chapter 6.

3

Application Procedures

The permit process is a small portion of the overall project process, which involves everything from project conception to completion. It is often unclear exactly at what point in the life of a project the permit process should be initiated; there are no guidelines to determine when, and with which agency, this process should commence. In general, however, it is advisable to seek permit information during the initial planning phases of a project. After the preliminary plans are prepared, a *preapplication meeting* with the permit-granting agency or agencies should take place. Application preparation, initial permit processing, and advanced design stages will usually be conducted simultaneously for environmentally sensitive projects to allow enough time for any modifications required to obtain the necessary permit(s). Projects whose environmental impact is relatively limited or small projects usually require few modifications; their application submission schedules should be adjusted accordingly.

The first step in the application process is to make telephone calls. Establish a list of potentially involved agencies, beginning with local agencies and eventually contacting state and federal agencies. Start with the telephone directory listing for specific city or county "environmental" departments. If there is no specific listing, call the general information number. Once you find the appropriate office, ask for the "permit processing officer." Although this title may vary from office to office, the operator will probably connect you with the right person. Do not forget to write down the names and direct phone numbers of everyone from whom you obtain information. Depending on the nature of the project, the telephone contacts may inform you that:

1. no permit is required;
2. a general permit exists for that type of project;
3. a letter of permission must be obtained;
4. the full permit procedure must be followed.

Following this initial advice, arrange a preapplication meeting(s). Depending on the nature of the project and the extent of prearranged intergovernmental coordination in any given area, the preapplication meeting may have to be repeated with various federal, state, regional, county, and/or city agencies.

PREAPPLICATION MEETING

A preapplication meeting with agency personnel is a very useful part of the permit process that applicants often overlook. It is an informal opportunity to meet with personnel from one or more agencies who will ultimately have permit authority over a project. The primary purpose of the preapplication meeting is to establish two-way communication and cooperation between agency personnel and soon-to-be applicants.

The applicants may come with representatives (architects, engineers, ecologists, and so forth) who can provide detailed information about the proposed project, identification of environmental issues, and their proposed method of treating each issue. The agency personnel should provide knowledgeable guidance about portions of the project that will require permits and portions that will not receive permits due to legal limitations; other agencies whose approvals or permits will be required (and their application procedures); application forms, fees, and supplemental information requirements (e.g., affidavits, supporting reports); timetables for all permit procedures; an agency review checklist; and information about appropriate routes of communication.

Depending on the individuals in the agency, the preapplication meeting can also create problems. In rare cases the overzealous agency individual (typically undertrained and overpowerful) may wish to rework an entire project to conform to some personal ideal or goal. This desire may stem from honest beliefs about what is best for the environment; however, such individuals are sometimes motivated by ego and/or inexperience. This situation calls for extreme patience and diplomacy. The applicant should make an effort to avoid confrontation: confrontations at any stage in the permit process can cause needless and costly delays and potential permit denials. If a conflict does arise try to deal with other individuals in the agency. When the overzealous individual must be dealt with, it is sometimes wise to bring in an outside consultant with experience and reputation who can quickly identify and control a situation within known agency guidelines and limitations.

The preapplication meeting should be held with the highest agency official available. This ensures maximum coordination of various agency personnel. Figures 3-1 and 3-2 illustrate some of the various ways environmental agencies are organized. It becomes obvious, when looking at these charts, why the highest-level official should be part of your preapplication meeting: often the permit-granting process requires numerous interdepartmental reviews. Working with someone aware of the entire process, and all of the personnel and timetables that will apply to a permit, will help create a realistic and informed dialogue that will facilitate the entire permit process.

The preapplication procedure may be summarized as follows (of course, this

(Text continued on page 30.)

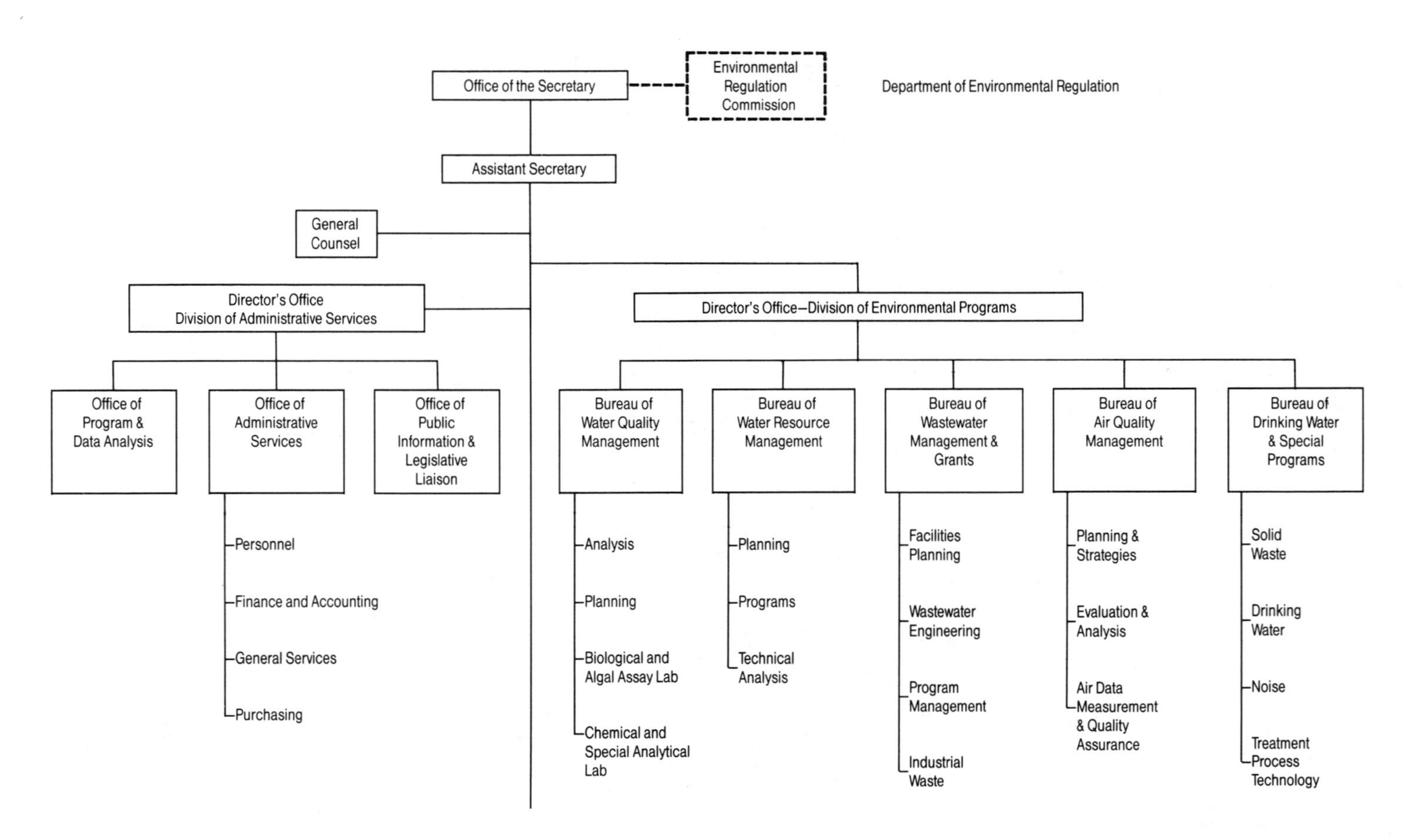
Office of the Secretary
Environmental Regulation Commission
Department of Environmental Regulation
Assistant Secretary
General Counsel
Director's Office Division of Administrative Services
Director's Office–Division of Environmental Programs
Office of Program & Data Analysis
Office of Administrative Services
Office of Public Information & Legislative Liaison
Bureau of Water Quality Management
Bureau of Water Resource Management
Bureau of Wastewater Management & Grants
Bureau of Air Quality Management
Bureau of Drinking Water & Special Programs
Personnel
Finance and Accounting
General Services
Purchasing
Analysis
Planning
Biological and Algal Assay Lab
Chemical and Special Analytical Lab
Planning
Programs
Technical Analysis
Facilities Planning
Wastewater Engineering
Program Management
Industrial Waste
Planning & Strategies
Evaluation & Analysis
Air Data Measurement & Quality Assurance
Solid Waste
Drinking Water
Noise
Treatment Process Technology

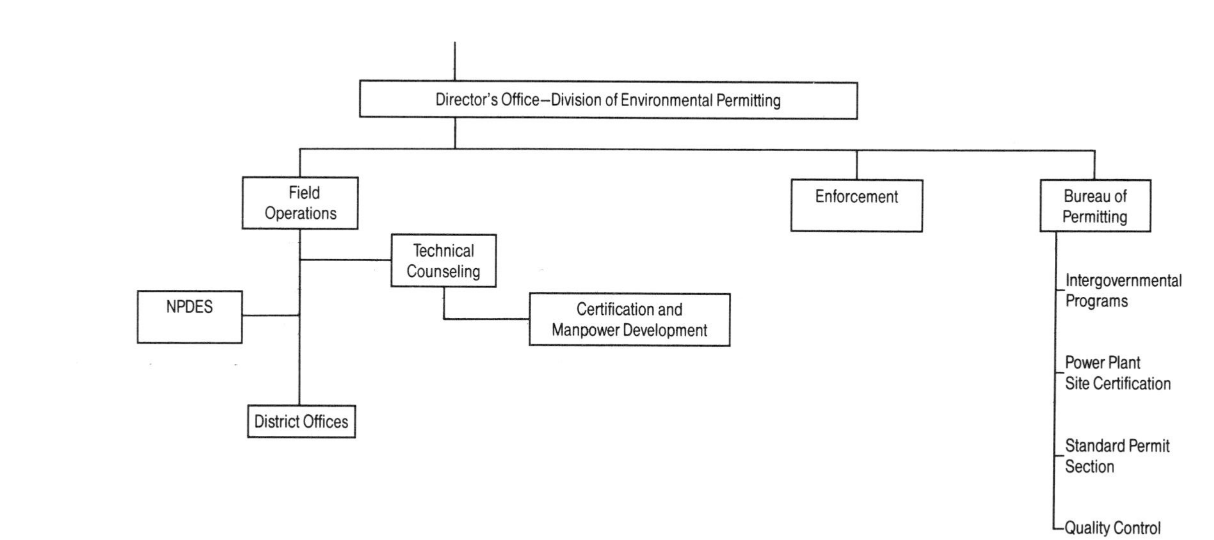

3-1. Example of the organization of a state environmental agency.

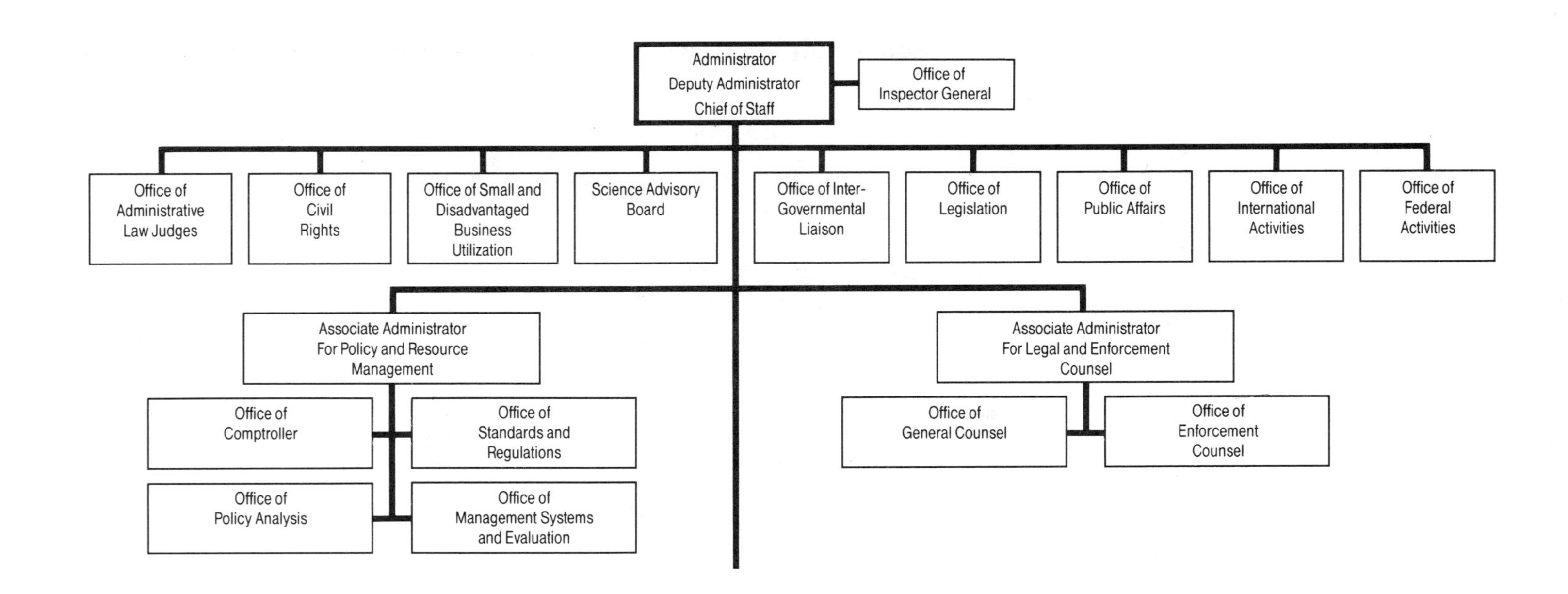

Administrator
Deputy Administrator
Chief of Staff
Office of
Inspector General
Office of
Administrative
Law Judges
Office of
Civil
Rights
Office of Small and
Disadvantaged
Business
Utilization
Science Advisory
Board
Office of Inter-
Governmental
Liaison
Office of
Legislation
Office of
Public Affairs
Office of
International
Activities
Office of
Federal
Activities
Associate Administrator
For Policy and Resource
Management
Office of
Comptroller
Office of
Standards and
Regulations
Office of
Policy Analysis
Office of
Management Systems
and Evaluation
Associate Administrator
For Legal and Enforcement
Counsel
Office of
General Counsel
Office of
Enforcement
Counsel

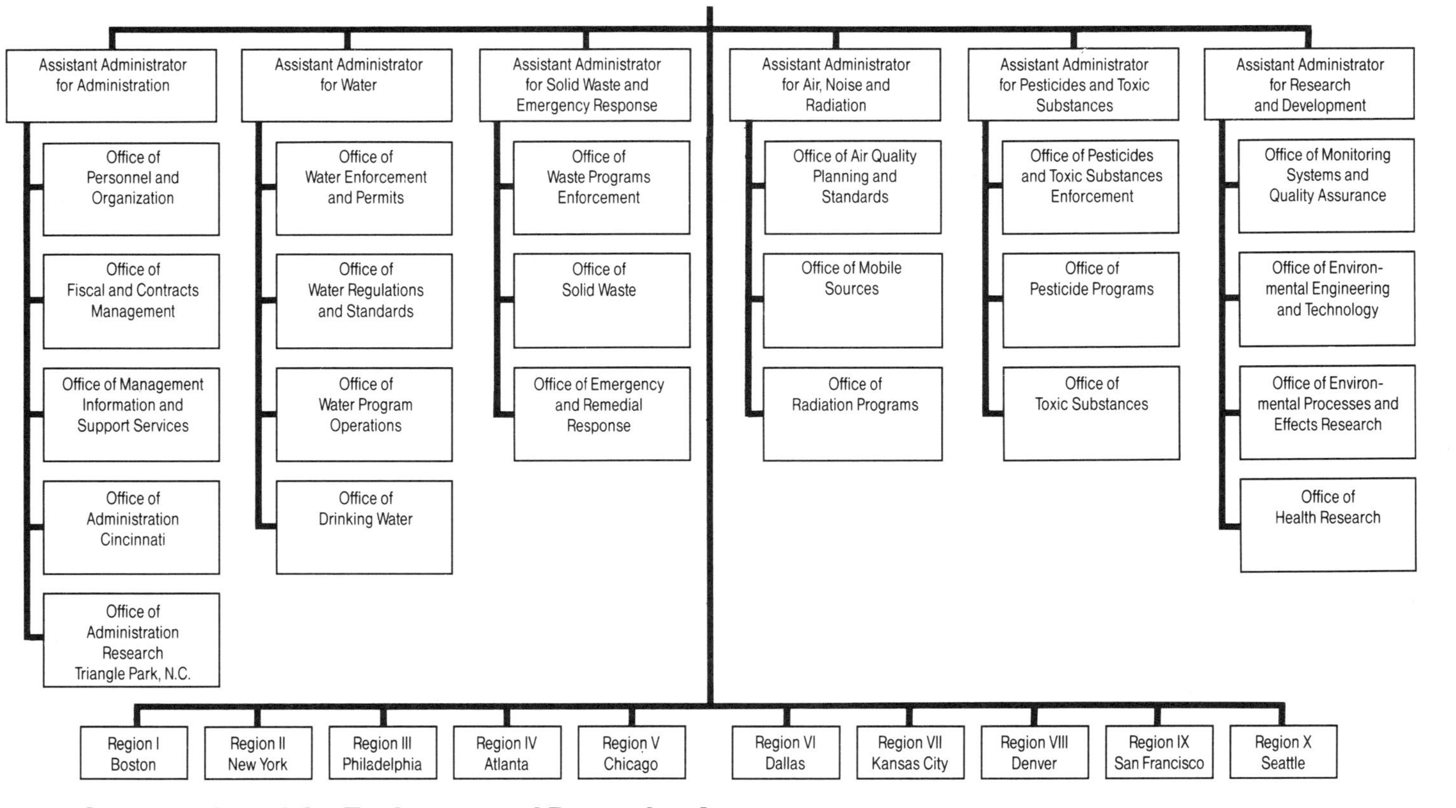

3-2. Organization of the Environmental Protection Agency.

procedure will be modified depending on the extent and nature of the proposed project):

PRIOR TO THE MEETING

1. Secure forms.
2. Arrange meeting (by phone).
   a. explain nature of project
   b. ask which agencies will probably be involved
   c. ask for meeting with highest agency representative available
   d. ask if multi-agency meeting may be arranged
3. Review forms and information obtained over the phone; make a list of specific questions.
4. Prepare package of materials to take to the meeting (plans, photographs, etc.).

AT THE MEETING

1. Establish a list of "players" (agency representatives); determine who does what and obtain their phone numbers.
2. Get a complete outline of the permit process, from preapplication through granting of the permit and any postpermit requirements (e.g., inspections, monitoring, "as-builts"). Establish a timetable for the entire process.
3. Make sure that the preapplication list of questions is answered.
4. Make a list of new questions/concerns; establish who will answer the questions and when.
5. Make a list of the permits that will be required and the agency contacts for each.

3-3. EIS process flow chart.

Notification of EIS Requirement	*Approximate time in days*
Environmental Assessment Initiated	0
↓	
Draft EIS/Public Notice	225–400
↓	
Public Hearing	300–600
↓	
EIS filed with CEQ	375–800
↓	
Final Action	450–900

FOLLOW-UP

1. Write a letter thanking agency officials for their time; include a meeting summary or confirmation of the procedures and contacts expected for the permit(s).
2. Prepare application and accompanying materials as required.

You may have to repeat the preapplication meeting three to five times, depending on the nature of the project, the agencies involved, and the degree of preexisting cooperation they have with one another. The entire permit process differs slightly from agency to agency because of the differing laws and rules that empower them (see figures 3-3 to 3-8).

As mentioned above, you should follow up the preapplication meeting with a letter to the agency supervisor (i.e., director or chairperson). This letter, aside from expressing your thanks for his/her time and cooperation and that of his/her staff (it is always wise to establish a good working relationship), should briefly describe the proposed project and the understood permit procedure (an attached conference summary is excellent). This will help crystallize the requirements in your mind, creating a realistic framework for preparation of the

3-4. U.S. Army Corps of Engineers review process flow chart.

	Approximate time in days
Submission of Application	
↓	
Application received	0
acknowledged	
processed	30
↓	
Public Notice	
↓	
Receipt of comments	
↓	60
Public hearing	
↓	
Evaluation	
↓	120
Approval ——— Denial	
↓ (Approval)	
Permit Issued	180

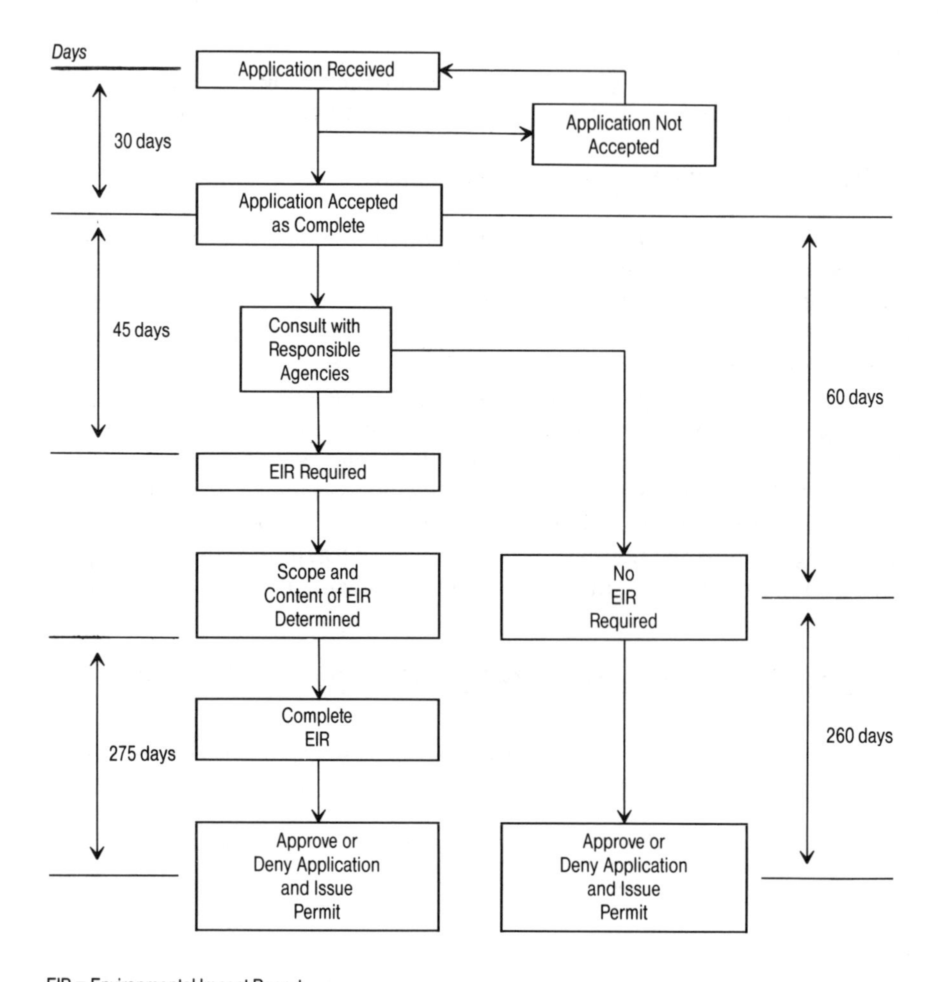

EIR = Environmental Impact Report
Source: Sample Guidelines for Processing Permits for Development Projects, State of California, 1984.

3-5. Example of a permit process flow chart (California).

application, and it will allow the agency to assign personnel as required. It will also serve as a final check that all requirements have been discussed, understood, and shall be met.

APPLICATION PREPARATION

Most environmental permits have standard application forms. These forms provide the basis on which the decision to grant or not grant a permit is made, so great care must be taken in their wording and presentation.

The application usually requires a clear, concise definition of the project, the

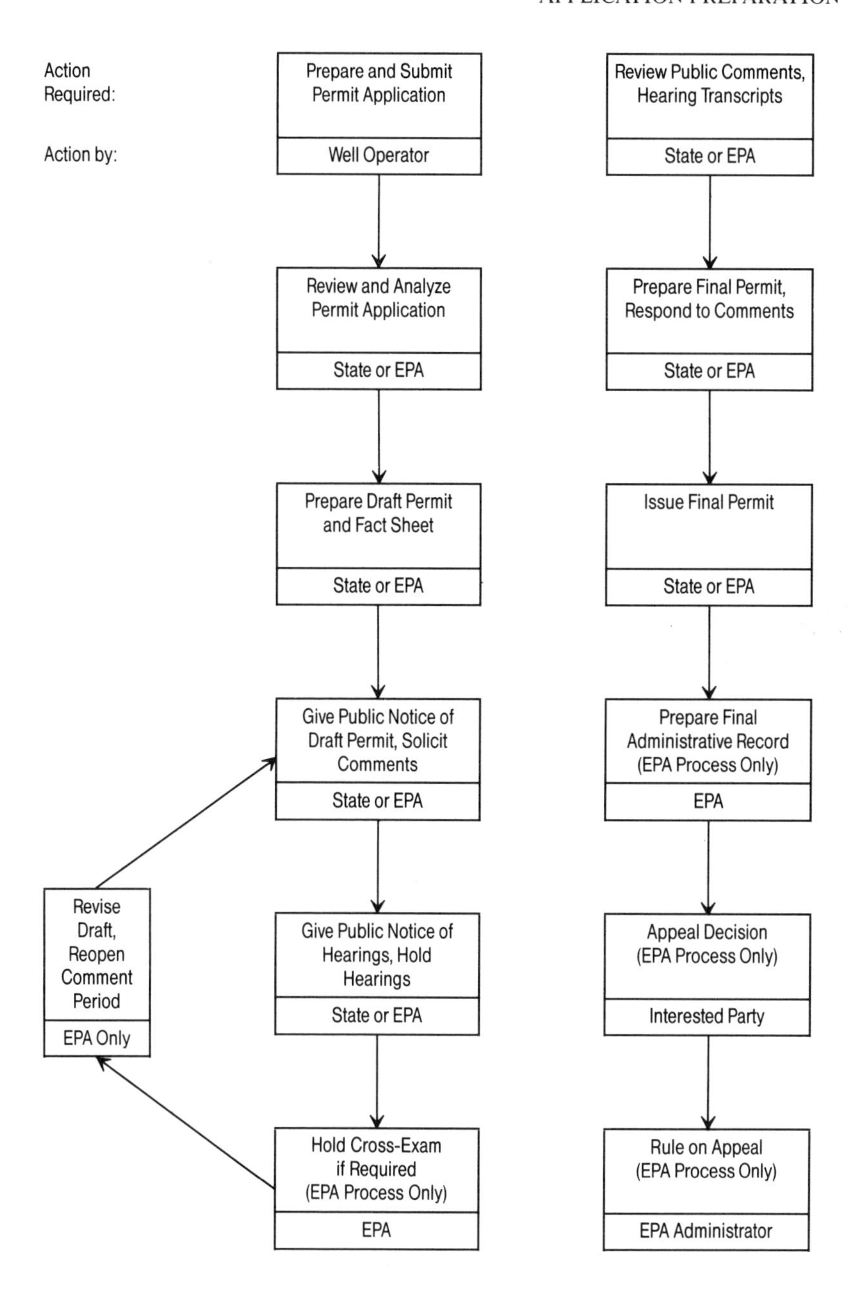

Source: A Guide to the Underground Injection Control Program, C–2, EPA Office of Drinking Water, June 1979.

3-6. The UIC permit process.

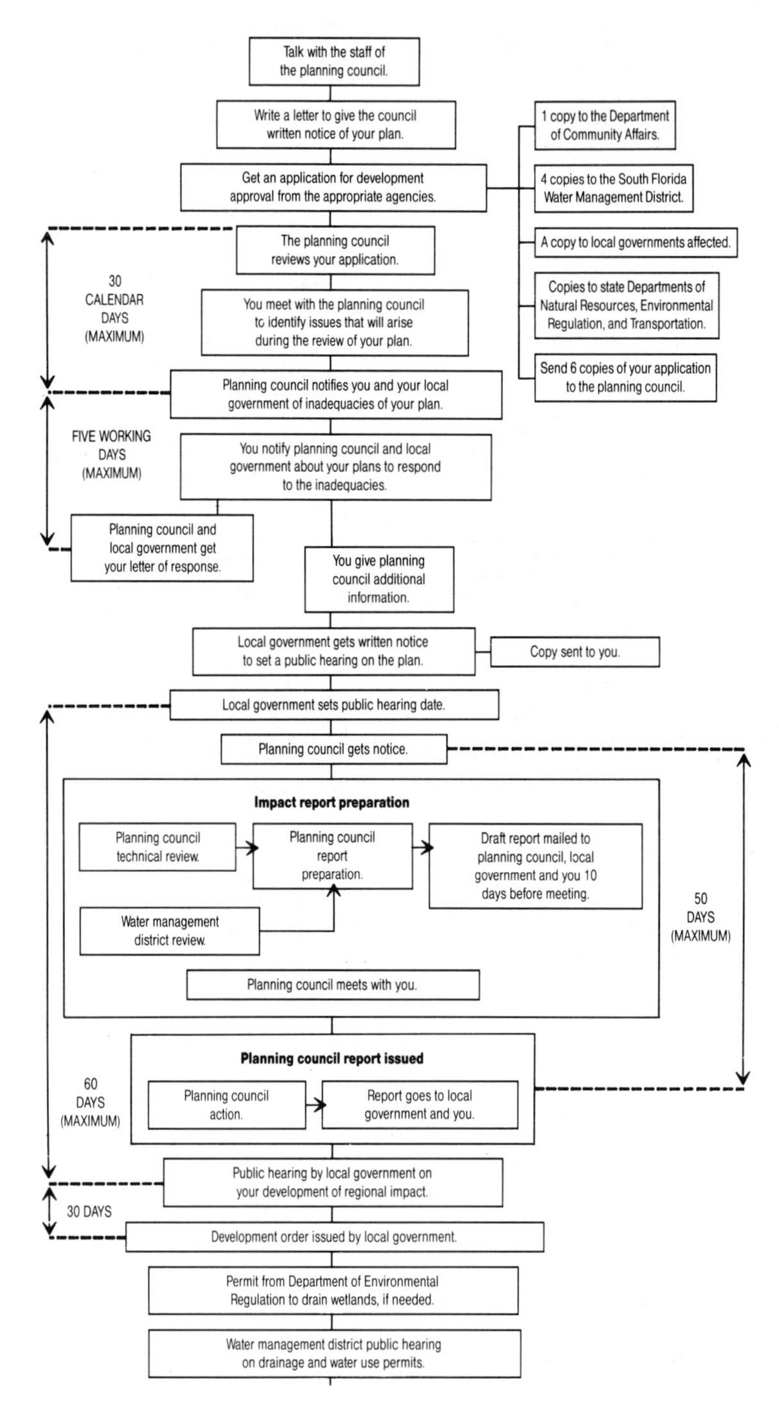

3-7. Example of a permit process flow chart (Florida).

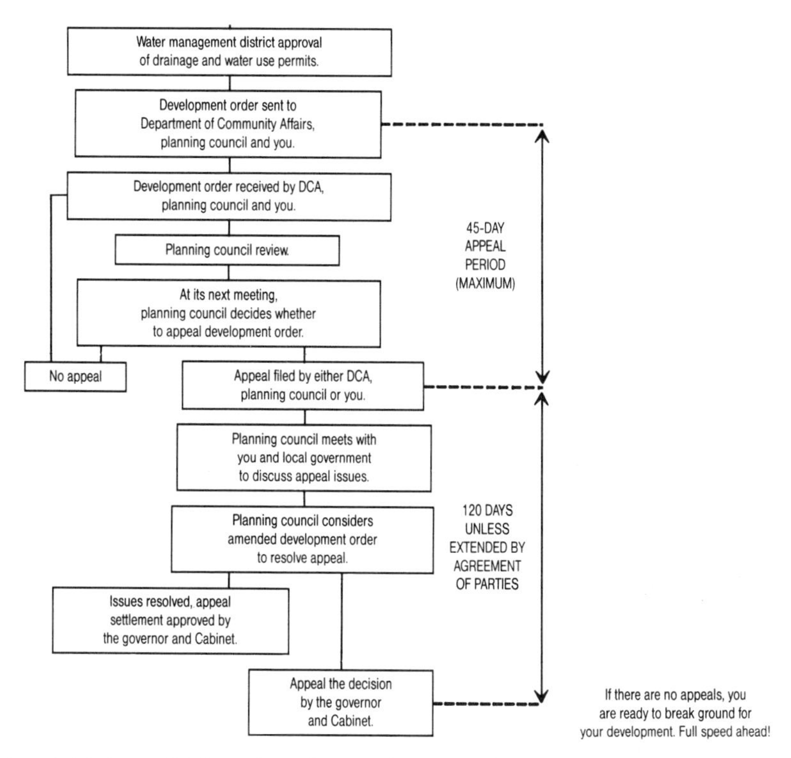
Water management district approval of drainage and water use permits.
Development order sent to Department of Community Affairs, planning council and you.
Development order received by DCA, planning council and you.
Planning council review.
At its next meeting, planning council decides whether to appeal development order.
No appeal
Appeal filed by either DCA, planning council or you.
45-DAY APPEAL PERIOD (MAXIMUM)
Planning council meets with you and local government to discuss appeal issues.
Planning council considers amended development order to resolve appeal.
120 DAYS UNLESS EXTENDED BY AGREEMENT OF PARTIES
Issues resolved, appeal settlement approved by the governor and Cabinet.
Appeal the decision by the governor and Cabinet.
If there are no appeals, you are ready to break ground for your development. Full speed ahead!

Source: From Paul Anderson, "Challenge: Not to Cope, but Control," Part Six of Shaping Progress and Policy Series, Miami Herald, *March 24, 1985.*

3-7. *Continued.*

project location, identification of the project owners and engineers, and sketches of each major aspect of the work. Application forms vary in length and detail. Figure 3-9 is an example of an application checklist. Figure 3-10 illustrates a fairly complex application form. Instructions for the completion of these forms also vary greatly. In general, the following rules apply to all forms:

1. Fill in forms completely, neatly, and precisely.
2. Use clear, concise, and objective language.
3. Provide all information in standard units and indicate units with every number given.
4. Check each form and sketch for accuracy and consistency before submission.
5. Keep a copy of all submitted forms, drawings, photos, and other documents.

Although these rules appear simple, they are often overlooked and can lead to substantial delays.

Most projects require federal, state, and local permit applications. It is imperative that the information be identical in all applications. A common error in applying for permits is to submit early drawings to one agency and

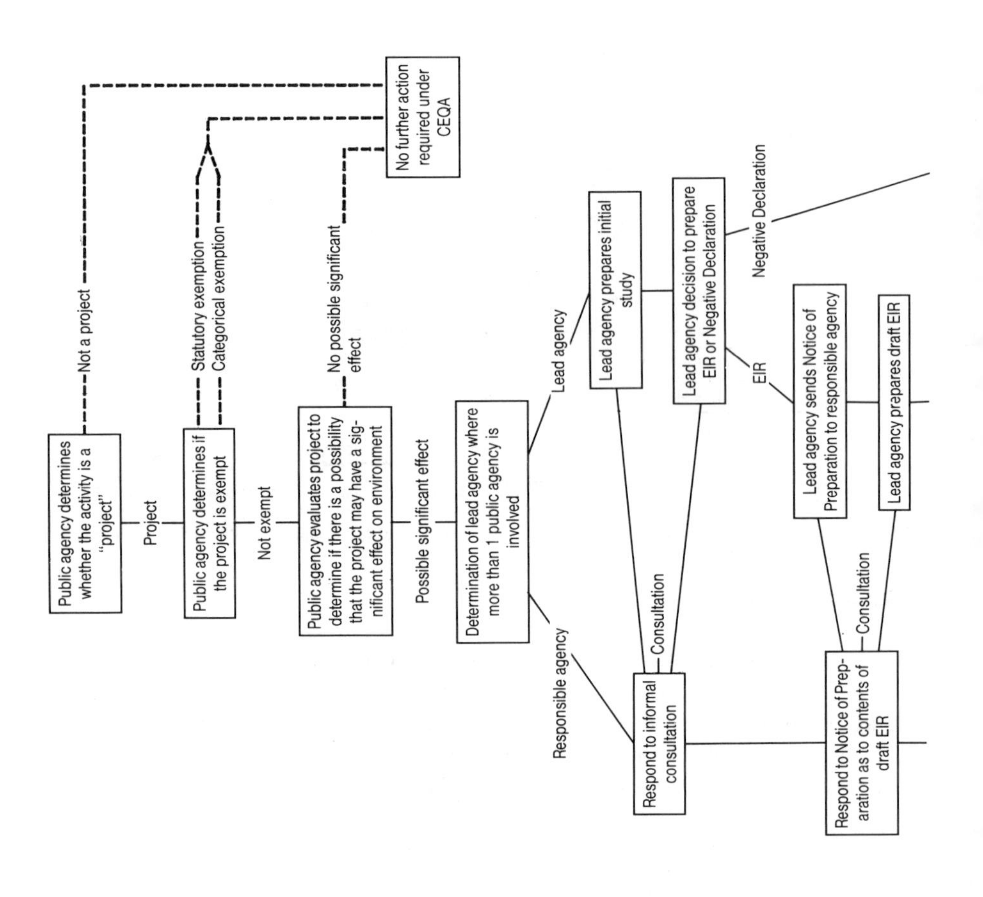
Public agency determines whether the activity is a "project"
Not a project
Project
No further action required under CEQA
Public agency determines if the project is exempt
Statutory exemption
Categorical exemption
Not exempt
Public agency evaluates project to determine if there is a possibility that the project may have a significant effect on environment
No possible significant effect
Possible significant effect
Determination of lead agency where more than 1 public agency is involved
Lead agency
Responsible agency
Lead agency prepares initial study
Consultation
Respond to informal consultation
Lead agency decision to prepare EIR or Negative Declaration
Negative Declaration
EIR
Lead agency sends Notice of Preparation to responsible agency
Lead agency prepares draft EIR
Consultation
Respond to Notice of Preparation as to contents of draft EIR

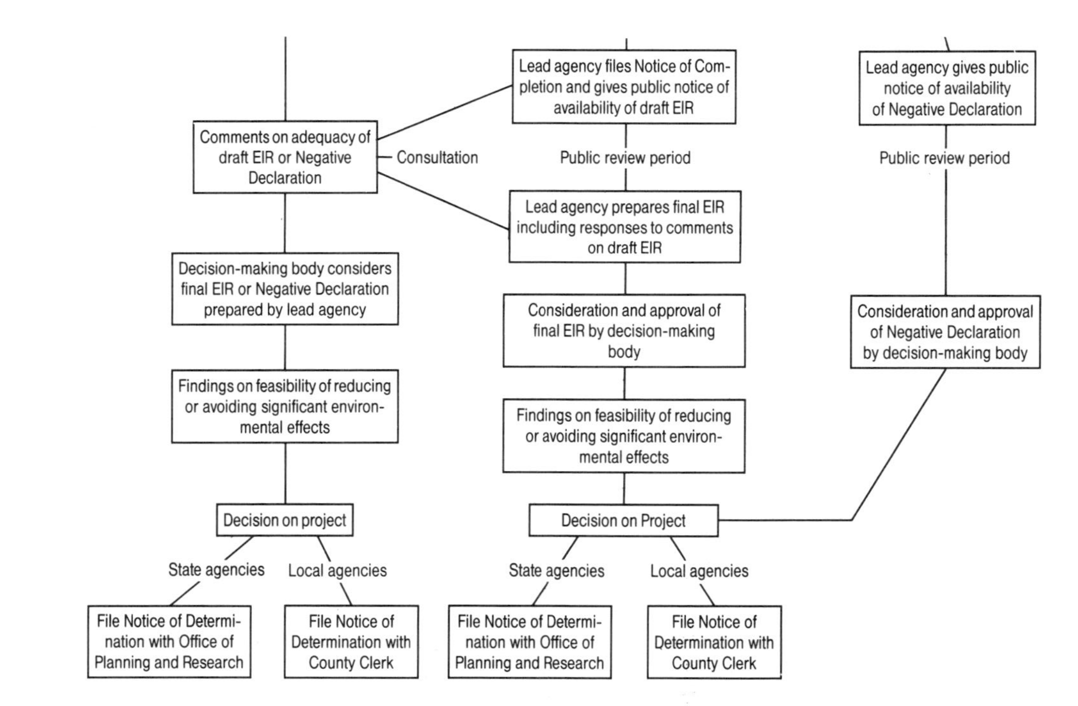

This flow chart is intended merely to illustrate the EIR process contemplated by these Guidelines. The language contained in the Guidelines controls in case of discrepancies.

3-8. Example of a clearinghouse agency flow chart (California).

3-9. U.S. Army Corps of Engineers application checklist.

General

- ☐ Submit one original or good-quality reproducible set of all drawings on 8″ × 10½″ tracing cloth, tracing film or paper. Submit the fewest number of sheets necessary to adequately show the proposed activity. Drawings should be prepared in accordance with the general format of the samples. Block-style lettering should be used.
- ☐ A 1-inch margin should be left at the top edge of each sheet for purposes of reproduction and binding. A ½-inch margin is required on the three other edges.
- ☐ Title block of each sheet submitted should identify the proposed activity and contain the name of the body of water; river mile, if applicable; name of county and state; name of applicant or agent; number of the sheet and total number of sheets in set; and date the drawing was prepared.
- ☐ Drawings should not reflect the approval, nonobjection, or action of other agencies.
- ☐ Since drawings must be reproduced photographically, color shading cannot be used. Drawings must show work as a dot shading, hatching, or similar graphic symbols.

Vicinity Map. Identify the map or chart from which the vicinity map was taken and show the following:

- ☐ Location of the activity site including latitude and longitude and river mile, if known.
- ☐ Name of waterway.
- ☐ All applicable political (county, borough, town, city, etc.) boundary lines.
- ☐ Name of and distance to local town, community, or other identifying location.
- ☐ Names of all roads in the vicinity of the site.
- ☐ Graphic scale.
- ☐ North arrow.

Plan View. The plan view of the proposed activity should show the following:

- ☐ Existing shorelines.
- ☐ Ebb and flood in tidal waters and direction of flow in rivers.
- ☐ North arrow.
- ☐ Graphic or numerical scale.
- ☐ Mean high and low water lines if the proposed activity is located in tidal areas on the Atlantic and Gulf coasts.
- ☐ Mean higher high water and mean lower low water lines if the proposed activity is located in tidal areas on the Pacific coast.
- ☐ Ordinary high water line and ordinary low water line if the proposed activity is on a lake or ordinary high water if on a stream.
- ☐ Water depths around the project.
- ☐ Principal dimensions of the structure or work and extent of encroachment beyond the applicable high water line.
- ☐ Waterward dimension from an existing permanent fixed structure or object.
- ☐ Distances to nearby federal projects, if applicable.
- ☐ Number of cubic yards, type of material, method of handling, and location of fill or spoil disposal area if applicable. If spoil material is to be placed in approved dumping grounds, a separate map showing the location of the dumping grounds should be attached. The drawing must indicate proposed retention levees, weirs, and/or other devices for retaining hydraulically placed materials.
- ☐ Distance between proposed activity and navigation channel, where applicable.
- ☐ Federal harbor lines, if established and if known.
- ☐ Location of structures, if any, in navigable waters immediately adjacent to the proposed activity, including permit numbers, if known. Identify purpose of all structures.
- ☐ Location of any wetlands, swamps, marshes, etc. Identify.

Elevation and/or Section View. The elevation and/or section view of the proposed project should show the following:

- ☐ Same water elevations as in the plan view.
- ☐ Depth at waterward face of proposed work, or if dredging is proposed, show dredging grade.
- ☐ Dimensions from applicable high water line for proposed fill, float, or pile supported platform. Identify any structures to be erected thereon.
- ☐ Graphic or numerical scale.
- ☐ Cross-section of excavation or fill, including approximate side slopes.
- ☐ Elevation of spoil areas.

Notes on Drawings

- ☐ List names of adjacent property owners whose property also adjoins the water and are not shown on plan view.
- ☐ State purpose (private use, commercial, etc.) of proposed activity.
- ☐ State datum used in plan and elevation views. Use mean low water, mean lower low water, National Ocean Survey Datum or National Geodetic Vertical Datum of 1929.

updated and altered drawings to another agency. If such errors are not caught before public hearings or permit approval, delays and loss of trust between the agency and the applicant may occur. If variations are found to exist from one permit to another, the project may be stopped by agency inspectors, who must call a halt to any work not performed strictly according to what is permitted by their agency. These changes, no matter how trivial, may result in work stoppage, which is seldom trivial.

Reviewing other project application forms may help you determine how much time to allow in preparing an application. As mentioned above, this can be done at the preapplication meeting. Preparing a simple list of rules for those who will prepare and draft the application (after examination of similar applications and discussions with agency representatives) should help simplify and reduce the cost of the preparation procedure. Such a list might include a standard project name to be placed on all documents, standard units and abbreviations, a standard reference system to additional materials (i.e., attachments, appendices, or exhibits), and a list of items that must be included on each drawing, such as title block, north arrow, and property lines. Figure 3-11 is a sample list of project drawing standards.

The importance of using clear language in permit applications cannot be overemphasized. Perhaps even more important is the use of *objective* language. Phrases such as "good plan," "well designed," "will not harm," and "in my experience" are subjective: that is, someone's judgment has been used in the description. Subjective statements are open to, and indeed invite, skepticism. It is always assumed that the applicant is an advocate of his or her own project and has a vested interest in its successful completion. Subjective statements support such assumptions by appearing to present the applicant's personal—and thus

(Text continued on page 47.)

3-10. Application form (Michigan). (Sample drawings are usually included with the application form.)

APPLICATION FOR PERMIT

FOR OFFICIAL USE

Corps of Engineers
Department of the Army

Corps Process No.

State of Michigan
Department of Natural Resources
Land Resource Programs

DNR File No.

PLEASE READ INSTRUCTIONS *BEFORE* FILLING OUT THIS APPLICATION—PRINT OR TYPE

1. APPLICANT (individual or corporate name)

ADDRESS

CITY STATE ZIP

TELEPHONE
(Work)
(Home)

SOC. SECURITY or FED I.D. NO.

AGENT/CONTRACTOR (firm name, if known)

ADDRESS

CITY STATE ZIP

TELEPHONE

2. If applicant is not owner of the property where the proposed activity will be conducted, provide name and address of owner and include letter of authorization from owner:

OWNER'S NAME **MAILING ADDRESS** **CITY** **STATE** **ZIP**

3. **PROJECT LOCATION** Street/Road Village/City BODY OF WATER (Lake, stream, creek, pond or drain)

County Township Town Range Section(s) Subdivision or Plat Lot No. Private Claim

4. **PROJECT INFORMATION**

(a) Describe proposed activity

(b) Attach drawings of the proposed activity prepared in accordance with the *DRAWING REQUIREMENTS* on pages 1 & 2 of Instructions.

(c) Check appropriate Project Type (below) **See Samples of Drawings Required**

1) ☐ Dredging, Filling, Draining or Construction Work in Inland Lakes or Streams, Great Lakes Bottomlands or Wetland Areas . . . 1, 2, 3, 4, or 5
2) ☐ Work in Riverine Flood Plain . . . (See SPECIAL INSTRUCTIONS, Section 1, on back of this form) . . . 6
3) ☐ New or Replacement Bridge or Culvert . . . (See SPECIAL INSTRUCTIONS, Section 2, on back of this form) . . . 7, 8, 9 and 10
4) ☐ Dam Construction or Reconstruction . . . (See SPECIAL INSTRUCTIONS, Section 3, on back of this form) . . . 11

NOTE: If boxes 2, 3, and/or 4 above, are checked provide appropriate additional information on the back under "SPECIAL INSTRUCTIONS"

(d) PROPOSED USE: (Check appropriate box)
1. ☐ Public; ☐ Private; ☐ Commercial; ☐ Other (specify) ______
2. Will the project site be served by a new on-site Sewage Disposal System (Septic Tank) ☐ No ☐ Yes

(e) Location of Source of Fill if more than 50 cubic yards are required for other than commercial source:

County	Township	Town	Range	Section	¼ Section

Futher Description (provide vicinity map of Source Site (Sample Drawing 5) if more than 50 cubic yards and source is other than commercial)

(f) Dredge Spoils Disposal Location Site (if required):

County	Township	Town	Range	Section	¼ Section

Further Description (provide vicinity map for Disposal Site (Sample Drawing 3)

(g) Describe any project alternatives considered.

If fill is required, is project water dependent? ☐ No ☐Yes

(h) Date activity will commence if permit is issued ______; be completed ______.

(i) Is any portion of the requested project now complete? ☐ No ☐ Yes. If yes, identify the completed portion on the drawings you submit and give the date activity was completed. ______

DO NOT WRITE IN THIS SPACE—FOR CASHIER USE ONLY

APPLICATION CONTINUED ON REVERSE SIDE.

REMOVE INSTRUCTIONS BEFORE MAILING.

DO NOT REMOVE THIS STUB

APPLICATION FOR PERMIT
LAND RESOURCE PROGRAMS

Continued

3-10. *Continued.*

(APPLICANT COMPLETE THE FOLLOWING)

NAME OF REMITTOR

ADDRESS

☐ 1972 P.A. 346 Pemit Application Fee ☐ 1979 P.A. 203 Permit Application Fee

5. State why you believe the project will not cause pollution, impair or destroy the water or any natural resources:

6. List all other Federal, State or local governmental agency permits or certifications required for proposed project.
   Specify permit approvals or denials already received. Explain reasons for denials.

Agency	Type Approval	Identification No.	Date Applied	Date Approved/Denied

State reasons if permit denied:

7. Is there any present litigation involving the subject property? ☐ No ☐ Yes If "Yes," explain:

8. Adjoining Riparian (Neighboring Waterfront Property Owner) Name and mailing address at which they may be reached.

Name of Riparian #1	Address	City	State	ZIP
Name of Riparian #2	Address	City	State	ZIP
Name and Address of Lake Association	Address	City	State	ZIP

READ CAREFULLY BEFORE SIGNING.

9. Application is hereby made for a permit or permits to authorize the activities described herein. I certify that I am familiar with the information contained in this application, and that to the best of my knowledge and belief such information is true and accurate and in compliance with the State Coastal Zone Management Program. I certify that I have the authority to undertake the activities proposed in the application. By signing this application, I understand to allow representatives of the Michigan Department of Natural Resources and the U.S. Corps of Engineers to enter upon said property in order to inspect the proposed project. I understand that the granting of other permits by local, county, state or federal agencies does not release me from the requirements of obtaining the permit requested hereon before commencing the project. I understand that the payment of fee does not guarantee permit.

SPECIAL INSTRUCTIONS

SECTION 1. FOR WORK IN FLOODWAY AREAS, a hydraulic engineering report prepared by a Registered Professional Engineer showing the impact of the proposal on flood stage or discharge characteristics may be needed.

SECTION 2. FOR NEW OR REPLACEMENT BRIDGES OR CULVERTS. To assist in the selection of an appropriate size structure, a design discharge may be requested from the Department of Natural Resources, Water Management Division. Requests should be accompanied by a location description giving the town, range, section, stream and road name. A location map as illustrated on page 4, Sample Drawing 11, should be included with the submission.

STRUCTURAL DATA:	**EXISTING**	**PROPOSED** (replacement)
Type	________	________
Entrance Design	________	________
Span, Rise	________	________
Length (width)	________	________
Waterway Area (total)	________	________
ELEVATIONS (Bench Mark Datum)	________	________)
Low Steel (culvert crown)	upstr ____ dnstr ____	upstr ____ dnstr ____
Invert	upstr ____ dnstr ____	upstr ____ dnstr ____
Highwater (observed or recorded)	________	________
ROAD GRADES	**EXISTING**	**PROPOSED**
At structure	________	________
Low Point of Approach	________	________

SECTION 3. DAM CONSTRUCTION OR RECONSTRUCTION, COMPLETE THE FOLLOWING:

Proposed head ________ ft. (Difference between normal pond level and stream water surface level below dam).

Proposed Impoundment size (flooded area) ________

If the proposed dam project has a head of five (5) feet or more or impounds five (5) or more surface acres, compliance with the Dam Construction Approval Act is required. Following a review of the environmental impacts of the proposed dam construction and clearance for the issuance of an Inland Lake and Streams Act Permit, you will be notified of the need to submit construction plans and specifications, prepared by a Registered Professional Engineer. Also, at that time you will be notified that an additional fee will be required in accordance with the following schedule:

Head less than five (5) feet . . . No Fee Head greater than five (5) feet but less than eight (8) feet . . . $200.00

Head greater than eight (8) feet but less than 20 feet . . . $400.00 Head greater than 20 feet . . . $600.00

Continued

3-10. *Continued.*

GENERAL INSTRUCTIONS

1. Please type or print legibly the application. Use black ink or pencil. (Do not use blue ink.)
2. Answer each question thoroughly. Incomplete applications will be returned.

FORM COMPLETION INSTRUCTIONS

1, 2 and 3 are self-explanatory.

4.(a) Describe proposed activity. EXAMPLE: Construct a steel seawall for shore protection. Dredge 50 cubic yards of material for fill behind the wall. Place 30 cubic yards of sand for beach construction.

4.(b) Drawings are required for each activity. Drawing instructions are included on pages 1 and 2 of these instructions. Sample drawings are on pages 3 and 4 of instructions.

4.(c) If Project Types 2), 3) and/or 4) are checked, additional information is required as specified under SPECIAL INSTRUCTIONS on the back of application form.

4.(d) 4(e) and 4(f) are self-explanatory.

4.(g) Describe any project alternatives considered. EXAMPLE: Placement of rock rip rap was determined "unacceptable" because of wave wash at the shoreline.

4.(h) and 4(i) are self-explanatory.

5. Reasons you believe project will not cause pollution, etc.; EXAMPLE: The project will stabilize shoreline and minimize erosion.

6, 7 and 8 are self-explanatory.

9. Read carefully before signing. Should you have any questions, please contact the Corps of Engineers at (313) 226–6812 or the Department of Natural Resources at (517) 373–9244.

FEES

(A) For all nongovernmental projects located on an inland lake or stream or any project located in a wetland, a state fee of $25.00 must accompany the application. The payment of the fee does not guarantee a permit. Projects located on the Great Lakes, unless they are under the Wetlands act, do not require a fee. Make checks payable to: "STATE OF MICHIGAN."

(B) The U.S. Army Corps of Engineers will notify you of the appropriate Federal filing fee when their permit application review has been completed and a preliminary determination has been made that a permit will be issued. Fees are assessed as follows: (1) Commercial or industrial users–$100.00 and (2) noncommercial uses–$10.00. DO NOT SUBMIT ANY FEE TO THE U.S. ARMY CORPS OF ENGINEERS UNTIL YOU ARE NOTIFIED OF THE REQUIRED AMOUNT.

(NOTE: The federal filing fee is in addition to any fee required by the State of Michigan.)

PLEASE READ THE PRIVACY ACT AND APPLICATION PENALTY STATEMENT ON PAGE 2 OF THESE INSTRUCTIONS.

MAIL APPLICATION AND STATE FILING FEE (if applicable) TO:

Department of Natural Resources
Land Resource Programs Division
Box 30028
Lansing, Michigan 48909

DRAWING REQUIREMENTS

PLAN SHEETS REQUIRED

SEE ATTACHED SAMPLE DRAWING. Drawings must be legible in black ink or pencil on standard weight paper of 8½ × 11″ size. Additional sheets may be required. In addition, each drawing must contain the following elements. If plans are engineering plans larger than 8½ × 11″, submit a minimum of five copies.

(1) Simple title block on lower right hand corner of each sheet with applicant name; name of water way; city; village or township; county; description of activity and scale of drawing (number of sheets, total number in set (sheet 1 of 4), date drawing was prepared).

(2) Location map stating source of the map, nearest main road, crossroad, and route to project site.

(3) Existing and proposed structures clearly identified, the dimensions of each showing existing shore features and man-made structures and the length of shore frontage, water depths and bottom configuration around project. If existing structures were previously authorized by federal/state permit, show corresponding permit number.

(4) Typical cross sections of existing and proposed shoreline, waterline, structures, dredge cuts and fills, including dimensions and elevations, location of wetlands.

(5) A description of construction materials such as: thickness, slope, type (stone, concrete, etc.) and size or weight of riprap material when placing a protective facing on earth retention structure. Include the total amount (volume) cubic yards of all fill or dredge material.

(6) Existing water depth and ground surface elevations referenced to (Low Water Datum (L.W.D.) on International Great Lakes) U.S.G.S. Quad sheets and areas flooded by water part of the year, existing shorelines and high water marks, flow and circulation patterns, if any.

(7) Show names and addresses of the owners of the adjoining property on both sides and if on a stream, address of opposite riparian, and the relative location of any structures which may exist along shoreline of adjacent properties. If there are no adjacent structures, show existing shorelines. Size relationship between proposed project and total area, i.e. one acre fill in 15 acre wetland.

(8) If proposed activity involves bulkhead construction, show the distance along both property lines from the face of the bulkhead to the centerline of a street or other definable reference point, i.e. NE corner of concrete patio, 12″ maple on W property line.

–CONTINUED ON BACK–

I.C. 2731
Rev. 10/82

DRAWING REQUIREMENTS (con't.)

(9) If activity involves dredging, furnish the following:
   a) If the dredged material is to be placed on-site, outline the disposal area on the drawing. If the dredged material is to be hauled away, provide a vicinity map showing the disposal area. If more than 50 cubic yards of fill is required, indicate the location or source of fill. This information will be used by the History Division to evaluate the impacts on historic or archaeologic resources.
   b) Show method of containing dredged material to prevent re-entry into any waterway or wetland. Describe all procedures by which applicant proposed to minimize adverse effects of construction.

(10) If your activity involves flood plain alterations, proposed and existing contours must be shown on a site development plan. Sample Drawing #6.

(11) If your activity involves a bridge or culvert crossing, furnish the following:
   a) Typical stream valley cross-section representative of the channel and flood plain area downstream of the proposal. Sample Drawing #7.
   b) Plan view of the proposal. Sample Drawing #9.
   c) Cross-sectional view of the proposed structure superimposed on the existing structure (if applicable). This view should include the existing and proposed road centerline profile for the width of the stream valley or a minimum of 300 feet either side of the structure. Sample Drawing #10.
   d) Photographs of the structure, channel and flood plain areas.
   e) A profile view of the proposed structure showing the proposed end treatment and bank stabilization. Sample Drawing #8.

(12) If the project activity involves dam construction, provide a sketch showing the head and approximate flooded area. Sample Drawing #11.

Continued

3-10. *Continued.*

DATA REQUIRED BY THE PRIVACY ACT OF 1974
(5 U.S.C. 552a)

TITLE OF FORM
APPLICATION FOR PERMIT (Form PR 2731) from the STATE OF MICHIGAN
DEPARTMENT OF NATURAL RESOURCES and DEPARTMENT of the ARMY, CORPS OF ENGINEERS.

PRESCRIBED DIRECTIVE
ER 1145-2-303

1. AUTHORITY

SECTION 10 River & Harbor Act 1899, Section 103 Marine Protection, Research & Sanctuaries Act of 1972, and Section 404 of the Clean Water Act Amendments.

2. PRINCIPAL PURPOSE(S)

Application form for permits authorizing structures and work in or affecting navigable waters of the United States, the discharge of dredged or fill material into navigable waters, and the transportation of dredged material for the purpose of dumping it into ocean waters.

3. ROUTINE USES

Describes the proposed activity, its purpose and intended use, including a description of the types of structures, if any, to be erected on fills, or pile float-supported platforms, and the type, composition and quantity of materials to be discharged or dumped and means of conveyance.

If the application is made at the Detroit District level, a copy will be furnished to the Michigan Department of Natural Resources, conversely if the application is submitted to the Michigan DNR, a copy will be furnished to the Detroit District, and subsequently the content is made a matter of public record through issuance of a public notice.

The application is made available to any requesting state and Federal agencies, dealing with the review of the application. The form itself is not made available; only that information which is pertinent to the evaluation of the permit request.

The form (or copies) could be kept on file at the Michigan DNR, Detroit District, Division or OCE level, depending on the details surrounding the case. The information could become a part of any record of a reviewing agency with a need to know; such as U.S. Fish & Wildlife; Environmental Protection Agency; etc.

The application will become a part of the record in any litigation action by the Department of Justice or the Michigan Attorney Generals Office involving the work or activity.

4. MANDATORY OR VOLUNTARY DISCLOSURE AND EFFECT ON INDIVIDUAL NOT PROVIDING INFORMATION

The disclosure of information is VOLUNTARY. Incomplete data precludes proper evaluation of the permit application. Without the necessary data (i.e., name, address and phone number), the permit application cannot be processed or a permit subsequently issued.

PENALTY

18 U.S.C. Section 1001 provides that: Whoever, in any manner within the jurisdiction of any department or agency of the United States knowingly and willfully falsifies, conceals, or covers up by any trick, scheme, or device a material fact or makes any false, fictitious or fraudulent statements or representations or makes or uses any false writing or document knowing same to contain any false fictitious or fradulent statement or entry, shall be fined not more than $10,000 or imprisoned not more than five years, or both.

I.C. 2731
Rev. 10/82

3-11. Sample standards for project drawings.

CALCULATIONS
1. Show all calculations
   a. source of equations
   b. source of input data
   c. any assumptions
   d. show units at each calculation
2. Date calculations
3. Identify who made the calculations
4. Show all update calculations

DRAWINGS
1. North arrow
2. Title box
   a. title of drawing
   b. date of drawing
   c. drawn by
3. Scale
4. Property line–identify adjacent property owners
5. Indicate direction of water flow–Ebb, flow, currents, etc.
6. Data
7. Space for permit file numbers

UNITS (note: this will vary with projects)
1. Length–feet, miles
2. Time–hours
3. Volume–cubic yards

biased—opinion of the project. Objective statements such as "standard design (technique)" or "according to documented evidence" imply that there is general concurrence with an idea—that it is not merely one individual's opinion. (Of course, you must provide backup data to substantiate such statements.) The use of objective language helps prevent exaggeration or misleading statements and should be part of every permit application.

Figure 3-12 is a sample application with accompanying sketches.

ACCOMPANYING DOCUMENTATION

The documents that accompany an application form as part of the application package are usually prescribed by the individual agencies. These documents are different from the "additional information" (reports and assessments) sometimes required by agencies, which will be discussed in chapter 6. Such accompanying documentation may include maps, aerial photographs, surveys, and proof of ownership (figure 3-13). Generally, these materials are existing and available, but occasionally they must be prepared for the application. When such is the case, an agency may initiate the processing of the permit application even though one item is missing, in order to avoid delays. However, the processing will not be completed without a missing item unless special variances are granted. The applicant, once asked for specific materials, should request a

(Text continued on page 52.)

JOINT APPLICATION
DEPARTMENT OF THE ARMY/FLORIDA DEPARTMENT OF ENVIRONMENTAL REGULATION
FOR
ACTIVITIES IN WATERS OF THE STATE OF FLORIDA

Refer to Instruction Pamphlet for explanation of numbered items and attachments required.

1. Application number (To be assigned)

2. Date

Day Mo. Yr.

3. For official use only

4. Name, address and zip code of applicant

Billy G. Wells
1641 S. Hampton Boulevard
Jacksonville, Florida 32211

Telephone Number 904/725-8386

5. Name, address, zip code and title of applicant's authorized agent for permit application coordination

N/A

Telephone Number

SAMPLE

6. Describe the proposed activity, its purpose and intended use, including a description of the type of structures, if any, to be erected on fills, or pipe or float-supported platforms, and the type, composition and quantity of materials to be discharged or dumped and means of conveyance.

Bulkhead and backfill with 880 cubic yards of clean sand obtained from upland areas. Construct pier for mooring private sailboat. No fuel pumps or toilet facilities to be constructed on pier. No structures are to be erected on fill.

	Dredged/Excavated		Filled/Deposited	
Volume of Material:	CY	880 CY	CY	CY
	Waterward of O.H.W. or M.H.W.	Landward of O.H.W. or M.H.W.	Waterward of O.H.W. or M.H.W.	Landward of O.H.W. or M.H.W.

7. Proposed use

Private [x] Public [ ] Commercial [ ] Other [ ] (Explain in remarks)

8. Name and address including zip code of adjoining property owners whose property also adjoins the waterway.

Robert L. Hampton
12467 Ridge Road
Jacksonville, Florida 32227

Cedric L. Clark
12571 Ridge Road
Jacksonville, Florida 32227

9. Location where proposed activity exists or will occur

Street address

Longitude Latitude (If known)

Sec. 31 Twp. 27S Rge. 41E

Florida	Duval	Jacksonville	
State	County	In City or Town	Near City or Town

10. Name of waterway at location of the activity Big Haw Creek

3-12. Sample application with sketches.

11. Date activity is proposed to commence 30 days after receipt of permit

Date activity is expected to be completed 6 months after receipt of permit

12. Is any portion of the activity for which authorization is sought now complete? Yes [ ] No [x]

If answer is "Yes" give reasons in the remarks section. Month and year the activity was completed ______

______. Indicate the existing work on the drawings.

13. List all approvals or certifications required by other Federal interstate, state or local agencies for any structures, construction, discharges, deposits or other activities described in this application, including whether the project is a Development of Regional impact.

Issuing Agency	Type of Approval	Identification No.	Date of Application	Date of Approval
DNR	Confirmation DF Title		29 Feb 75	———
City of Jacksonville	Permit	7J123	15 Jan 75	31 Jan 75

14. Has any agency denied approval for the activity described herein or for any activity directly related to the activity described herein?

Yes [ ] No [x] (If "Yes" explain in remarks)

15. Remarks (see Instruction Pamphlet for additional information required for certain activities)

SAMPLE

16. Application is hereby made for a permit or permits to authorize the activities described herein. I agree to provide any additional information/data that may be necessary to provide reasonable assurance or evidence to show that the proposed project will comply with the applicable State Water Quality Standards or other environmental protection standards both during construction and after the project is completed. I also agree to provide entry to the project site for inspectors from the environmental protection agencies for the purpose of making preliminary analyses of the site and monitoring permitted works, if permit is granted. I certify that I am familiar with the information contained in this application, and that to the best of my knowledge and belief such information is true, complete, and accurate. I further certify that I possess the authority to undertake the proposed activities.

Billy G. Wells ______ 6-20-77 ______

Signature of Applicant Date

18 U.S.C. Section 1001 provides that: Whoever, in any manner within the jurisdiction of any department or agency of the United States knowingly and willfully falsifies, conceals, or covers up by any trick, scheme, or device a material fact or makes any false, fictitious or fraudulent statements or representations or makes or uses any false writing or document knowing same to contain any false, fictitious or fraudulent statement or entry, shall be fined not more than $10,000 or imprisoned not more than five years, or both.

The application must be signed by the person who desires to undertake the proposed activity; however, the application may be signed by a duly authorized agent if accompanied by a statement by that person designating the agent and agreeing to furnish upon request, supplemental information in support of the application.

FEE: Attach Checks/Money Orders on front

Payable to Department of Environmental Regulation

$200 Standard form projects

$20 Short forms and Chapter 403 projects only

Continued

EXAMPLE – PIER

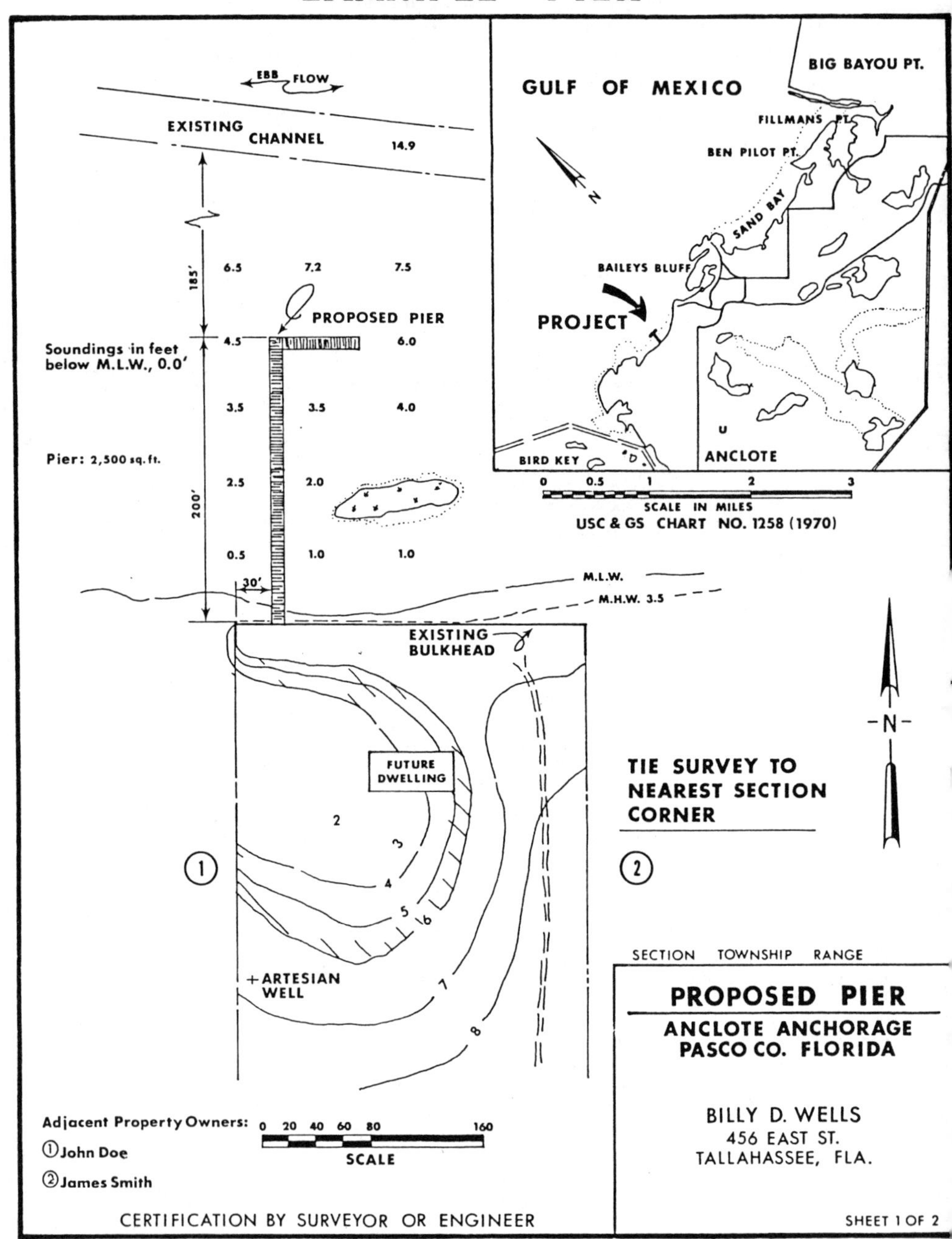

3-12. *Continued.*

EXAMPLE PLAN

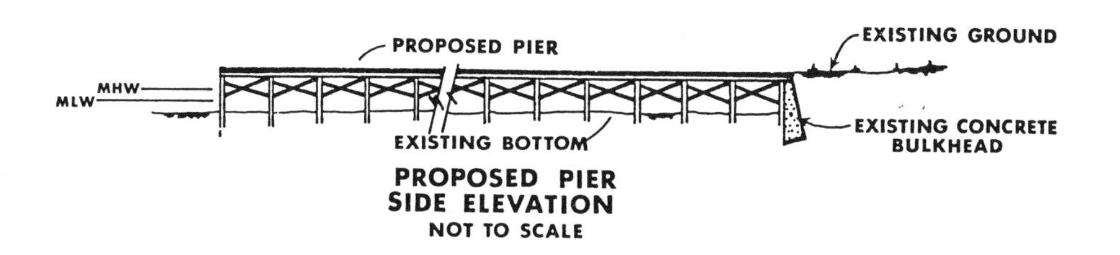

PROPOSED PIER
SIDE ELEVATION
NOT TO SCALE

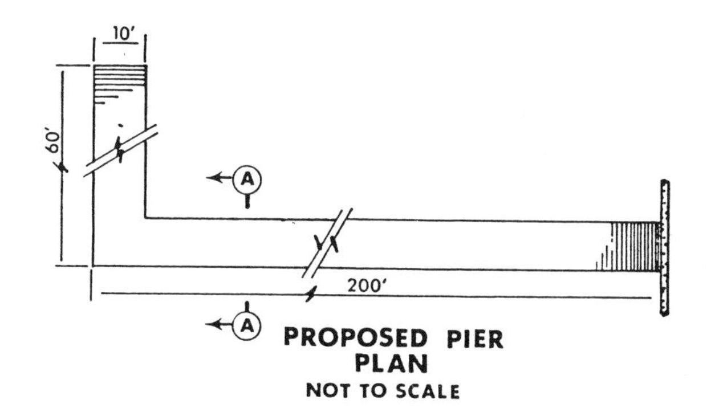

PROPOSED PIER
PLAN
NOT TO SCALE

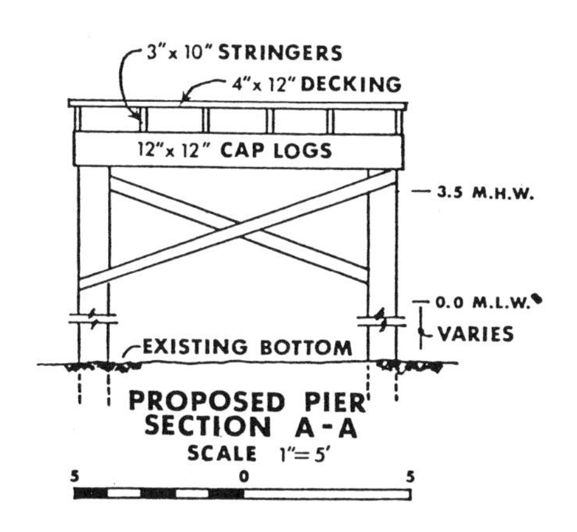

PROPOSED PIER
SECTION A-A
SCALE 1"=5'

SHEET 2 OF 2

3-13. Sample affidavit of ownership.

INFORMATION NEEDED ON AFFIDAVIT OF OWNERSHIP OR CONTROL

1. The affidavit of ownership or control should be signed by at least one owner of the property, lessee, or easement holder.
2. If subdivided, the property should be described by lot number, block number (if any), name of subdivision, and plat book number and page where recorded.
3. If unplatted, the property should be described by section, township, and range, and subdivision of section.
4. If the description is based on courses and distances not following the plane coordinates, the point of beginning should be identified.

Please contact this office should clarification or further instructions be necessary.

AFFIDAVIT OF OWNERSHIP OR CONTROL
TO THE DEPARTMENT OF ENVIRONMENTAL REGULATION:

I hereby certify that I am the (check one):

record owner ______________________

lessee ______________________

record easement holder ______________________

applicant to record owner for easement ______________________

of the below described property situated in County, Florida; and that said property is all the property that is contiguous to and landward of the area in which the work proposed in the permit application is to be conducted. Furthermore, I certify that as record owner, lessee, or record easement holder I have or will have prior to undertaking the work all necessary approvals or permission from all other persons with a legal interest in said property to conduct the work proposed in the permit application.

LEGAL DESCRIPTION

____

____

Sworn to and subscribed before me at ______________________

______________________ County, ____________, this ____________ day

of ____________, 19______.

____

NOTARY PUBLIC

My commission expires:

sample of similar materials submitted with similar applications if it will facilitate production of acceptable materials. If materials must be prepared in support of a permit application, ask the agency to describe clearly all required aspects, in writing, before the material is prepared. If a new aerial photograph is required, and it must be to a certain scale, or if a map must contain certain information, find out *before* it must be done twice!

Occasionally an applicant may provide nonrequired documentation in support of an application. This might include a historian's report on the area, old photographs, a biologist's statement, or a report on the environmental effects of similar activities. Extraneous, unsolicited information may or may not be used in the evaluation of an application. Only material that is relevant to the agency's authority to grant or deny permits should be submitted (and again, the applicant must remember to keep copies).

SUBMISSION STRATEGY

The timing of the submission of the application, accompanying documentation, and additional information required by an agency may affect the cost of producing these materials. Two basic submission strategies exist: (1) submit everything at once; and (2) submit material in stages. There are advantages and disadvantages to both methods. The keys to determining which strategy to follow are project size and agency personnel. Generally, the larger the project, the more economical the submit-by-stages strategy. Also, the more picky and difficult the agency personnel, the more economical the submit-by-stages strategy.

Submitting all information at one time is no problem for small-to-intermediate-size projects. If a project will have limited environmental impact (e.g., adding an additional building to an already developed site, or duplication of a project already in place next door), it is also advisable to submit all information immediately. This will help create an atmosphere of total cooperation with the agency and also ensure that the agency evaluation "time clock" for permit processing (discussed in chapter 4) will begin immediately. Generally, any questions, clarifications, or alterations that arise as part of the review process for such projects may be answered by telephone or letter and do not require major rewriting of reports or redrawing of plans.

This method does, however, leave the applicant open to complete revision of all application materials if major difficulties arise, an eventuality that can double the preparation costs.

Submitting by stages allows the agency to put its requests in writing before an applicant hires experts to write reports or prepare assessments. For large projects that might entail evaluation or testing procedures (such as a flushing evaluation of a boat basin or a soil percolation test), or that might require significant alterations (such as the need for a wastewater treatment plant on site), it is often more economical to perform the required evaluations or tests *after* initial agency review.

Example: An applicant has planned a 12-foot-deep boat basin for a project site. At the preapplication meeting he learns that the local environmental agency will require a hydrographic evaluation of the flushing times of the basin. If he pays for preparation of the flushing evaluation *before* initial review, he will have to pay for a second one when he learns that the state will not permit a basin deeper than 6 feet in that area. By waiting for initial comments from various agencies, the applicant could save the cost of preparing and distributing a second report. Hopefully, the major issues (such as basin depth) would be caught in the preapplication meeting. This would depend on the completeness of the plans available at the meeting and the personnel (and the review time they have) attending the meeting.

You must evaluate your submission strategy in terms of preparation and distribution costs versus potential time costs. In some cases time is the most critical element. You should also bear in mind any previous experience(s) with similar projects when deciding on a submission strategy. Remember, however, that you are not engaged in a fight with the agency or agencies; your submission strategy should be determined in cooperation with agency representatives. Their advice and experience with similar projects should be sought out. This advice must then be evaluated in terms of your own knowledge of the project, personnel, and economics involved.

PRACTICAL SUGGESTIONS

Drafting

Drawings communicate a tremendous amount of information about a project in a very compact manner. They are a vital component in all phases of project planning, evaluation, permit application, and construction. Drafting can also be used as a tool in communicating various components of a project. As with any form of communication, simplicity, clarity, consistency, and completeness are crucial.

With adequate planning and an understanding of the various components of the permit process, drafting (and redrafting) can be kept to a minimum. In designing the drafting program, the most stringent requirements should be made the standard, to which all drafting must conform. It is far easier to simplify graphics for specific needs than it is to add information constantly. Imposing stringent requirements throughout will yield uniform illustrations and will minimize the amount of redrawing required for each alteration to a project.

All of the individuals working on graphics for a project must employ the same standards, formats, and units. This is best achieved by establishing a written set of instructions standardizing every aspect of preparation and production, including size of drafting paper, type styles, scales, aspects, and internal referencing (e.g., insets for details). All aspects of drafting should be placed under the control of one individual so that consistent style and review may be established.

The importance of review for accuracy cannot be overstated. Missing or erroneous decimals, missing units, and simple typographic errors are very difficult for a draftsman, involved in looking at drawings for hours on end, to pick up. New computerized drafting programs can aid in redrafting when errors are found.

Record Keeping

During the initial stages of the permit-application process, the novice applicant often tends to be remiss in record keeping. An informational phone call, a letter, even forms are sometimes unrecorded, lost, or destroyed. In some instances these can be very important later in the permit process, so it is wise to begin thorough record keeping from the beginning.

Omissions or inaccuracy in project information often stem from the fact that months or even years may have elapsed between project conception and permit application. During that period, numerous changes and revisions often take place. It is not uncommon for a permit applicant to forget that some detail has changed when the project has gone through a process of revision, rerevision, and readoption of the original (or a modified) plan on any number of components. Here too, excellent record keeping along with constant redrafting and uniform drafting standards (more on this above) is the best (although not foolproof) way to avoid these omissions.

Records should consist of a *master file* with the original of every letter received, a copy of every letter and form sent, and records of all conferences (figure 3-14), phone conversations (figure 3-15), interoffice memos, and any other correspondence. The *working file* should contain a copy of any information necessary for an individual to perform his or her daily work. For example, a project engineer might keep copies of all correspondence concerning design aspects of the permit process and of all drawings sent; a biologist's working file would contain materials of his or her specific interests in the project. Under no circumstances should material be "borrowed" from the master file for anything other than copying. You cannot afford to lose such material, most of which is irreplaceable.

3-14. Sample conference report.

CONFERENCE REPORT **(company logo)**

Project:
Subject:
Date:
Location:

File numbers:

Attendees:

Name	Representing	Phone

Notes:

Concluding Action:

Date	Person Called	Person Calling	Firm/Organization	Subject

3-15. Sample telephone log.

You should develop a system of recording and referencing all materials sent to agencies (photos, surveys, reports, etc.). It should include date of submission as well as any agency identification numbers. A rubber stamp such as the one shown in figure 3-16 may prove useful if a large number of documents (or permits) is necessary.

Conference notes should clearly identify all parties to the conference: who they represented, date, subject of conference, all pertinent comments, and all conclusions, including a clear list of who is to do what next. Such conference notes should be standardized for easy reference and for simplicity of preparation. The same recording procedure should be followed for all phone conversations in which a project is discussed.

All record keeping should follow several basic rules:

1. Keep it simple: Make your record-keeping system easy to use, easy to store, and easy to refer to after weeks or months of inactivity. Do not clutter your files with extraneous materials.
2. Keep it clear: Eliminate all unnecessary clutter from each piece of correspondence and from all official forms. If voluminous materials or detailed discussions need to be conveyed, present them as accompanying reports. Use clear, simple language and leave no doubt as to the purpose of the letter and the desired response and/or timing required, if appropriate.
3. Keep it professional: Remember that your records may be subpoenaed at any time. Think of any sentence of any letter being read aloud in a court (perhaps out of context). This is the best single editorial tool when reviewing any written material.
4. Make no personal remarks: Personal remarks do not belong in professional records. Again, think of any statement being read aloud in court. In a letter

DOCUMENT:	______	
	______	
Specifics:	______	clear identification of the document
	______	
Date:	______	
Copy ______ of ______		
TRANSMITTAL		
To:	______	
	______	record of transmittal
Date:	______	
Project Name:	______	
Project ID:	______	
	______	
SUBMITTED BY:		
JOHN Q. DOE		applicant's name
address		
telephone		

3-16. Rubber stamp for accompanying documents.

to a client or an agency superior, do not refer to Mr. Smith as a "turkey," "idiot," or by any other such term. Simply state, for example, that Mr. Smith has been unavailable by telephone for more than two weeks, or that Mr. Smith has caused delays by repeatedly asking for answers to the same question.

5. Keep it: Do not throw away any vital part of your records. (Duplicate copies may be eliminated.) If in doubt—keep it.

Correspondence

All project correspondence should follow formal business format (see figure 3-17 for an example). The following points are particularly important:

1. A clear, simple project reference (e.g., XYZ Project, or Banyon Harbor Expansion) should be used consistently in all correspondence, along with all known agency file numbers. This information should be noted in the subject line, or reference line, of the letter. (The letter will usually be filed by this line.)
2. A simple, opening statement of the objective of the letter should be kept short. If the letter is a response to a call or letter, this should be stated, along with any information necessary to maintain continuity (e.g., "In response to your letter of 18 May, 1986, asking for additional air-quality data . . .").
3. The body of the letter should convey the desired material in as complete a manner as possible. If part of a question cannot be answered immediately, state the part, the reason for delay, and the expected answer date. Do not just

(company letterhead)

18 January, 1984

Mr. H. J. Smith
Deputy Director
State Environmental Agency
1111 State Street
Capital, State 00000

Re: ABC Redevelopment Project
State Environmental Agency file #00-00-0000-0
Corps of Engineers file # 000-0-0000

Dear Mr. Smith:

Following our telephone conversation of 12 January, I am enclosing three copies of the aerial photos you requested of the proposed ABC Redevelopment Project area.

I understand that the permit application package is now complete and that your agency will respond within 45 days.

Thank you for your assistance.

Sincerely,

A. B. Jones
Project Coordinator

enclosures

cc. with enclosure
Col. C. D. Doe
U.S. Army Corps of Engineers

3-17. Sample letter to an environmental agency.

ignore portions. The letter should leave no questions unanswered, unless requesting the answer is the object of the letter.

4. The letter should close with a clear statement of the desired result or response from the letter and any deadlines for this response.

Since the various components of a project are directed (through an agency coordinator) to different departments within an agency, it is advisable to keep each letter to one subject, the exception being a response to a multiple-question letter. In answering a letter that itself posed many questions, answer each question individually; do not combine answers. Try to refer to any sequencing given in the requesting letter—i.e., questions *a* through *j*, or *1* through *8*. If a multi-question letter requires facts to be repeated from one answer to another, repeat the information if it is short; if it is longer, simply refer to the previous answer. The key is to answer *every* question clearly. If any question appears unanswered, the permit processing may be stopped.

Consultants

A consultant is usually brought into a project by an applicant to perform a specific task. Consultants provide expertise. They may assist in administrative or technical matters, solve a complex problem, provide an unbiased opinion, supplement staff expertise, introduce new technology, or provide expert witness. Sometimes an entire consulting team, experienced in complete EIS preparation and procedures, is required (figure 3-18). The expertise and reputation consultants can bring to a project can significantly increase the chances that a permit will be granted and decrease permit preparation time in major projects. A consultant should be selected only after careful review of his or her technical qualifications, reputation, and experience with similar projects. (Technical qualifications are, of course, the first, but not necessarily the deciding, selection criterion.)

Before a consultant or consulting firm is hired, the project's objectives, scope, budget, and schedule requirements should be established (figure 3-19). All existing information, plans, and so forth should be assembled. Information about previous permits (secured or attempted) should be included in this background package.

The consultant, in addition to functioning as a technical expert, must function as part of the project team in the preparation of reports, presentations, and testimony. He or she must also agree to conform to budgets, schedules, and the level of technical standards required to assist in a cost-effective environmental evaluation and permit process for the project.

Once the consultant is selected, it is vital that open communication be maintained. The project's background, and any previous permit applications should be discussed, as should management philosophies. A clear definition of the desired "product" or end result should be established, with schedules for completion. Progress meetings with all key personnel should be scheduled. One staff member, designated as *consultant coordinator*, can integrate the work of all consultants into compatible reports and presentations.

Once a consultant's services are contracted, the consultant should be kept abreast of all project activities. The consultant should be a source of previous experiences with similar projects, and should be utilized to ensure cost-effective productivity of all activities in project permitting (Rona 1984).

Consultants are generally hired under one of three categories of contracts: per diem, multiplier, or lump sum. A *per diem* contract stipulates a per day fee,

3-18. Partial list of professional expertise commonly used in preparation or review of EIS.

Characteristics under Investigation	*Field of Expertise*
Physical Characteristics	
	Agricultural Engineering
	Air Pollution Engineering
	Air Quality Chemistry
	Biology
	Botany
	Ecology
	Geography
	Geology
	Hydrology
	Land Use Planning
	Meteorology
	Oceanography
	Pedology
	Sanitary Engineering
	Seismology
	Soils Engineering
	Water Quality Biology
	Water Quality Chemistry
	Zoology
Socioeconomic Characteristics	
	Acoustical Engineering
	Archaeology
	Civil Engineering
	Demography
	Economics
	Educational Planning
	Electrical Engineering
	Health Planning
	History
	Hydroelectric Engineering
	Land Use Design
	Land Use Planning
	Landscape Architecture
	Paleontology
	Political Science
	Recreational Planning
	Sanitary Engineering
	Sociology
	Solid Waste Management
	Structural Engineering
	Traffic Engineering
	Transportation Planning

Sources: Warden and Dagodag, 1976; and Cross and Hennigan, 1974.

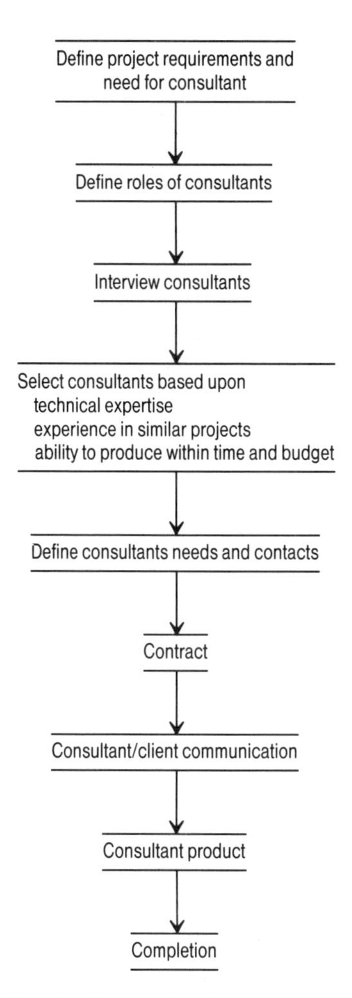

3-19. Using a consultant–flow chart.

which varies according to the type of work being performed. An expert witness, for example, usually earns a higher per diem than an individual who prepares reports. A per diem contract is generally used for individual consultants, when the scope of the project or need for the consultant is not well defined, or for short-term requirements.

Under a multiplier contract, an individual's salary is multiplied by a fixed factor to cover overhead and profit. This method is frequently used by institutions such as universities and consulting firms, where several people might be involved in a project. This method allows for hours of "technician time" or "supervisor time" to be charged to the client as they occur, rather than one per diem being charged for all personnel.

A lump sum is a fixed, negotiated fee for all defined services. It is only used when the scope of the required services is completely defined.

A consultant is used as an expert witness in a hearing or court case, to provide information and opinions based on his or her specialized skill, knowledge, and experience. The judge determines the qualification of the individual to serve as an expert witness on the basis of his or her education, fellowships and research, teaching experience, publication record, membership in learned societies, and the totality of experience in the area(s) of specialization. The precise combination of education and experience required for "expert witness" status will vary greatly with the subject matter and the judge (Rose 1983).

SOURCES OF HELP

The first source of information about any permit is the agency that issues that permit. In some municipalities (and some states) a central information bureau or clearinghouse can assist you in defining every permit you will need. In some cases these bureaus can even provide you with all of the necessary application forms and assist you in setting up preapplication meetings. This form of centralized information service is not yet available everywhere.

If you have requested information from a regulatory agency and have waited beyond a reasonable amount of time for a response, your elected representative can help you. His or her office can call the agency on the applicant's behalf and expedite the permit process. Although you as a taxpayer ultimately pay the salaries of all bureaucrats, your congressperson actually pulls the purse strings (Lesko 1983).

You can reach your congressperson by contacting the local district office (see the white pages of your phone book under "U.S. Government"), or you can contact: Coordinator, Federal Information Center, 7th and D Streets, S.W., Washington, DC 20024; telephone (202)-224-3121. If you do not know the name of your congressperson, call your local newspaper. Your state, county, and city officials can assist you in local matters.

4

Agency Review Procedures

When an application is received by an environmental agency, a logging-in procedure records the date of receipt and the items received (i.e., application forms, maps, photographs, reports, and so forth) and assigns an agency number to the project. It is important for the applicant to create a file to be used in following the progress of the application. This file should contain all of the above-mentioned information plus the names and phone numbers (with extensions) of each individual who will be involved in overseeing the permit through the review process. The applicant should ask each agency for a listing of each element of the review process and periodically check the review status. See figure 4-1 for an example of an agency checklist.

In most agencies one individual is assigned to oversee the application process. This individual may be referred to as the *coordinator*. The preapplication meeting can alert the agency to assign an individual with the appropriate area(s) of expertise to this task. However, more often the coordinator is an administrator or is assigned by geographic region as a regional coordinator. This is particularly true in federal agencies.

COMPLETENESS REVIEW

Following the logging-in procedure, the assigned agency coordinator goes through a cursory completeness review and informs the applicant of any deficiencies in the application in a letter of receipt. (It is usually this letter that first informs the applicant of the agency file number assigned to the project.) This review usually takes the form of a checklist (figure 4-2). As mentioned in chapter 3, the applicant should obtain a copy of this checklist at the preapplication meeting. This review will check that all of the necessary forms are complete and have been submitted in proper quantity, that the correct units of measurement have been used, and so forth. Certain omissions (e.g., submission of four copies of the application instead of five) will not delay the processing of the application in most agencies. More significant omissions, however, will result in delays.

Once a permit application is deemed complete, most agencies have a legislated "time clock" that requires action on the permit application in a certain

File # ______________________________

Submission date: ______________________________

Project Title (or type): ______________________________

Applicant: ______________________________

Address: ______________________________

____

Phone: ( ) ______ – ______ Ext. ______

Agent: ______________________________

Address: ______________________________

____

Phone: ( ) ______ – ______ Ext. ______

Contact: ______________________________

Submission package:

Date		Complete
______	Application forms (5 copies)	______
______	Evidence of ownership	______
______	Plans (5 copies)	______
______	Engineers' certification	______
______		______
______		______
______		______
______		______

Other Permitting Agencies

Local	Application/Permit #
______	______
______	______
______	______
State	
______	______
______	______
______	______
Federal	
______	______
______	______
______	______

4-1. Sample agency checklist for application completeness.

number of days. This "time clock" does not begin until the permit application is complete. The applicant must consider it his or her responsibility to maintain close contact with the coordinator to prevent minor gaps or omissions in paperwork from halting the review process completely. Since most environmental agencies are understaffed and overburdened, an incomplete application can be buried by a pile of new applications. Thus, the "clock" is effectively delayed from starting, and the applicant is at fault.

The requirements for a complete permit application package will vary, depending on the nature and size of the project. An example of the review criteria for a major project is given in figure 4-3. The criteria for a less complex project would be correspondingly less detailed. Figure 4-4 shows the review criteria of the U.S. Army Corps of Engineers.

COMPETENCE REVIEW

The agency coordinator distributes copies of the application to the relevant departments within the agency and sometimes to other agencies (figure 4-5). An application for canal dredging work, for instance, might go to biologists, geologists, hydrologists, road engineers, and utility specialists (to ensure that no local power, telephone, or water crossings would be disturbed). Local agencies often lack the in-house, multi-expertise review staff necessary for large or complex projects and rely on state or federal agency review capabilities. In most states the "burden of proof" of "no environmental degradation" from a project rests with the applicant. Thus, the agency review personnel do not have to prove that a project will degrade the environment: the applicant must prove it will *not*. Since proof implies factual evidence with little or no latitude for interpretation, and since there are few if any absolutes in environmental issues, this difficult task is often best performed by technical experts, whom agency personnel will respect. Of course, the magnitude of the project and the potential environmental degradation are the key issues.

The competence review takes place on two levels. First, the materials provided are evaluated objectively. Figure 4-6 presents the competence review criteria for a major project. The second level is more subjective and includes factors from the reviewers' experiences. A developer's reputation, his or her past efforts in developing adjacent properties, or a reviewer's fondness for a particular wilderness site can influence the decisions made about a project. For this reason open and continuous communication, and/or the use of environmental permit specialists as representatives of the applicant, can help prevent or identify delays caused by subjective reviews.

LEGAL REVIEW

Most agencies are empowered by legislation that specifies the nature of the work for which permits can be granted by that agency. The law may also specify activities for which permits cannot be issued unless specific conditions are met

(Text continued on page 74.)

4-2. Agency review checklist.

Submitted	Completed	Not Submitted	*Application Reference* *Subject Requirement & 40 CFR Section Nos.*	*Comments*	*Location in Application*	*Due Date*
			PART XIII—GROUNDWATER PROTECTION			
			A. Subpart F:			
			A-1 Interim status period groundwater monitoring data summary: 265.92 122.25(c)(1)			
			• EPA interim primary drinking-water standards			
			• Parameters establishing ground-water quality (chloride, iron, manganese, phenols, sodium, and sulfate)			
			• Indicator parameters for ground-water contamination (pH, specific conductance, TOC and TOH).			
			A-2 Hydrogeological survey 122.25(c)(2)			
			• Location and identification of uppermost and hydraulically interconnected aquifers under facility			
			• Depth, porosity, horizontal and vertical permeability for the aquifers and the depth to, and lithology of, the first confining bed(s)			
			• Vertical permeability, thickness, and extent of confining beds			
			• Direction and rate of ground-water flow			
			• Depth to ground water and seasonal variations			
			• Estimate of worst-case seepage rate			
			• Basis for identification (survey, published information, etc.).			
			A-3 Topographic map (general topo required under Part I-B-3 264.95 122.25(c)(3) 122.25(a)(19)			
			• Delineation of property boundary			
			• Delineation of waste management area			
			• Delineation of proposed point of compliance			
			• Proposed or existing location of ground-water monitoring wells			
			• Delineation of existing contamination plume.			

			Requirement	Citation			
			A-4 Description of existing contamination • Concentrations of each Appendix VIII constituent throughout the plume, or • Maximum concentrations of each Appendix VIII constituent in the plume.	122.25(c)(4)			
			A-5 Detailed plans and engineering reports (existing or proposed) of Ground Water Monitoring Program	264.97 122.25(c)(5)			
			Monitoring Wells				
			• Number of wells	264.97(a)			
			• Location of wells 1. Test boring location and depth 2. Evaluation of test boring 3. Represent water unaffected by leakage 4. Represent water quality at point of compliance				
			• Description of wells and casing 1. Depth 2. Backfilling test hole with impermeable material 3. Installation –casing/liner –well screen –gravel pack –other 4. Grouting or filling of annular space 5. Well development	264.97(c)			
			Sampling Methods & Procedures • Sample collection methods including purging of well prior to sampling • Sample preservation/shipment • Chain of custody control • Description of analytical procedures (Appendix III of Part 261 or appropriate procedure) including referenced documentation number • Procedure for determination of ground-water elevation with each sample • Procedure for annual determination of uppermost aquifer flow rate and direction.	264.97(d)-(f)			

Continued

4-2. Continued.

Submitted	Completed	Not Submitted	Application Reference Subject Requirement & 40 CFR Section Nos.		Comments	Location in Application	Due Date
			A-6 Description of Detection Monitoring Program	264.91(a)(4) 264.98			
			If hazardous constituents have not been detected:	122.25(c)(6)			
			• Proposed list of indicator parameters, waste constituents, and reaction products to be monitored	264.93			
			• Basis for proposed parameters: 1. Type, quantities, concentrations of constituents expected in waste 2. Mobility, stability, persistence in unsaturated zone 3. Detectability in ground water	264.98(a)			
			• Background ground-water concentrations or values, coefficients of variation of constituents in background, and procedures used to determine background values	264.98(c)			
			• If no background values are available, procedures proposed to be used to determine background and calculate values.				

			A-7 Description of Compliance Monitoring Program If hazardous constituents have been detected: • List of waste previously handled at facility • Characterization of contaminated ground water 1. hazardous constituents identified 2. hazardous constituents concentrations • List of hazardous constituents to be compliance-monitored • Proposed compliance period • Proposed concentration limits for constituents with justification.	264.91(a)(1) 264.99 122.25(c)(7)			
			A-8 Description of Corrective Action Program If the concentrations of hazardous constituents exceed established limits or 265 monitoring indicated concentrations of hazardous constituents over background: • Characterization of contaminated ground water 1. Hazardous constituents identified 2. Hazardous constituents concentrations • Concentration limit for each hazardous constituent • Detailed plans and an engineering report describing the corrective action to be taken • Time period necessary to implement the corrective action program • Description of how the monitoring program will assess the adequacy of the corrective action • Description of corrective action to be taken for constituents in ground water beyond compliance point.	264.100 122.25(c)(8)			

4-3. Completeness review criteria of a major project.

I. COMPLETENESS
- A. Project Description
  - 1. Purpose of action
    1. Identify all purposes of the project.
    2. Describe each purpose in detail.
    3. Establish criteria for measuring achievement of each purpose.
    4. Quantify the extent to which each purpose is achieved.
    5. Use innovative quantification approaches.
    6. Develop a purpose achievement profile.
  - 2. Description of action
    1. Identify and list all project activities.
    2. Describe all activities clearly and in detail.
    3. Relate each activity to the project's purposes.
    4. Provide a location map on an appropriate scale.
    5. Provide maps and sketches of all layouts and complete engineering drawings.
    6. Provide detailed engineering plans.
    7. Describe structures to be built.
    8. Provide construction details about:
       - Timing
       - Material flow
       - Equipment use
       - Construction monitoring plans
       - Safety plans
       - Labor needed
       - Services required
       - Pollution abatement procedures and hardware
    9. Give the area covered by each structure.
    10. Discuss important design features.
    11. Explain construction practices.
    12. Develop contingency plans for rare events which might occur during implementation of the project.
    13. Assess in detail the methods for shipping of materials and equipment.
  - 3. Environmental setting
    1. Describe the environmental quality prior to proposed action based upon:
       - Air
         - Particulates
         - Sulfur dioxide
         - Hydrocarbons
         - Carbon monoxide
         - Nitrogen oxide
         - Oxidants
       - Water
         - Physical/chemical quality
           - Flow pattern
           - Turbidity
         - Dissolved oxygen
         - pH
         - Phosphates
         - Hazardous substances
       - Ecology
         - Habitat
         - Diversity
         - Endangered species

I. COMPLETENESS
A. Project Description
3. Environmental setting *(continued)*

Aesthetic quality
- Appearance
- Land/water interface
- Floating materials
- Wooded shoreline

Land
- Specific parameters of land-use patterns
- Soil erosion, aesthetic quality
- Noise conditions
- Ecology

2. Describe existing federal activities and their impact upon the above parameters.
3. Discuss existing state and local activities and their impacts upon the above parameters.

B. Land-Use Relationships

1. Conformity and conflicts

1. Identify existing federal, state, regional, and local plans, policies, and controls.
2. Identify potential plans, policies, and controls.
3. Define quantitative requirements for conformity and conflict criteria.
4. Identify various conflicts and conformities using a matrix approach or project characteristics, based upon the existing and potential criteria defined above.
5. Quantify, on the basis of the above criteria, the extent of the project's conformity and conflict with:
   - Air-quality plan
   - Water-quality plan
   - Noise levels and impact on land uses
   - Transportation plan
   - Regional land-use plan
   - Community services plan
   - Other plans
6. Discuss conflicts and conformities in detail.
7. Specifically state the conflicts that need to be reconciled.

2. Conflict assessment

1. Define alternatives for conflict reconciliation.
2. Quantify the extent of conflict reconciliation for each alternative for the proposed project.
3. Systematically compare the overall conflict reconciliation for the various alternatives.
4. Select and justify the choice of alternative reconciliation.
5. Quantify the extent of incomplete reconciliation.
6. Justify incomplete conflict reconciliation.

C. Impacts

1. Positive and negative effects

1. Quantify, as well as possible, national and regional impacts of the project upon:
   - Air
     - Particulates
     - Sulfur dioxide
     - Hydrocarbons
     - Carbon monoxide
     - Nitrogen oxide
     - Oxidants

Continued

4-3. *Continued.*

I. COMPLETENESS

C. Impacts

1. Positive and negative effects *(continued)*

- Water
  - Physical/chemical quality
    - Flow pattern
    - Turbidity
    - Dissolved oxygen
    - pH
    - Phosphates
    - Hazardous substances
  - Ecology
    - Habitat
    - Diversity
    - Endangered species
    - Aesthetic quality
      - Appearance
      - Land/water interface
      - Floating materials
      - Wooded shoreline
  - Land
    - Specific parameters of land-use patterns
    - Soil erosion
    - Aesthetic quality
    - Noise
    - Ecology

4. Quantify as much as possible the potential international impacts upon the above factors.
5. Assess, evaluate, and document the impacts listed above.

2. Direct/indirect consequences

1. Quantify as much as possible, in terms of specific, measurable parameters, all primary effects upon:
   - Human health
   - Ecology
   - Vegetation
   - Materials
2. Quantify, as much as possible, in terms of specific, measurable parameters, all secondary effects upon:
   - Labor productivity
   - Food chain
   - Unemployment
   - Alienation, etc.
3. Develop specific profiles for each impact.

D. Alternatives

1. Identify a set of possible alternatives in terms of:
   - Location
   - Equipment
   - Operation procedure
   - Engineering design
2. Quantify costs of each alternative.
3. Quantify benefits of each alternative.
4. Compare benefits with costs.
5. Establish the most desirable alternatives.
6. Develop a profile for the desirable alternatives, their benefits, and their costs.

E. Unavoidable Effects

1. Identify unavoidable impacts.
2. Quantify the above impacts as well as possible.
3. Develop a profile of unavoidable impacts.

I. COMPLETENESS

E. Unavoidable Effects *(continued)*

4. Justify why the unavoidable impacts cannot be eliminated.
5. Identify mitigation alternatives.
6. Quantify, as well as possible, the costs and benefits of these alternatives.
7. Compare the benefits and costs of these alternatives.
8. Select a set of feasible alternatives.
9. Develop a profile for the unavoidable effects of these alternatives.

F. Short-Term Uses Versus Long-Term Productivity

1. Identify short-term gains and losses from the project.
2. Quantify the above gains and losses as well as possible.
3. Identify long-term gains and losses from the project.
4. Quantify the above gains and losses as well as possible.
5. Assess trade-offs between short-term gains and long-term losses.
6. Assess trade-offs between short-term losses and long-term gains.
7. Develop a profile for trade-offs.
8. Identify and describe future options without the proposed project.
9. Identify and describe options that will be foreclosed by the project.

G. Commitment of Resources

1. Identify all resources needed to construct the project.
2. Quantify all resources that will be used.
3. Quantify all resources that will cease to exist when the project is completed:
   - Labor
   - Material
   - Natural features
   - Cultural objects
4. Classify the above resources into major categories.
5. Develop a profile for irreversible commitment of resources.

H. Other Interests

1. Quantify all benefits of the proposed action based upon terms of the present and projected needs of the proponent agency.
2. Quantify all benefits of alternatives based upon the above mentioned criteria.
3. Discuss in detail the mitigation effect of major benefits.
4. Establish superiority of proposed action.

I. Discussion of Review

1. Summarize comments of other agencies.
2. Identify all points of disagreement.
3. Indicate agency's response to the issues raised and comments made by the reviewers.

II. Accuracy

A. Readability

1. Write clearly.
2. Remove all ambiguities.
3. Avoid use of technical jargon; all technical terms should be clearly explained.

Continued

4-3. *Continued.*

II. Accuracy

B. Flavor and Focus

1. Do not slant or misinterpret findings.
2. Avoid use of value-imparting adjectives or phrases.
3. Avoid confusion or mix-up between economic, environmental, and ecological impacts and productivity.
4. Avoid unsubstantiated generalities.

C. Quantification

1. Use well-defined, acceptable qualitative terms.
2. Quantify factors, effects, uses, and activities that are readily amenable to qualification.

D. Data

1. Identify all sources.
2. Use up-to-date data.
3. Use field data collection programs as necessary.
4. Use technically approved data collection procedures.
5. Give reasons for use of unofficial data.

E. Methods and Procedures

1. Use quantitative estimation procedures, techniques, and models for arrival at the best estimates.
2. Identify and describe all procedures and models used.
3. Identify sources of all judgments.
4. Use procedures and models acceptable by professional standards.

F. Professional Expertise

1. Use data collection and analysis procedures designed by recognized professional experts.
2. Identify scientists participating in the EIS preparation.
3. State qualifications of these scientists.
4. Use expert judgment for all high-level impact areas where there is uncertainty.

G. Interpretation of Findings

1. Consider and discuss all impact areas before any are dismissed as not applicable.
2. Give thorough treatment to all controversial issues, and discuss the implications of all results.
3. Consider the implications for each area of a range of outcomes having significant uncertainty.
4. Analyze each alternative in detail and give reasons for not selecting it.
5. Scrutinize and justify all interpretations, procedures, and findings that must stand up under expert professional scrutiny.

Source: Jain, Urban, and Stacey, 1977.

(e.g., "no structures shall be constructed seaward of the dune line"). The applicant's authority to develop the property is assured by some form of proof of ownership, which is usually part of the application package. These two points—ability of the agency, under the law, to issue a permit, and ownership—are the extent of the legal review typically required for permits. Occasionally, additional legal review is required where conflicting legislation, pending legislation, conflicting interpretations of legislation, or court precedents are involved.

4-4. U.S. Army Corps of Engineers permit review evaluation.

The following statement appears on the back of every public notice issued by the U.S. Army Corps of Engineers during the permit processing.

Evaluation

The decision whether to issue a permit will be based on an evaluation of the probable impact including cumulative impact of the proposed activity on the public interest. That decision will reflect the national concern for both protecting and utilization of important resources. The benefit which reasonably may be expected to accrue from the proposal must be balanced against its reasonably foreseeable detriments. All factors which may be relevant to the proposal will be considered, including cumulative impacts thereof; among those are conservation, economics, esthetics, general environmental concerns, cultural values, fish and wildlife values, flood hazards, flood plain values, land use, navigation, shoreline erosion and accretion, recreation, water supply and conservation, water quality, energy needs, safety, food production, and in general, the needs and welfare of the people. Evaluation of the impact of the activity on the public interest will also include application of the guidelines promulgated by the Administrator, EPA, under authority of Section 404(b) of the Clean Water Act or of the criteria established under authority of Section 102(a) of the Marine Protection, Research, and Sanctuaries Act of 1972 as appropriate. A permit will not be granted unless its issuance is found to be in the public interest. If this project is located within the State's coastal zone, it will have to be approved by the appropriate State agency in accordance with the State's Coastal Zone Management Plan.

Request for Public Hearing

Any person may request a public hearing. The request must be submitted in writing to the District Engineer within the designated comment period of this notice and must state the specific reasons for requesting the public hearing.

ENVIRONMENTAL CONSIDERATIONS

The evaluation of the potential effects of a project on the environment is the primary purpose of the environmental permit procedure. The review process should establish and present to the decision makers of the agency a concise summary of the project's potential positive and negative impacts. To accomplish this, agency personnel must first review the applicant's statements of potential impacts, which are based on the information provided in the application and supporting information (e.g., data reports or technical reviews), if applicable. Then, based on their own experience and the published experience of others, they must determine the potential positive and negative impacts *independent* of the applicant's statements. These impacts are considered in specific areas, such as air, land, water, biology, and aesthetics. The review should be straightforward and objective with regard to any projected physical alterations to the land, water, air, or habitat. Unfortunately, as explained in chapter 1, there are seldom clear "if *a* then *b*"-type arguments where the environment is concerned. Environmental systems are complex, and biological adaptability often depends on immeasurable factors. Biologists rely on three basic strategies in forecasting environmental effects: comparative analysis, monitoring, and limited experiment (Ortolano 1984). The comparative analysis method takes similar environmental settings and/or species and compares the effects of proposed changes. Monitoring involves taking measurements and observations over a long period in order to define characteristics of the system. The limited-experiment technique utilizes laboratory and/or field alteration of environmental factors to observe potential outcomes. These three strategies

4-5. Sample clearinghouse coordination checklist.

1. Project Title: ____

2. Lead Agency: ____ 3. Contact Person: ____

3a. Street Address: ____ 3b. City: ____

3c. County: ____ 3d. Zip: ____ 3e. Phone: ____

PROJECT LOCATION 4. County: ____ 4a. City/Community: ____

4b. Assessor's Parcel No. ____ 4c. Section ____ Twp. ____ Range ____

5a. Cross Streets: ____ 5b. For Rural, Nearest Community: ____

6. Within 2 miles: a. State Hwy # ____ b. Airports ____ c. Railways ____ d. Waterways ____

7. *DOCUMENT TYPE*

CEQA

01. ____ NCP 06. ____ NOE
02. ____ Early Cons 07. ____ NOC
03. ____ Neg Dec 08. ____ NOD
04. ____ Draft EIR
05. ____ Supplement/ Subsequent EIR
(Prior SCH No.: ____)

NEPA

09. ____ NOI 11. ____ Draft EIS
10. ____ FONSI 12. ____ EA

OTHER

13. ____ Joint Document
14. ____ Final Document
15. ____ Other ____

8. *LOCAL ACTION TYPE*

01. ____ General Plan Update
02. ____ New Element
03. ____ General Plan Amendment
04. ____ Master Plan
05. ____ Annexation
06. ____ Specific Plan
07. ____ Community Plan
08. ____ Redevelopment
09. ____ Rezone
10. ____ Land Division (Subdivision, Parcel Map, Tract Map, etc.)
11. ____ Use Permit
12. ____ Waste Management Plan
13. ____ Cancel Ag Preserve
14. ____ Other ____

9. *DEVELOPMENT TYPE*

01. ____ Residential: Units ____ Acres ____
02. ____ Office: Sq. Ft. ____ Acres ____ Employees ____
03. ____ Shopping/Commercial: Sq. Ft. ____ Acres ____ Employees ____
04. ____ Industrial: Sq. Ft. ____ Acres ____ Employees ____
05. ____ Water Facilities: MGD ____
06. ____ Transportation: Type ____
07. ____ Mining: Mineral ____
08. ____ Power: Type ____ Watts ____
09. ____ Waste Treatment: Type ____
10. OCS Related
11. ____ Other: ____

10. *TOTAL ACRES:* ____ 11. *TOTAL JOBS CREATED:* ____

12. *PROJECT ISSUES DISCUSSED IN DOCUMENT*

01. ____ Aesthetic/Visual
02. ____ Agricultural Land
03. ____ Air Quality
04. ____ Archaeological/Historical
05. ____ Coastal Zone
06. ____ Economic
07. ____ Fire Hazard
08. ____ Flooding/Drainage
09. ____ Geologic/Seismic
10. ____ Jobs/Housing Balance
11. ____ Minerals
12. ____ Noise
13. ____ Public Services
14. ____ Schools
15. ____ Septic Systems
16. ____ Sewer Capacity
17. ____ Social
18. ____ Soil Erosion
19. ____ Solid Waste
20. ____ Toxic/Hazardous
21. ____ Traffic/Circulation
22. ____ Vegetation
23. ____ Water Quality
24. ____ Water Supply
25. ____ Wetland/Riparian
26. ____ Wildlife
27. ____ Growth Inducing
28. ____ Incompatible Landuse
29. ____ Cumulative Effects
30. ____ Other ____

13. *FUNDING* (approx) Federal $ ____ State $ ____ Total $ ____

14. *PRESENT LAND USE AND ZONING:*

15. *PROJECT DESCRIPTION:*

16. SIGNATURE OF LEAD AGENCY REPRESENTATIVE: ____ DATE: ____

NOTE: Clearinghouse will assign identification numbers for all new projects. If a SCH number already exists for a project (e.g. from a Notice of Preparation or previous draft document) please fill it in.

REVIEWING AGENCIES

_______ Resources Agency
_______ Boating/Waterways
_______ Conservation
_______ Fish and Game
_______ Forestry
_______ Colorado River Board
_______ Dept. Water Resources
_______ Reclamation
_______ Parks and Rec
_______ Office of Historic Preservation
_______ Native American Heritage Comm
_______ S.F. Bay Cons. & Dev't. Comm
_______ Coastal Comm
_______ Energy Comm
_______ State Lands Comm
_______ Air Resources Board
_______ Solid Waste Mgmt Board
_______ SWRCB: Sacto
_______ RWQCB: Region #_______
_______ Water Rights
_______ Water Quality

_______ Caltrans District _______
_______ Dept. of Transportation Plng
_______ Aeronautics
_______ CA Highway Patrol
_______ Housing & Community Dev't
_______ Statewide Health Plng
_______ Health
_______ Food & Agriculture
_______ Public Utilities Comm
_______ Public Works
_______ Corrections
_______ General Services
_______ OLA
_______ Santa Monica Mtns
_______ TRPA-CALTRPA
_______ OPR–OLGA
_______ OPR–Coastal
_______ Bureau of Land Management
_______ Forest Service
_______ Other: _______________
_______ Other: _______________

FOR SCH USE ONLY

Date Received at SCH _______________
Date Review Starts _______________
Date to Agencies _______________
Date to SCH _______________
Clearance Date _______________

Catalog Number _______________
Applicant _______________
Consultant _______________
Contact _______________ Phone _______________
Address _______________

Notes: ____

____

____

____

____

____

4-6. Competence review criteria.

1. Presentation
   - Clearly written
   - Clearly organized
   - Data sources defined
   - Separation of impacts
     - environmental
     - economic

2. Contents
   - Data
     - current
     - obtained by approved procedure
     - sources identified
   - Analysis
     - current
     - models and techniques
       - best currently available
       - tested and verified
       - clearly identified and described
   - Objectivity
   - Areas of uncertainty identified
   - Alternatives presented clearly

3. Expertise
   - Identification of all participants
     - their role
     - their qualifications
     - their contribution to the final product

4. Interpretation
   - Areas of judgment in analysis clearly identified
   - No generalities
   - No unsubstantiated statements
   - Clear presentation of interpretation

5. General
   - Thorough treatment of all controversial issues

are not mutually exclusive and may be used in combination, depending on the situation. None of these techniques, alone or in combination, provides unambiguous, unquestionable results.

This is the point at which inexperienced agency personnel can cause undue delays by asking for more information. It is certainly desirable to have indisputable proof that no negative effects will result from a project or that the effects will be fully predictable; again, there are seldom absolutes. Judgments must be made on the basis of the available information, experience in similar projects, and available current scientific knowledge.

5

Agency/Applicant Interactions

In dealing with agency personnel it is important to understand their frame of reference. First, the education they bring to the review process can vary widely both in degree and kind. Many possess degrees in science (anything from anthropology to zoology), and some have degrees in law or business; some have graduate degrees. Their working backgrounds may vary as well, from none to many years in industry, although outside experience is rare owing to the fact that agency pay scales are usually lower than those in industry. Because of this pay-scale discrepancy, a large percentage of permit-agency personnel are usually fresh out of school and stay in an agency only long enough to develop sufficient experience to move on to industry: hence, the common conception of permit agencies as having a "revolving door." Their permit processing experience may range from small, simple projects to large, complex ones. This wide variety in training and experience may affect their attitudes toward development and developers. Inexperienced personnel may have personal biases and preconceived notions that are not based on facts. More experienced personnel may harbor memories of similar projects and the "headaches" that went with them. All too often agency personnel lose sight of the fact that the ultimate objective of all permits, and the laws that govern them, is to protect the environment while allowing reasonable development and growth to continue, and allowing an individual the right to utilize privately owned property.

Agency personnel in review positions are typically biologists or engineers. Decision-level personnel frequently have legal or administrative backgrounds. Thus, early communication with high-level agency personnel may give no indication of the viewpoints, attitudes, or experience of review personnel, and vice versa. When this form of interagency background difference is suspected, it is advisable to request early meetings with multilevel agency staff present. Although review criteria should be unambiguous, it cannot be overstated that environmental issues are seldom black and white. The background and experience of agency review personnel will have a significant effect on the review process. This is quite different from the building code, electrical codes, etc., which have created workable (albeit bureaucratic bogs) that carry unambiguous technical security of a project's permits.

Applicants vary too, from inexperienced to experienced: from homeowners seeking to repair a backyard dock, to developers seeking to create an entire community, to an industry seeking to build a power plant. Projects more complex than the simple individual application for small development or property improvement often employ an agent to obtain permits. While the individual applicant is usually a novice, the agent should be expert at the preparation of permit documents and have an understanding of agency requirements and experience in dealing with agency procedures and personnel. While the bulk of agency correspondence is directed to the novice applicant, it is the agent applicant, handling larger, more complex projects, that consumes the most agency-personnel time.

DEALING WITH REQUESTS FOR ADDITIONAL INFORMATION

Once a permit application has been received and reviewed, various agency personnel may request additional information from the applicant. These requests should be consolidated and funneled through the agency-applicant contact (or coordinator) assigned at the preapplication meeting. (For smaller projects, the contact should be established when the application is submitted.) If these requests come directly through the various review personnel, a chaotic chain of communications, miscommunications, and even rumors may ensue. Once the agency personnel determine that more information is required, the "time clock" which runs for most applications is stopped. It is therefore in the applicant's best interest to respond to the request immediately.

The information requested varies, from additional copies of submitted materials such as aerial photographs, to detailed studies and reports on particular aspects of the project. Study and report requests are usually preliminarily defined in the preapplication process and may already be initiated, and occasionally prepared, when the request is received.

Applicants' responses to such requests also vary, depending on the request, from total compliance to refusal. The latter stalemates the process, requiring the intervention of high-level agency personnel, and sometimes lawyers. If requests seem unreasonable, it is advisable to hold additional meetings with agency personnel. Often, in the process of standardizing or consolidating communications, the information request is distorted. Before refusing to comply with a request, the applicant should obtain clarification of the request and the reasoning behind it *in writing,* to prevent time and money from being wasted. Close communication can often prevent misunderstandings and delays.

If, after all of the requested information has been submitted, secondary requests are generated, this is clearly a time to meet with agency administration to determine the cause. Sometimes such secondary requests result from project revisions made during the permit process. These revisions may be applicant- or agency-generated; nevertheless, any changes that alter the basic information already submitted will require updated data to all agencies involved in the permit-granting process.

Repeated requests for additional information indicate serious problems on the part of the applicant and/or the agency. Either the applicant does not understand the nature of the request, or the agency is not communicating its needs properly, or both. Occasionally, repeated requests are caused by agency personnel who are using applicants to perform their research for advance degrees or publications. On rare occasions, agency personnel make these requests to avoid decision making. Remember, the "time clock" is stopped while the agency awaits applicant fulfillment of requests. Under any circumstances, these requests cause delays both in the permit process and in getting the project under way. What is more, they often cost the applicant considerable sums in consulting fees. They should thus be avoided if at all possible.

It is not uncommon for an applicant to feel that he or she is being manipulated by an agency. Neither is it uncommon for agency personnel to feel that an applicant is trying to "get away with something." In fact, each party has an obligation to communicate clearly and effectively to the other. If either or both parties fail to meet this obligation, delays will be inevitable.

DEALING WITH THE INTENT TO DENY

Most permit-granting procedures include a process for issuing an *Intent to Deny.* This is a letter issued to the applicant before the permit application is actually denied. In effect, it means that while the agency has formally declared that it will not issue the permit as submitted, the files are not yet closed. Once an Intent to Deny is issued, it becomes the applicant's responsibility to change the agency personnel's opinion of the project or alter the proposed plans to conform to agency standards. In reality, however, once issued, it is seldom possible to have the decision reversed. The Intent to Deny usually indicates that the agency personnel do not consider the project to be within the agency's standards for air quality, water quality, or other habitat considerations. Occasionally, the difficulties are legal ones, concerning such matters as mandated preserve areas or erosion control lines. To reverse an Intent to Deny, the applicant must show that the agency's conclusions were based on incorrect premises, or in some way present a strong argument as to why an exception should be granted for a particular project.

An Intent to Deny generally contains a statement as to why it was issued (e.g., the applicant will exceed federal air pollution standards, or the project will have a detrimental effect on the downstream water quality). If the applicant wishes to appeal the decision, it is then his or her responsibility to answer these specific concerns. Direct technical refutation usually requires the use of experts.

The applicant may proceed to an appeal process. The appeal for many environmental agencies is a technical hearing in which the case is presented to a judge, who will listen to both agency and applicant presentations and arguments and render a decision based on fact and law. This decision will be given to the agency as a recommended action, and the agency head will then make the final decision to issue or deny the permit.

A technical hearing is useful if the applicant feels the technical merits of the application have been inadequately reviewed or assessed by the agency personnel, or if special technical considerations or controversies are present which merit special evaluation and consideration. A project may employ new technologies, methodologies, or theories that agency personnel are reluctant to accept. If the applicant feels his or her experts can successfully argue before the hearing officer, a technical hearing may remove the stumbling block in the permit process. The technical hearing will be discussed in greater detail in chapter 7.

BARGAINING

Once an agency indicates that denial of the application in its present form is imminent, a bargaining process usually begins in earnest. Bargaining may take two forms: alteration of the proposed project to eliminate the offensive portions, or mitigation (e.g., agreeing to replace a damaged habitat).

At this time the economics of the alternatives must be carefully considered. The project is usually far enough along in planning and investment of both time and money, that giving up the project would be costly. However, alternatives are usually costly too. It is easy for the applicant, pressured by the threat of permit denials, to agree to any and all proposed alternatives. This is a dangerous route. If the alternative chosen is later found to be economically or technically infeasible, then a great deal of time and money will have been irretrievably lost. If the special requirements are not met, costly and time-consuming court battles may ensue. Thus, great care must be taken in the bargaining sessions to consider alternatives. They must be evaluated and accepted or rejected on the basis of economic and logistic feasibility.

The personalities of agency personnel are evident at this time more than at any other. Certain individuals will use this opportunity to exert their "power" over developers. Others will work closely to find acceptable solutions. This is the area where agency personnel's typical lack of industrial experience will cause the greatest conflicts. Their inexperience in dealing with economically feasible solutions can place a strain on communication and working relations between applicant and agency.

As at all other times, the applicant should strive to maintain open communication with agency personnel. During bargaining sessions, work with the highest-level personnel in the agency to ensure that the options under consideration can indeed be approved by the agency. Keep records of bargaining meetings, showing which options have been considered and the reasoning behind the dismissal of any options. If more than one option is available as a solution to a deadlock, establish a priority and work with the agency personnel to achieve a workable solution.

6

Preparation of Reports and Assessments

Reports and assessments are documents produced in support of an environmental permit application. They may take numerous forms and be produced at any time during the permit process (figure 6-1). Reports are typically produced after the preapplication meeting or the initial agency review addressing specific issues. They are usually written to accompany or complement permit applications, elaborating on specific points or giving background data. Assessments are more formal than reports, with formats and/or contents specified by legislation. The National Environmental Policy Act (NEPA) identifies four key elements of an environmental impact assessment: impact indentification, measurement, interpretation, and communication (Canter 1977). The term *assessment* will here be used exclusively to describe such well-defined documents as the federal Environmental Impact Statement (EIS) and the regional Development of Regional Impact/Application for Development Approval (DRI/ADA).

Both the EIS and the DRI/ADA are all-inclusive assessments of a project's impact. The economic, social, and environmental impacts are all presented for evaluation. Neither the EIS nor the DRI/ADA is a permit application, but either may be required by governmental agencies before permits or approvals are granted or other actions (such as zoning changes or funding) are taken. Their format and contents are largely defined by legislation. They are typically expensive and time consuming to produce because of their broad technical range and the detail (and hence research) required. These are often unwieldy documents, requiring several pounds of paper per copy. Their efficient production usually requires the talents of a knowledgeable coordinator and numerous specialists.

Report formats vary, depending on the ultimate use to which the information will be put. The purpose of the report—a clear presentation of the information to be conveyed and the prospective audience—must be precisely defined before report writing begins. The ultimate uses and audiences must be known to prevent costly rewrites. If it is anticipated that an EIS or DRI/ADA will be required on a project, preliminary reports, or reports required for state and

6-1. Sample supplemental reports by project type.

PROJECT	*A*	*B*	*C*	*D*	*E*	*F*	*G*	*H*	*I*	*J*
Airports, heliports, and landing strips	X	X	X	X	X	X	X	X	?	
Batch plants	X	X	X	X	X	X	X	X	?	
Bulk storage	X	X	X	X	X	X			?	
Commercial developments on three or more acres	X	X	X	X		X	X		?	
Commercial forest-products removal		X	X				X	?	?	
Construction in stream channels	X	X	X	X		X			?	
Developed campgrounds	X	X	X	X	X	X	X		?	
Educational facilities	X	X	X	X	X	X	X	?	?	
Electrical substations and power plants	X	X	X	X	X		X		?	
Fish and wildlife	X		X						?	
Forest management programs	X	X	X					?		
Golf courses	X	X	X	X	X			X	?	
Government buildings	X	X	X	X	X	X	X	?	?	
Harbors and marinas	X	X	?				X		X	
Highways, roads, and structures	X	X	X	X	X	X		X	?	
Medical facilities	X	X	X	X	X	X	X	?	?	
Mobile-home parks	X	X	X	X	X	X	X	X	?	
Motels and apartments with 5 or more units	X	X	X	X	X		X	X	?	
Multiperson dwellings	X	X	X	X	X	X	X		?	
Organized recreation camp	X	X	X	X	X		X	?	?	
Overhead or underground utilities	X	X	X	X						
Parking lots	X	X	X	X	?	X		X	?	
Piers and shoreline construction	X	X							X	
Pumping stations	X	X	X	X			X		?	
Quarries	X	X	X	X	X	X		X	?	
Radio, TV, and telephone relay stations	X	X	X	X						
Sewage and water-treatment plants	X	X	X	X	X	X	X	X	?	
Skiing facilities	X	X	X	X	X	X		X		
Solid-waste transfer stations	X	X	X	X	X		X		?	
Stations and transmission lines and structures	X	X	X	X			X		?	
Stream crossing	X	X	X	X		X				
Subdivisions	X		X	X	X	X	X	X	?	
Variances	X	X	?	?	X			?	?	
Water storage tanks and reservoirs	X	X	X	X	X	X		X		
Wrecking yards	X	X	X	X	X	X	X		?	

Source: California Office of Planning and Research, 1984.

A. Environmental Impact Statement
B. Plot Plan
C. Grading and Slope Stabilization Report
D. Vegetation Preservation and Protection Report
E. Slope Analysis Map
F. Storm Drainage and Hydrology Report
G. Fire Protection Report
H. Soil and Geology Report
I. Shoreline Report
J. Land Capability Report
X = required
? = requirement depends on project

local agencies, should conform in content, if not in style, to the EIS format. If it is anticipated that a less rigorous format will ultimately be required, time and funds should not be wasted preparing a report that meets the strict requirements of an EIS. Instead, the focus should be on communicating specific environmental issues.

Professionalism in the preparation and presentation of a document is essential. Each portion must be technically impeccable, since poor quality in any portion, no matter how small or seemingly insignificant, casts a shadow of doubt upon the entire document. The integrity and objectivity of all material will always be questioned. An applicant hires a consultant or expert to prepare these documents, and it is always assumed (even if incorrectly) that the benefit of any doubt is given to whomever pays for the work. Charges of "environmental prostitution" are common, and the technical experts who contribute to documentation must therefore be carefully chosen for their expertise, reputation, and integrity.

Document statements concerning probable environmental impacts (positive and negative) must be authoritative. Rash statements made by businessmen and lawyers have been known to create additional problems when dealing with agency personnel and environmental "watchdog" groups.

Simplicity and brevity are the fundamental characteristics of a readable, usable document. For simplicity, the use of appendices to convey technical information affords the opportunity to use referenced, summary-type statements in the document proper. This method also permits alteration or updating of individual parts of the document without the major effort and expense of total rewrites, reproduction, and redistribution.

The clarity of a report or assessment and its appendices is a crucial ingredient to successful communication. Clearly organized, well-presented, well-documented reports can provide sufficient supporting information for reviewing-agency personnel to make decisions regarding a project. Excessive use of jargon creates confusion unless a glossary is included. The inclusion of a glossary is typically beyond the scope of such reports, however, and is usually unproductive.

TYPES OF REPORTS/ASSESSMENTS

Reports and assessments required as part of the environmental permit process generally fall into three categories: data reports; technical literature and technical expert reviews; and prescribed format/content statements (e.g., EISs).

Data Reports

Data reports are collections of various data concerning the site or project in question, with minimal prose. Data reports are frequently composed of site maps, aerial photographs, and information such as lists of plants found on a site, animals observed over a given time period at the site, water parameters (such as salinity and dissolved oxygen) at various locations (and possibly depths) in a water body, or air-quality measurements over a period of time at a location. An example of a table of contents for a data report is presented in figure 6-2. The writing contained in the data report should be restricted to statements about data collection and/or correlation techniques and the conditions under which the data were collected. It is important that interpretation *not* be included in a data report: it is an objective document that can provide the basis for decision making, and the inclusion of any interpretation compromises the integrity of the entire document.

Technical accuracy and clarity are essential to the integrity of a data report.

6-2. Sample data report table of contents.

Testing must be done by standard methods and preferably by an independent laboratory. Each testing method should be outlined or referenced in the report. Species lists should be compiled by qualified individuals. (They are frequently compiled by individuals with minimal training, under the supervision of experts.) If there is any doubt as to the acceptability of a testing technique or an individual's qualification to perform as a technical expert, the environmental agencies can usually assist. They cannot, however, be expected to do the testing or to recommend a particular individual, except in very rare cases.

Technical Literature Reviews

Both technical literature and technical expert reviews are subjective and should therefore be presented as separate reports, independent of background documentation or data reports. Technical literature reviews are made subjective by the selection of literature that supports the viewpoint of the reviewer. This selectiveness may be subconscious, but it is a factor nonetheless. Even when the review is done by an independent organization, it is generally assumed that the opinions of the party paying for the review will affect the review. It is for this reason that the technical literature reviews should be written and compiled in a report independent of a data report.

Literature surveys try to serve as proving grounds for efforts to be included as part of a permit application. For example, if a garage is planned as part of a development, a survey of air-pollution literature might provide the following information: (a) the level of air pollution anticipated from a given number of automobiles; (b) the effect of climate on that level; (c) design effects on pollutant dispersion; (d) documentation of existing similar facilities; (e) standard pre- and postconstruction testing techniques.

Literature surveys take several forms. The simplest form is a bibliography. This lists references on a specific topic, giving title, author, and publication information. The annotated bibliography, presenting a short synopsis (often the author's abstract or summary) of each work, is a more elaborate form of literature survey. The literature survey is most often presented in the form of a paper (a readable report, not a list) organized by topics, summarizing the technical state of the art on a subject, with appropriate references. The bibliography becomes the reference source for confirmation by agency reviewers. Since most published technical works are reviewed prior to publication (thus attesting to the technical soundness of the relevant techniques or results), the literature survey, reporting the work of numerous professionals in the field, can be a useful method of bringing multiple expertise to a project.

Technical Expert Reviews

Expert reviews are written by professionals who are engaged in a given area of work and are recognized as experts in their fields. The experts' observations and evaluations of factors pertaining to the project can serve as both an education and evaluation tool for the agency reviewers. Using an expert combines the benefits of the literature and data surveys, since the expert's opinion is based on intimate knowledge of his/her field and a review of all the applicable data on the project. The expert can draw on his or her experience with similar projects and offer professional judgments about the project. While the criteria for expert status vary, academic degrees, years of experience in the field, and publications are the primary indicators.

Technical experts typically work in one field of study and often specialize in a very narrow branch of that study. For example, while technical experts on marine ecology are available, most have concentrated in such specific areas as marine invertebrates or algae. Some even specialize in a single species. Generally, the permit applicant who begins a search for technical expertise begins with a generalist, and specialists are added as needed.

If you need to locate specific technical experts, an excellent starting point is often professional societies. They can assist in locating one of their members with the special expertise needed. Experts are sometimes found in the environmental departments of corporations. Universities can also assist in locating experts; however, they will have a vested interest in using their staff, and they may not necessarily direct you to the best person for your needs. The best reference to a technical expert is a referral from someone who has gone through a similar permit process to the one in which you are engaged. Agencies cannot specify an expert to assist an applicant. They can, however, provide a list of two or more experts they have worked with, or have heard of, who might be contacted for further information.

An expert's review of a given set of data or an environmental issue can be of immense assistance in the permit review process. Agency personnel are seldom experts in all fields, and they respect the judgments of knowledgeable professionals. Like technical literature reviews, however, these judgments are subjective and should be separated from data reports, which should be strictly objective.

When a project involves more than one field of expertise, it is advantageous to have separate experts prepare free-standing reports. These reports can all reference the same data report as a source of field data on the site. They can also be consolidated into final reports or assessments as appendices. This technique allows independent reviews and independent rewrites or updates should they prove necessary. The "separate-report" technique is often preferred by reviewing agencies who can then assign staff "experts" to review particular sections. Also, each expert's contribution is more easily acknowledged, thus adding to the credibility of each review.

Development of Regional Impact (DRI)

The Development of Regional Impact Application for Development Approval (DRI/ADA) (this may go under different names but the concept will remain similar) is a regionally required assessment that varies greatly in scope and format from location to location. In some areas no DRI/ADAs or their equivalents are required. This assessment is prepared to support a project in a regional review. Often, such reviews are required on major projects prior to the granting of local permits or even zoning approvals. The DRI/ADA is usually required by a multi-city or multi-county agency set up to oversee the growth and development of a region such as a valley, drainage basin, bay region, or watershed.

The DRI is a review of all aspects of planning and land use for a project. It attempts to forecast the overall impacts of a development (i.e., on traffic, schools, drainage, water supply, sewerage, fire service, and so on) in terms of the region's needs and capacities. Thus, it considers the rules, regulations, and needs of more than one municipality. It usually considers a physical region with unique environmental considerations (e.g., a watershed or bay) in a more complete manner, and sometimes according to stricter environmental evaluation criteria, than could be offered by a single municipality, but with more site-specific criteria than could be offered by the state. The general content of a DRI/ADA is outlined in figure 6-3. The DRI/ADA usually employs so-called threshold criteria (figure 6-4). Any project with portions that exceed these criteria (for example, dock space for over 99 boats; parking for more than 150 cars) is automatically required to submit to the DRI/ADA process.

Planned Unit Development (PUD)

The planned unit development (PUD) is a review process many communities use in evaluating large projects such as housing tracts and shopping centers.

6-3. Sample outline of a DRI/ADA.

PART I

Provide the following general information about the proposed development. Any part of this information which is not relevant to a particular type of development may be omitted. A brief statement of the reasons for omitting any of the following information must be included.

A. Identify the Development.
   1. Name of development, developer, and authorized agent.
   2. Previous DRI development name(s) and developer(s), of project site.
   3. Distance of the development to the nearest adjacent county.
   4. Legal description (section, township, and range).
   5. Type of development pursuant to state code.
   6. Present ownership of property.
   7. Property owned, optioned or leased by the developer adjacent to or within one-half mile of the proposed site (location and acreage).

B. Describe the Project.
   1. Total acreage and proposed breakdown by land use (e.g. commercial, residential, open space, water bodies, etc.).
   2. Present and proposed land use and zoning.
   3. Number of units, gross square footage, parking spaces, etc. by development type pursuant to state code.
   4. Gross and net residential density.
   5. Proposed phasing of project by time increments, including the amount of development per phase.
   6. Project population at buildout and date of buildout.

C. Describe the Regulatory Status.
   1. Types of permits applied for, received, required or denied (including zoning).
   2. Permitting agency(ies) and status of pending permits.

D. Provide the Following Visual Exhibits:
   1. General location map.
   2. Proposed site plan/preliminary layout of the site.
   3. Existing land use map of the area.

E. Provide any other information which you believe should be considered in this determination.

PART II

The Department will request the following information, or portions thereof, if it is needed in order to complete the binding letter of interpretation. To expedite the binding letter review process, it is recommended that the applicant submit information pertaining to those resources identified in Part II which could be substantially affected by the proposed development.

A. Describe the Potential Impacts on Natural Resources.
   1. Air Quality:
      a. Source, amount and type of pollution.
      b. Project location in relation to attainment or non-attainment air quality areas.

Continued

6-3. *Continued.*

2. Water Quality:
   a. Potential impacts to existing water bodies receiving discharge from the project and their current state water quality classification.
   b. Potential impacts to ground water from pollution, run-off, or salt water intrusion.
3. Site Clearance–Type and amount of vegetation to be disturbed.
4. Impervious Surfaces–Type and amount (e.g. paved surfaces, structural, etc.)
5. Ground and Surface Water:
   a. Estimated depth to aquifer.
   b. Identify the location of existing surface water bodies contained on site or adjacent to the site.
6. Surface Drainage:
   a. Potential impacts to predevelopment drainage conditions.
   b. Identify the type of storm event that will be controlled at post development, impacts to any regional drainage system (e.g. canals, river basins, etc.), the point(s) of discharge, and the level of treatment to be provided prior to discharge.
7. Wetlands–The location and size of wetlands on site or affected by the proposed development and method or source of wetland identification. Provide a site plan indicating the boundary of wetland areas.
8. 100-Year Flood Plain–Percent of land area in flood plain and specify the percent of project in a velocity zone. Provide a site plan indicating the boundary of flood plain areas.
9. Dredge and Fill–Amount required and location, including spoil disposal areas. Provide a site plan indicating the boundary of dredge and/or fill areas.
10. Natural Vegetation–Specific types found on site.
11. Endangered or Threatened Plant and Animal Species–Number and species, and indicate if species have been found on site or if probable habitat exists.

B. Describe the Potential Impacts on Historical or Archaeological Sites–Indicate the number and type either on-site or in close proximity. Identify which portions of the development will impact such sites and what actions will be used to mitigate their disturbance. The state Department of State, Division of Archives, History and Records Management may be contacted by the applicant to assist in providing this information.

C. Describe the Potential Impacts on Public Facilities.
1. Sewage Flow:
   a. Gallons per day by land use.
   b. Level and type of treatment.
   c. On-site or off-site discharge location.
   d. Available capacity of regional or local treatment facility (if applicable).
2. a. Potable Water Supply–gallons per day by land use, source, and the available capacity of the source.
   b. Non-Potable Water Supply–gallons per day by land use, source, and availability of supply.
3. Solid Waste Generated:
   a. Pounds per day by land use, disposal method.
   b. The available capacity of a sanitary land fill or other disposal facility.
4. Transportation
   a. Provide trip generation rates, trip distribution, and source of transportation assumptions.
   b. Provide an analysis of traffic assumptions by completing Table 1 and/or Table II. (See Attachments)
   c. Describe any proposed improvements to the existing road system and provide legally binding documentation that these improvements are scheduled for completion at buildout of the proposed development.
   d. In preparing the traffic analysis please adhere to the following assumptions:
      - Traffic impact including level of service, average daily trips or peakhour trips.
      - Traffic impact for 100% occupancy, at buildout, and phased should the development schedule extend five years or more.
      - The traffic analysis should cover an area within a radius of not less than a half mile and not greater than five miles from the project site.
      - Service volumes or roadway capacities must be derived using acceptable DOT methodology or definitions.

e. Expected increases in water borne traffic, aircraft operations, or other transportation considerations (if applicable).

5. Daily Energy Demand–Source of supply and describe any alternative energy plans.
6. Available Fire and Police Protection including response time to the site.
7. Available Health and Education Facilities including proximity to site.
8. Disaster Preparedness, Civil Defense, and Evacuation Plans.

D. Describe the Potential Impacts on the Economy of the Region.

1. Subsidiary development and/or capital improvements expected to be generated by this proposal.
2. Employment opportunities generated or affected by this proposal.
3. Other economic impacts of the proposed plan.

Source: Adapted from instructions for DRI application from the South Florida Regional Planning Council.

Passed by local ordinances, each PUD process has different criteria for evaluation. PUDs, like DRIs, attempt to review all aspects of a development. They are "total package" reviews that include master plans, community land-use plans, subdivision regulations, sign and landscape codes, and municipal services.

Environmental considerations are part of the PUD "package"; however, the form and degree of the environmental portion of the PUD vary widely. Environmental considerations do not originate PUD ordinances as they do DRIs. As with any legislation, the factors that bring about the legislation most often determine the areas of greatest review emphasis.

Environmental Impact Statement (EIS)

In 1969 the National Environmental Policy Act (NEPA) was passed to require all federal government agencies to integrate environmental concerns into their planning and actions. NEPA created the Council on Environmental Quality (CEQ) and the Environmental Impact Statement (EIS). A breakdown of the contents of the NEPA is given in figure 6-5. The purpose of an EIS is to provide a consistent process for and record of an agency's efforts to formulate and evaluate actions in terms of their environmental consequences (Ortolano 1984). The EIS itself is a document designed to provide information to the personnel who make decisions about or issue approvals for a project. It describes the project, outlines its probable effects, and identifies alternatives and potential mitigation measures (figures 6-6 and 6-7).

The CEQ is in the Executive Office of the President and is charged with implementation of national environmental policy. The CEQ guidelines are policy statements about EIS preparation, not specific details. (The guidelines for EIS preparation are included in Appendix F.) Each federal agency must create its own detailed EIS preparation specifications for all EISs prepared within its jurisdiction. An example of an EIS guide from EPA is found in the *Federal Register* (Volume 40, Number 70, 14 April, 1975; see Appendix H).

The CEQ first designed its guidelines as nondiscretionary standards for

6-4. Development of regional impact–threshold criteria.

Type of Development	*Development of Regional Impact Threshold*	*Dade County Development of County Impact Threshold*
Airports	Any new airport, runway, or new runway extension	–
Attraction and Recreation	Expansion or Construction: 1. Single-performance facilities a. 2,500 parking spaces, *or* b. 10,000 permanent seats 2. Serial-performance facilities a. 1,000 parking spaces, *or* b. 4,000 permanent seats Any new parimutuel wagering facility *or* an expansion which increases parking or seats by 10%	1,000 parking spaces *or* 20 acres
Electrical Generation Facilities and Transmission Lines	Deleted in 1980	–
Hospitals	600 beds	–
Industrial Plants and Industrial Parks	1,500 parking spaces, *or* 1 sq. mile (640 acres)	500 parking spaces, *or* 50 acres
Mining Operations	Removal of or disturbance of over-burden over an area, whether or not contiguous, greater than 100 acres or whose consumption of water would exceed 3 million gallons per day	–
Office Parks	30 acres *or* 300,000 sq. ft. of gross floor area	10 acres *or* 125,000 sq. ft. gross *or* 1,000 parking spaces
Petroleum Storage Facilities	Located within 1,000 ft. of any navigable waterway with a storage capacity of 50,000 barrels, *or* storage capacity for 200,000 barrels	–
Port Facilities	Any water port and any marina over 99 slips	–
Residential	County Population No. Units (×1,000) <25 250 25–50 500 50–100 750 100–250 1,000 250–500 2,000 >500 3,000	250 units *or* 30 acres
Schools	Public or private postsecondary campus for a design population of more than 3,000 full-time equivalent students, *or* Expansion of a facility having a 3,000-FTE design population by 20%	–
Shopping Centers	40+ acres, *or* 400,000 sq. ft. of gross floor area, *or* 2,500 parking spaces	10 acres, *or* 100,000 sq. ft. of gross floor area
Mixed Use	–	160 acres *or* 250 dwelling units
Areawide Impact Development	–	Any development application or zoning request that has a significant impact on the natural or human resources of Dade County

6-5. Principal sections of the NEPA.

Declaration of Policy
  Section 101(A) Broad Policy Statement
  Section 101(B) Federal Government's Responsibilities

Provisions of Policy
  Section 102(2)(A) Interdisciplinary Approach
  Section 102(2)(B) Development of Methods and Procedures
  Section 102(2)(C) Prepare Environmental Impact Statements

Conformation of All Government Agencies
  Section 103 Mandates Review and Compliance of All Agencies with the Intent of This Law

Creation of the Council on Environmental Quality
  Section 201 Defines the President's Annual Report
  Section 202 Creates the Council on Environmental Quality
  Section 203 Defines the Council's Employees
  Section 204 Defines the Duties of the Council
  Section 205 Defines the Council Interactions
  Sections 206–209 Define Internal Logistics for the Council

6-6. EIS preparation process.

Action	*Product*
Organize	
Scope	
Make assignments	
Collect information	Data bank
Collate information and write draft	Draft EIS
Environmental setting	1. Environmental setting
Plan description	2. Plan description
Analyze and Write	
Environmental impact of project	3. Environmental impacts
Impacts of "no action"	4. "No action" impacts
Determine and Write	
Mitigation needs/techniques	5. Mitigation measures
Answer topical questions	6. Topical questions
Document sources and references	7. Sources and references
Coordinate review	
Incorporate review comments and prepare final EIS	Final EIS

Source: After Corwin and Heffernan, 1975.

agency decision making. Agencies and courts viewed these guidelines as being subject to a wide range of interpretation; indeed, more than seventy different sets of agency regulations were generated. The resulting inconsistencies in agency policies and practices impeded federal coordination and caused delays, duplication, and excess paperwork.

The present CEQ regulations have been refined and are binding on all federal agencies and provide uniform standards for conducting environmental reviews.

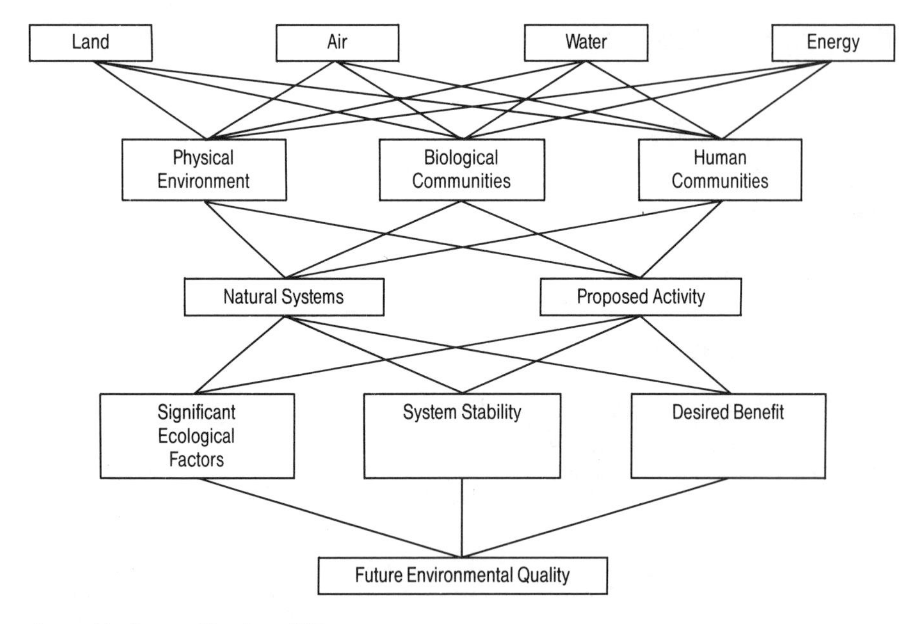

Source: After Cross and Hennigan, 1974.

6-7. EIS interactions.

Individual agency requirements still exist, but the degree of variation has been greatly reduced.

The CEQ regulations view the treatment of alternative action evaluations as the "heart" of the environmental impact statement. A comparative analysis of the consequences of the proposed action and the "no-action" alternative must be performed and reported in detail. The no-action evaluation is unique to the EIS process. The effect of the project *not* being undertaken is evaluated in as much detail as the potential effects of the project. If done properly, this is the most important phase in understanding a given environment and the human/environment interaction.

An EIS must include the following information:

1. Detailed description of the project and the existing environmental conditions
2. Environmental impacts of proposed action
3. Alternatives to proposed action
4. Unavoidable adverse impacts of proposed action
5. Evaluation of short- and long-term consequences
6. Evaluation of irreversible and irretrievable commitments of resources involved

The EIS is an intragovernmental process required of federal agencies to evaluate major or environmentally significant activities such as:

1. Agency proposals for legislation
2. Decisions of policy, procedure, or regulation

3. Projects and continuing activities such as federal permits, leases, licenses, certification; federal agency projects; and federal agency-funded projects

Thus, nongovernmental permit applications sometimes entail an EIS process. An agency's decision to issue or deny permits may have environmental consequences that must be evaluated under legislated mandate by an EIS.

Some states have adopted EIS processes for state actions which virtually ensure that an applicant for any major action requiring a permit in those states will be subject to the EIS process. State-level programs are often referred to as *mini-EIS* programs.

In general, the contents of an EIS may be outlined as in figure 6-8. The specific requirements vary and are found in the guidelines of each agency. The

6-8. Contents of an EIS.

1. PROJECT DESCRIPTION
   a. Purpose of action
   b. Description of action
      (1) Name
      (2) Summary of activities
   c. Environmental setting
      (1) Environment prior to proposed action
      (2) Future environment without the proposed action
      (3) Other related federal activities

2. DOCUMENTATION

3. LAND-USE RELATIONSHIPS
   a. Conformity or conflict with other land-use plans, policies, and controls
      (1) Federal
      (2) State
      (4) Regional
      (3) Local
   b. Conflicts and/or inconsistent land-use plans
      (1) Extent of reconciliation
      (2) Reasons for proceeding with action

4. PROBABLE IMPACT OF THE PROPOSED ACTION ON THE ENVIRONMENT
   a. Positive and negative effects
      (1) National and international environment
      (2) Environmental factors
      (3) Impact of proposed action
   b. Direct and indirect consequences
      (1) Primary effects
      (2) Secondary effects

5. ALTERNATIVES TO THE PROPOSED ACTION
   a. Alternatives available to the applicant
   b. Alternatives available to EPA
   c. Alternatives available to other permitting agencies
   d. Identification of reasonable alternative actions
      (1) Those that might enhance environmental quality
      (2) Those that might avoid some or all adverse effects

Continued

6-8. *Continued.*

e. Analysis of alternatives
   (1) Benefits
   (2) Costs
   (3) Risks
f. Possible mitigation

6. PROBABLE ADVERSE ENVIRONMENTAL EFFECTS THAT CANNOT BE AVOIDED
   a. Adverse and unavoidable impacts
   b. How adverse impacts will be mitigated

7. RELATIONSHIP BETWEEN LOCAL SHORT-TERM USES OF MAN'S ENVIRONMENT AND THE MAINTENANCE AND ENHANCEMENT OF LONG-TERM PRODUCTIVITY
   a. Trade-off between short-term environmental gains at expense of long-term losses
   b. Trade-off between long-term environmental gains at expense of short-term losses
   c. Extent to which proposed action forecloses future options

8. IRREVERSIBLE AND IRRETRIEVABLE COMMITMENTS OF RESOURCES
   a. Unavoidable impacts irreversibly curtailing the range of potential uses of the environment
      (1) Labor
      (2) Materials
      (3) Natural
      (4) Cultural

9. OTHER INTERESTS AND CONSIDERATIONS OF FEDERAL POLICY THAT OFFSET THE ADVERSE ENVIRONMENTAL EFFECTS OF THE PROPOSED ACTION
   a. Benefits of proposed action
   b. Benefits of alternatives

10. COORDINATION
   a. Agency objections and suggestions
      (1) Federal agencies
      (2) State agencies
      (3) Regional agencies
      (4) Local agencies
   b. Public objections and suggestions

11. LIST OF PREPARERS

Source: After Federal Register *Vol. 44, No. 216; and Jain, et al., 1977.*

details of such trivia as the cover sheet are specified in the *Federal Register* (see, for example, figures 6-9 through 6-11). Since the environment is divided into physical-chemical, biological, cultural, and socioeconomic parameters, the contents of an EIS must be prepared by an interdisciplinary team.

The NEPA states that the EIS must be prepared by the responsible federal agency involved, called the *lead agency.* This lead agency supervises all aspects of EIS preparation, from the original survey of the potential topics to be reviewed (commonly called "scoping"), through intragovernmental comment and distribution channels, to completion. Originally an agency could have an environmental impact statement prepared by a contractor. The lead agency selected and oversaw the contractor in all aspects of EIS production. The

ENVIRONMENTAL IMPACT
STATEMENT

____

____

(Describe title of project plan and give identifying number)

Prepared by: ____________________
(Responsible Agency Office)

Approved by: ____________________
(Responsible Agency Official)

____
(Date)

Source: EPA Final Regulations, Federal Register, *Vol. 40, No. 72, April 14, 1975.*

6-9. Cover sheet format for an EIS (EPA).

selection of a contractor was made with technical as well as "conflict of interest" considerations. Now, if a private (nongovernmental) applicant is preparing an EIS (as with a permit application), the applicant may obtain information from any source. The agency will then independently evaluate the information, since it is the agency that is still ultimately responsible for its accuracy. The use of professional environmental management teams can greatly reduce the cost of producing an EIS since it is usually a large document that lends itself to

A. Identify Project.

Name of Applicant: ______

Address: ______

Project Number: ______

B. Summarize Assessment.

1. Brief description of project: ______

2. Probable impact of the project on the environment: ______

3. Any probable adverse environmental effects that cannot be avoided: ______

4. Alternatives considered, with evaluation of each: ______

5. Relationship between local short-term uses of man's environment and maintenance and enhancement of long-term productivity: ______

6. Steps to minimize harm to the environment: ______

7. Any irreversible and irretrievable commitment of resources: ______

8. Public objections to project, if any, and their resolution: ______

9. Agencies consulted about the project: ______

State representative's name: ______

Local representative's name: ______

Other: ______

6-10. Environment impact appraisal—suggested format (EPA).

C. Reasons for concluding there will be no significant impacts: ________________

____

____
(Signature of appropriate official)
(Date)

Source: EPA Final Regulations, Federal Register, *Vol. 40, No. 72, April 14, 1975.*

"boilerplate" production techniques. (Professional environmental teams can combine the expertise of many professionals, including engineers, biologists, geologists, hydrologists, and atmospheric specialists.) An EIS must take multidisciplinary considerations into account. This requires not only technical expertise but the interfacing of that expertise with a system of reporting, analyzing, and determining alternatives and their relative effects. This makes the use of professional teams, who have worked together before, very economical. A section of NEPA (1502.17) requires the listing of all preparers of an EIS. The list must include a description of the qualifications and professional disciplines of each preparer. This ensures accountability and documents interdisciplinary participation.

Both the EIS and the DRI/ADA require public hearings and comments from various agencies prior to the final evaluation and approvals. (Public hearings will be discussed in detail in chapter 7.) Since these assessments are prepared only for larger projects, or those with significant probable impacts, these hearings can be significant events. For this reason, public forums and meetings with local environmental and neighborhood groups are often part of the preparation process: they serve to create a spirit of cooperation and can help avoid confrontation at public hearings.

SCOPING

Scoping is a relatively new concept in the environmental permit procedure. The scoping process identifies public and agency concerns in a proposed action, defines environmental issues and alternatives to be examined in the EIS process, and also defines nonsignificant issues. An effective scoping process will focus the EIS, making it a valuable document that efficiently covers relevant, significant issues and reduces the possibility that such issues will be overlooked. Scoping also identifies all relevant procedures (such as state and local permit interactions). The goal of the scoping process is to make preparation of the EIS more efficient and economical. Scoping helps agencies decide what the central issues are, how long the EIS will be, and how the responsibility for preparing the EIS will be allocated among the lead agencies and cooperating agencies.

(Check one)
( ) Draft
( ) Final

ENVIRONMENTAL PROTECTION AGENCY

____

(Responsible Agency Office)

1. Name of action. (Check one)
   ( ) Administrative action
   ( ) Legislative action
2. Brief description of action indicating what states (and counties) are particularly affected.
3. Summary of environmental impact and adverse environmental effects.
4. List alternatives considered.
5. a. (for draft statements) List all Federal, State, and local agencies and other sources from which comments have been requested.
   b. (for final statements) List all Federal, State, and local agencies and other sources from which written comments have been received.
6. Dates draft statement and final statement made available to Council on Environmental Quality and public.

Source: EPA Final Regulations, Federal Register, *Vol. 40, No. 72, April 14, 1975.*

6-11. Summary sheet for an EIS (EPA).

While scoping can facilitate any application for a permit, it is particularly useful (and well defined) in the EIS procedure, because of the latter's cumbersome, all-encompassing mandates. The formal scoping mechanism for agencies is established in section 1501.7 of NEPA. In developing this section, the Council on Environmental Quality tried to balance the benefits of scoping in federal decision making against potential significant disruptions to existing procedures. (Disruptions were sometimes caused because the scoping process did not interface with other agency "time tables." Also, formal scoping can involve public meetings at too early in the planning stages: this is costly and time-consuming.) The section leaves important elements of scoping to agency discretion.

Some object to the formal scoping process, since compliance with the provision in every case would be time-consuming and would inevitably lead to legal challenges by citizens and private organizations with objections to the agency's methods of conducting the process. As with other aspects of the environmental permit process, the policies and procedures are evolving with each new set of guidelines issued in response to experiences and court decisions.

TIERING

Tiering is a process of preparing overview portions of an environmental impact statement. The agency decides when tiering will be used. When done properly, tiering ensures that broad program-level environmental analysis is not duplicated for site-specific project reviews. The process is particularly useful for projects that are repeated frequently, such as housing projects on military bases.

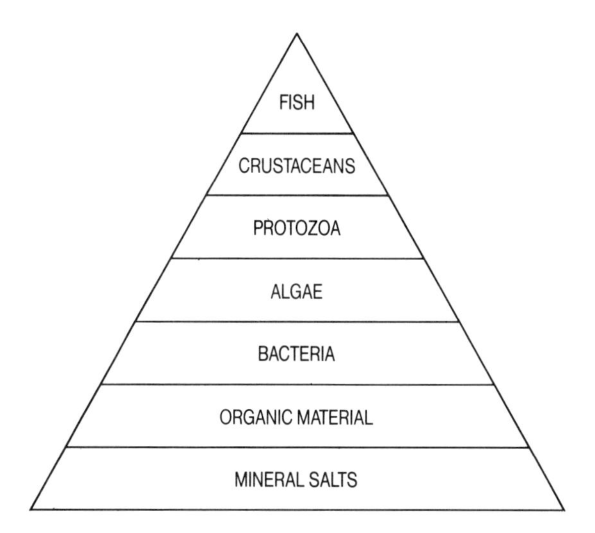

Source: After Sarnoff, 1971.

6-12. Example of an ecological community.

METHODOLOGIES

Environmental impact analysis is considered by some to be more an art than a science (Warner et al. 1974). Since there are no universally applicable procedures for conducting an unambiguous, complete analysis, various methodologies have been developed to assist in organizing and establishing priorities for the environmental assessment process.

The methodology used must relate the multiple levels within an ecological community (figure 6-12) to a wide variety of considerations—environmental, social, economic, and so on (figure 6-13). It must predict and present for assessment the impacts of a project and of the various alternatives under consideration. The following factors must be addressed:

Definition of Impact
Impact Determination
Impact Allocation
Measurement of Impact
Transformation of Scale
Rank Ordering/Weighing
Impact Integration
Sensitivity/Validation
Treating Uncertainty (Golden et al. 1974)

The methodologies used in impact statement preparation have been developed by borrowing assessment tools from other disciplines. No single methodology is best for all environmental assessments. Each has advantages and shortcomings, and each situation must be evaluated in terms of the needs to be fulfilled and

6-13. Assessment techniques.

Techniques	*Model Elements*								
	1	2	3	4	5	6	7	8	9
Averaging				X		X	X		
Checklist	X								
Clustering		X							
Consensus				X	X	X			
Cost/Benefit		X				X			
Decision Tree									X
Delphi Technique				X	X	X			
Demand Level				X					
Direct Measurement				X					
Dominance							X		
Equivalence				X					
Expert Judgment	X	X	X	X	X				
Factor Analysis					X			X	
Goal Structuring	X	X							
Imputed Worth				X		X			
Indices				X	X		X		
Lattice Theory		X							
Lexicographic Ranking							X		
Mapping							X		
Matrix, Allocation			X						
Matrix, Utility						X			
Moot Courts	X								
Network	X								
Objective Function							X		
Overlay							X		
Pairwise Comparison				X		X			
Parametric Variation								X	
Probability Distribution									X
Public Meetings	X								
Regression Techniques								X	X
Relevance Techniques			X			X			
Relevance Trees	X	X							
Scenarios									X
Simulation								X	
Survey Techniques	X	X	X	X		X			
Threshold Testing								X	
Trend Extrapolation				X					

Source: After Golden, Ovellette, Saari, and Cheremisinoff, 1979.

Elements:
1 Definition of Impact
2 Impact Determination
3 Impact Allocation
4 Measurement of Impact
5 Transformation of Scale
6 Rank Ordering/Weighing
7 Impact Integration
8 Sensitivity/Validation
9 Treating Uncertainty

the potential impacts to be evaluated. The criteria for selecting the appropriate methodology include the following:

Comprehensiveness
Specificity
Flexibility

Workability
Portrayability
Expandability
Explicitness of Criteria
Holism
Separation of Effects
Commensurability
Availability of Data
General Acceptance of Methodology
Public Involvement
Alternative Evaluation
Timing and Duration
Magnitude of Impacts
Objectivity
Risk Analysis (Golden et al. 1974; Warner et al. 1974; and Canter 1977)

The methodologies employed in environmental impact analysis include:

The *overlays.* A map of the evaluated characteristics of environmental consideration (biology, water, economics, etc.) is prepared on transparent materials, forming a composite that enables decision makers to view the overall situation.

The *matrix.* A list of project activities and a list of environmental attributes/impacts are related in a matrix to identify cause-and-effect relationships within a project.

The *network.* A cause-and-effect line is established for segments of project activities. Often, a set of cause-and-effect lines is produced to identify the effect of alternatives.

The *checklist.* This is a listing of environmental parameters to be investigated for potential impacts. This technique is used as an early guide for preliminary evaluations.

The *ad hoc method.* This is a general evaluation of areas of impact with no specific parameters defined or investigated.

Impact analysis methodologies should provide a standardized reference for evaluating both the absolute and relative impacts of alternatives. The National Environmental Policy Act (NEPA) identifies four key elements of an environmental impact assessment: impact identification, measurement, interpretation, and communication to information users. Within these four criteria, Warner et al. (1974) have created a set of twenty criteria for a methodology, which is presented in figure 6-14. Figures 6-15 through 6-19 show examples of methodologies.

The process of modeling any environmental system involves eight major elements:

Definition of systems: The system is described in terms of its component parts, each of which must be set forth in sufficient detail to allow for the process that follows.

Identification of impacts: Each potential impact that can be anticipated is identified and categorized. This includes individuals and/or classes of events or

6-14. Criteria for evaluating a methodology.

IMPACT IDENTIFICATION
1. Comprehensiveness
2. Specificity
3. Isolate Project Impacts
4. Timing and Duration
5. Data Sources

IMPACT MEASUREMENT
6. Explicit Indicators
7. Magnitude
8. Objectivity

IMPACT INTERPRETATION
9. Significance
10. Explicit Criteria
11. Uncertainty
12. Risk
13. Alternatives Comparison
14. Aggregation
15. Public Involvement

IMPACT COMMUNICATIONS
16. Affected Parties
17. Setting Description
18. Summary Format
19. Key Issues
20. NEPA Compliance

Source: Warner and Preston, 1974.

changes and areas of activities. The impacts, both positive and negative, are then determined on a component-by-component basis so that various alternatives can be compared.

Measurement of impact: The potential impacts are made more concrete through research into and comparison with the impacts of similar projects. This process entails the selection of specific, measurable characteristics to be used as indices of impact.

Uniform scale: Impacts must be reduced to common terms. Physical scales of time, distance, volume, and the like are readily determined. The critical *transformation* to common scales of values, cost equivalents, imputed worth, and so on is more difficult and is often the crux of debate as to the worth of a model.

Rank ordering: The relative weight to be attached to each portion of the impact analysis, including values such as health versus growth, is determined.

Impact integration: The weighted impacts are integrated in summary form, yielding a decision that assists output, usually supported by statistics or graphic techniques.

Validation: The model is compared with real projects and/or environments to check its accuracy and sensitivity to aspects of real (not projected) impacts of real projects.

Summary: The results of the model (and its limitations) are documented.

INSTRUCTIONS

1. Identify all actions (located across the top of the matrix) that are part of the proposed project.
2. Under each of the proposed actions, place a slash at the intersection with each item on the side of the matrix if an impact is possible.
3. Having completed the matrix, in the upper left-hand corner of each box with a slash, place a number from 1 to 10, which indicates the MAGNITUDE of the possible impact; 10 represents the greatest magnitude of impact and 1, the least (no zeroes). Before each number place + if the impact would be beneficial. In the lower right-hand corner of the box place a number from 1 to 10, which indicates the IMPORTANCE of the possible impact (e.g. regional vs. local); 10 represents the greatest importance and 1, the least (no zeroes).
4. The text which accompanies the matrix should be a discussion of the significant impacts, those columns and rows with large numbers of boxes marked and individual boxes with the larger numbers.

SAMPLE MATRIX

	a	b	c	d	e
a		2/1			8/5
b		7/2	8/8	3/1	9/7

PROPOSED ACTIONS CHEMICAL CHARACTERISTICS	A. MODIFICATION					
	a. Exotic flora or fauna introduction	b. Biological controls	c. Modification of habitat	d. Alteration of ground cover	e. Alteration of ground-water hydrology	f. Alteration of drainage
1. Earth a. Mineral resources						
b. Construction material						
c. Soils						
d. Land form						
e. Force fields and background radiation						
f. Unique physical features						
2. Water a. Surface						
b. Ocean						
c. Underground						
d. Quality						
e. Temperature						
f. Recharge						
g. Snow, ice, and permafrost						

Source: United States Geological Survey.

6-15. Methodology example.

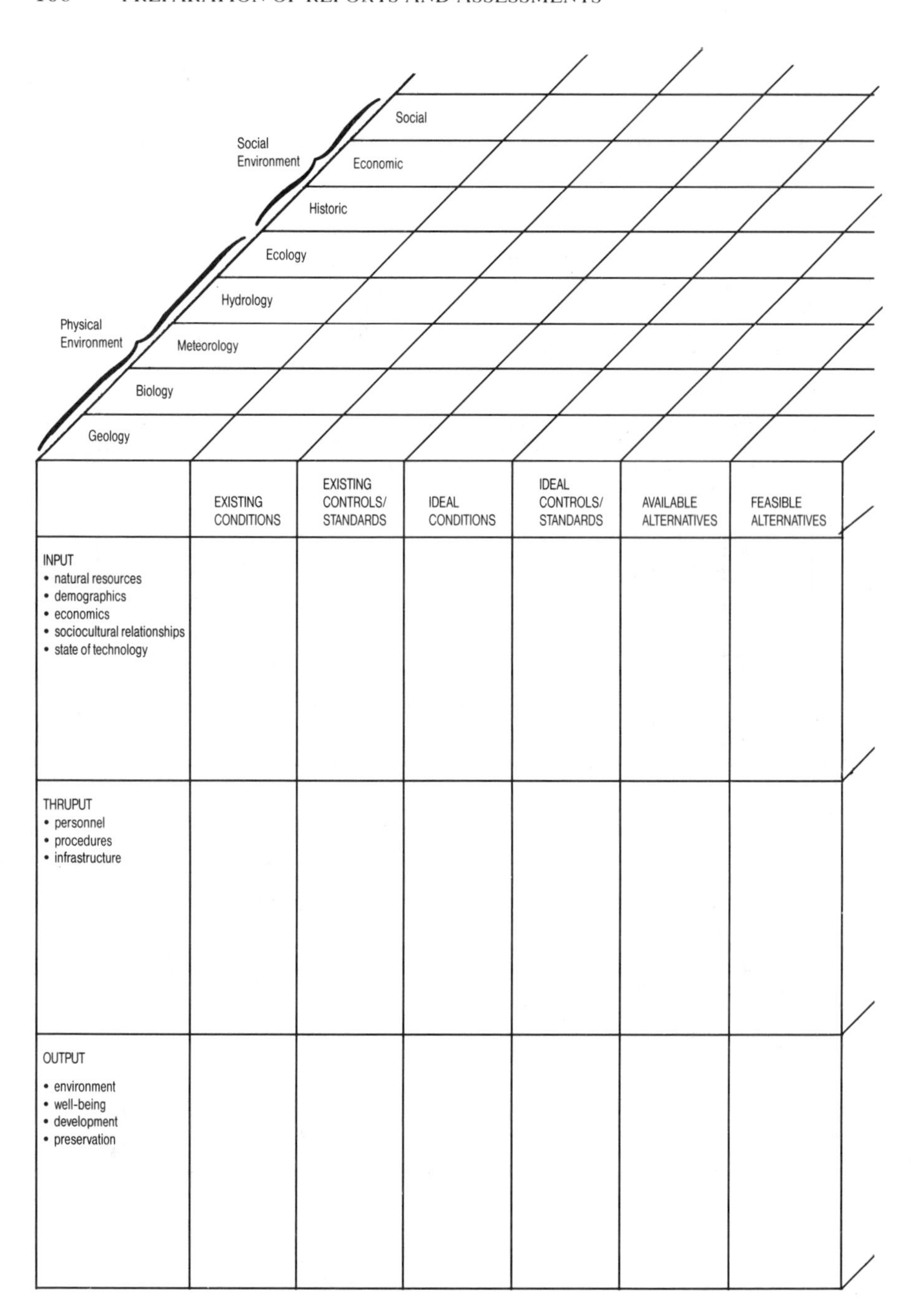

Source: Adapted from Hendricks et al., 1975.

6-16. Evaluation matrix for systematic analysis.

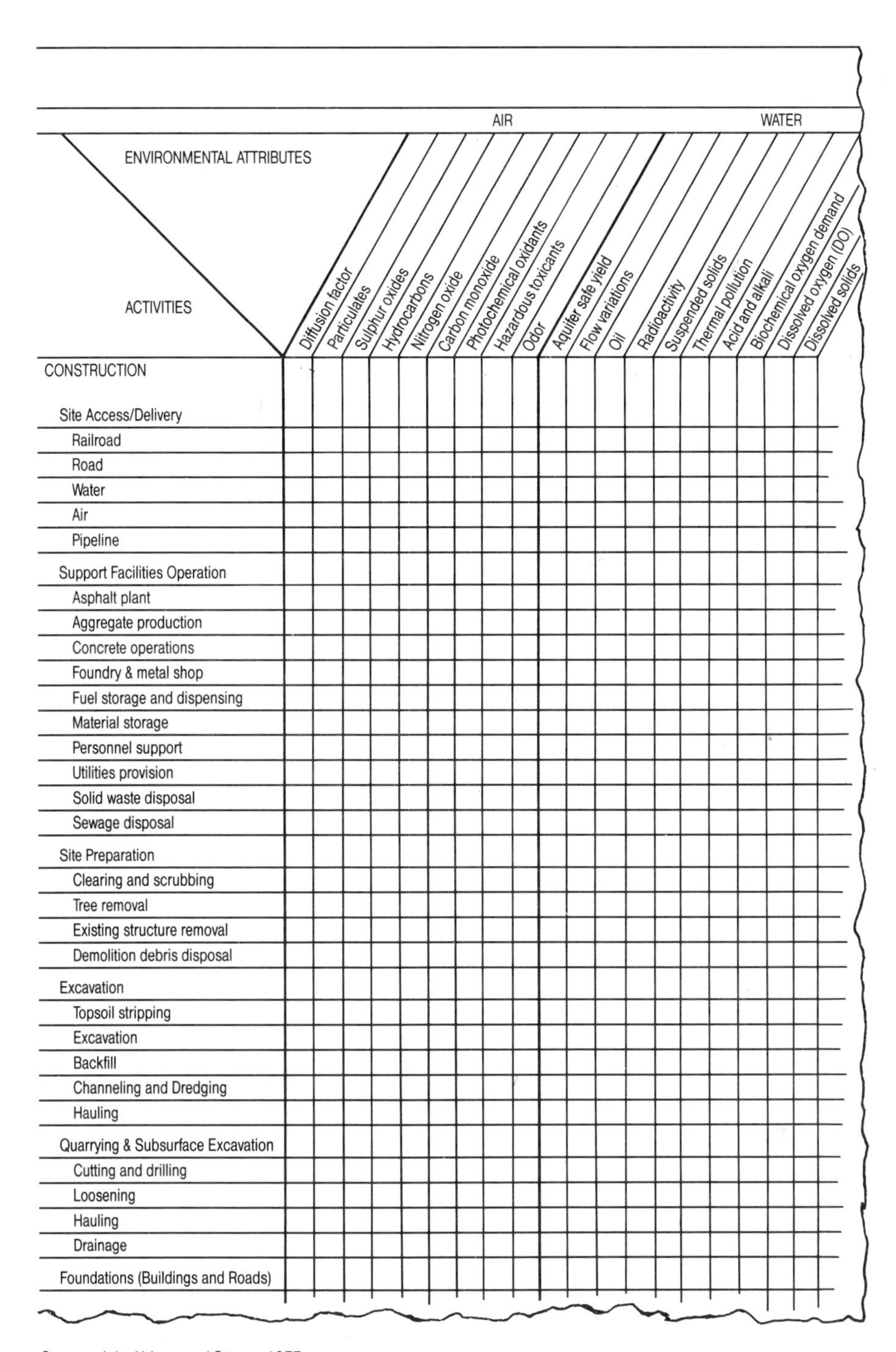

ENVIRONMENTAL ATTRIBUTES	AIR									WATER									
ACTIVITIES	Diffusion factor	Particulates	Sulphur oxides	Hydrocarbons	Nitrogen oxide	Carbon monoxide	Photochemical oxidants	Hazardous toxicants	Odor	Aquiter safe yield	Flow variations	Oil	Radioactivity	Suspended solids	Thermal pollution	Acid and alkali	Biochemical oxygen demand	Dissolved oxygen (DO)	Dissolved solids
CONSTRUCTION																			
Site Access/Delivery																			
Railroad																			
Road																			
Water																			
Air																			
Pipeline																			
Support Facilities Operation																			
Asphalt plant																			
Aggregate production																			
Concrete operations																			
Foundry & metal shop																			
Fuel storage and dispensing																			
Material storage																			
Personnel support																			
Utilities provision																			
Solid waste disposal																			
Sewage disposal																			
Site Preparation																			
Clearing and scrubbing																			
Tree removal																			
Existing structure removal																			
Demolition debris disposal																			
Excavation																			
Topsoil stripping																			
Excavation																			
Backfill																			
Channeling and Dredging																			
Hauling																			
Quarrying & Subsurface Excavation																			
Cutting and drilling																			
Loosening																			
Hauling																			
Drainage																			
Foundations (Buildings and Roads)																			

Source: Jain, Urban, and Stacey, 1977.

6-17. Example of a construction impact evaluation matrix.

6-18. Modeling of water systems.

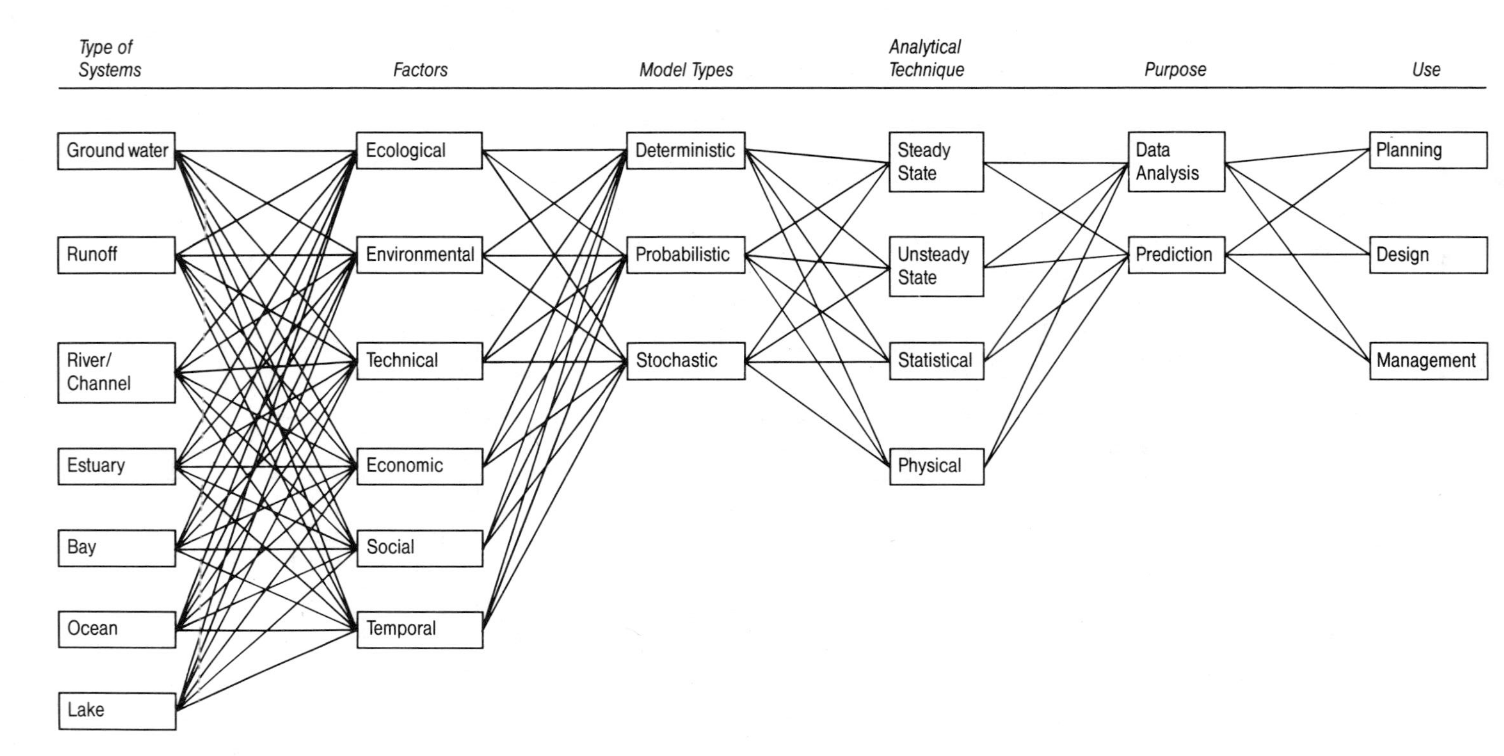

Sources: After Golden et al., 1979; and Shahane, 1976.

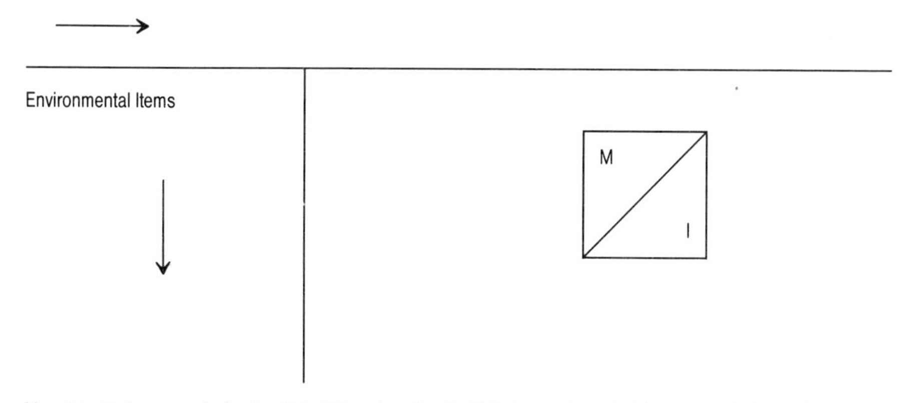

M = magnitude–numerical value (1 to 10) assigned, with 10 being an impact of large magnitude, and 1, of small magnitude.
I = importance–numerical value (1 to 10) assigned, with 10 being an important interaction, and 1 an interaction of relatively low importance.

Source: Canter, 1977.

6-19. Leopold Interaction Matrix.

All models have shortcomings. The most severe of these are the inability to treat uncertainties and the difficulties involved in creating a uniform system of assigning values to such intangibles as clean air (or a degree of pollution within the air).

The final result of an evaluation is an assessment, usually organized as shown in figure 6-20.

TIMING

The timing of report and assessment preparation and distribution must be carefully considered. Those materials that represent relatively unchanging factors, such as geology, hydrology, or historical background, may be prepared as soon as the project is conceived. However, the need for this information as well as the ultimate presentation format will vary greatly, depending on the complexity of the final product. Detailed portions of a document require all phases of the project to be clearly defined and all plans to be made final. However, the permit process, at least the preapplication meetings, should be initiated long before the final project drawings are produced. This will allow alterations based on environmental considerations to be incorporated into final designs.

This need to integrate design and permit timing can cause considerable difficulties in producing time and cost projections for large, environmentally sensitive projects.

6-20. Components of an environmental assessment.

Description
- The proposed action
- The existing environmental system
- Delineation of the environment to be affected
- Relevant environmental impact areas
- Conflicts with state, regional, or local plans and programs
- Existing controversy regarding the action

Prediction
- Significant adverse impacts that cannot be avoided
- Relationship between local short-term uses and long-term productivity of the area
- Irreversible and irretrievable commitments of resources

Evaluation
- The potential environmental impacts
- Specification and comparison of alternatives to proposed action

Source: After Golden, Ovellette, Saari, and Cheremisinoff, 1979; and Canter, 1977.

MYTHS ABOUT ENVIRONMENTAL ASSESSMENTS

In concluding this discussion of environmental assessments, it should be noted that several misconceptions about the environmental assessment process have developed into myths. These myths are held by permit applicants and administrators alike. Like other myths, they are founded in half-truths or beliefs and are held uncritically. While they may even contradict each other, these myths are hard to dispel.

Myth 1: Assessments and Evaluations Eliminate Uncertainty about the Impact of the Project

Because of the very nature of the environment, with complex variables and interdependencies that we do not fully understand, an assessment or evaluation can never eliminate all uncertainties. Certainty implies proven, fixed, dependable, reliable, indisputable, and inevitable knowledge. Few features of the physical and biological world are so well defined that total certainty can be achieved; thus, "degrees of certainty" are evoked. The degrees of certainty are improved with evaluation and research directed to answer specific questions. The circular argument remains, however: the answers can only be as good as the questions.

Myth 2: Any Environmental System Can Be Defined Completely if Enough Research Is Done

Using this argument, research could be supported forever, and all planning, development, and implementation of projects would be postponed indefinitely. Since all environmental systems are interdependent, the complete definition of any one depends on the complete definition of all. New factors are constantly being introduced into all environmental systems, however. Research performed to answer specific questions may (and often does) raise new questions and uncover influences that may require additional research. To facilitate timely solutions, lines must be drawn somewhere and the best available information used to make decisions. In an easily measured factor, such as noise pollution, the levels change as technology changes, and our understanding of effects becomes more sophisticated every year.

Myth 3: Any Scientific Study Contributes to Better Decision Making

Scientific studies address specific questions of interest to the scientific disciplines involved in the study. Most scientific studies yield results that are inherently biased as a result of the methodology used in the study. The information required for the planning and management of a given region, habitat, or species involves specific spatial and temporal scales (Holling 1978). Unless a scientific study is undertaken for management purposes, it is unlikely that it will produce results that can be directly integrated into the decision-making process. Therefore, the results of a study may or may not provide directly applicable data. In some cases the analysis of data can lead different reviewers to different conclusions. While each scientific study may add to the information pool that may eventually resolve questions as yet unasked, the applicability of the existing data should be evaluated at each step of the decision-making process.

Myth 4: Each Assessment Is Unique

Each assessment is unique in its integration of information, but each factor within the assessment (e.g., species, geology, economic interactions, watershed character) has probably been previously identified, studied, modeled, and managed. In some cases similar environmental types, habitats, or associations may have been dealt with in great detail in prior assessments. Identifying and utilizing appropriate existing information and experience will yield timely and useful assessments, minimizing the requirements for data gathering and analysis or tailoring data-gathering procedures for maximum effectiveness.

Myth 5: All Possible Impacts Must Be Fully Considered in an Environmental Study

Every impact is an effect, be it positive, negative, long-term, short-term, large, small, of known or unknown cause, that can be associated with the performance (or nonperformance) of an activity or event. The sheer magnitude of numbers involved in relating all physical, biological, and socioeconomic factors is staggering. Evaluating all of the foreseeable impacts of a project thus becomes an art form: one must consider only those aspects that are known to have impacts in a form that has measurable value. However, evaluating known impacts, however complex, is simple compared to the challenge of evaluating the unforeseen impacts that always exist. "The interesting question is rather: What does the fact that it is impossible to foresee all (or even most) of the impacts imply for the structure of the basic development plan and assessment research?" (Holling 1978).

Myth 6: Pollutant Levels or Standards Are Inviolate Numbers that Define a System

Any measurement reflects the technology available to perform it. The relative merit of that measurement within a natural environment is dependent on the level of understanding of the system and its response to changes (natural and manmade) over time and the sensitivity of the measurement to monitor those changes. It becomes clear that since we are in the primitive stages of understanding and measuring environmental parameters, standards must be set and altered as our understanding and measurement technology improve.

Myth 7: Systems Analysis Will Allow Effective Selection of the Best Alternative from Proposed Plans

Even if models could predict accurately the results of every alternative (modeling is not yet that sophisticated), assigning values to the result of each alternative would be a highly subjective process. The value of a resource within a given environment cannot be uniformly specified, nor has anyone yet come up with a method to do so. What is the value of a clam? A sixty-year-old oak tree? A pond? A job? The oxygen produced by a field? What are the costs of lung cancer? Aesthetic compensation? Loss of a species? Loss of a job?

7

Hearings

PUBLIC HEARINGS

Public hearings are a vital part of most environmental permit application procedures. They are required in the statutes that empower the environmental agencies. In most cases a public hearing is a standard part of the permit procedure. An exception, however, is the U.S. Army Corps of Engineers' public notice process. The district engineer can set a hearing date immediately and state it in a public notice, or he can send a notice and wait to see if public demand warrants a hearing. The public notice with a short description of the project and drawings is circulated to agencies and individuals on the mailing list (anyone can request to be put on the district's mailing list). Unless the notice specifies that a public hearing will be held, the public has a specified amount of time (the comment period) to request a hearing, with specific reasons, in writing. The district engineer will determine if the request represents a valid interest and set a hearing date as required.

While public hearings are a method of informing the public about a project, their primary objective is to acquire information or evidence that will be considered in the permit application, evaluation, and review. However, prehearing notices and files, available in public offices prior to the hearing, ensure that the interested public can obtain details of the proposed project and be prepared to make presentations at hearings.

The public hearing provides an opportunity to review the application status (i.e., whether the application is complete, whether an intent to approve or deny has been established by agency personnel, etc.) for the general public and to receive the public's views, opinions, and information on the project for consideration prior to final agency action.

Public hearings have a relatively standard cast of characters and a "script" required by law. The degree of formality of the hearing may alter the titles of the cast. Every hearing, however, will include a hearing officer, a recorder, agency representatives, applicant representatives, and members of the general public.

The hearing officer is usually a staff member of an environmental or related agency. The hearing officer directs the hearing and should show no bias. It is his or her responsibility to allow for all comments to be presented in an orderly manner, such that they may be used in the final agency decision-making process. The recorder varies from an official court reporter to a tape recorder, depending on the formality of the hearing. Representatives from other permit-granting agencies, and those that only comment to permit-granting agencies, may or may not be present, depending on the nature of the project and their agency's degree of interest. The permit applicant will usually be present or represented by an agent. If technical environmental questions are anticipated, the applicant will sometimes bring technical experts to a hearing. The general public that attend a hearing may be neighbors of a proposed project site or concerned citizens from any location. Local environmental organizations often send representatives to attend all public hearings in their area, to act as "environmental watchdogs." For projects of significant environmental impact, national environmental organizations may send representatives. The public may also bring "expert witnesses" if it wishes.

Procedures vary for the conduct of public hearings, but they generally include an introduction by the hearing officer, statements by the applicant and agency representatives, and public comments. Occasionally, the applicant is permitted to respond to questions from the public in some portion of the hearing. Agencies usually schedule hearings for many projects in a given area on the same day. While this grouping of hearings reduces agency expenses such as travel and recorder expenses, it may impose strict time limitations on each project's presentation and subsequent discussion. Multiple-agency hearings are also held when the purposes, projects, and required formats are amenable.

The introduction by the hearing officer will identify the project being considered, outline the permit procedure, and explain how the input from the hearing will be used. The program of the hearing will then be defined. Time limits are usually specified to allow everyone who wishes to make comments to do so. Also, the process of identifying each speaker for the record (orally or in writing) will be specified. Some hearings will require any speaker to rise and state his or her full name; others require submission of full name and address in writing before making comments.

Applicant statements at public hearings are optional and are specified by agency policy. A short statement that describes the project, outlines applicant responsibilities, and expresses concern for the local environment may be accompanied by an explanation of the procedure(s) proposed to prevent or mitigate environmental degradation (if any). This is done to anticipate and lessen public apprehension and negative comments about a project.

Representatives of other agencies are typically given time in the hearing program to present statements about the project, their evaluations, recommendations, and information about their permit procedures and status, if applicable.

The comments elicited in public hearings typically range from emotional recollections of how pristine the environment was, and how it will never be so again if the proposed project is approved, to factually based descriptions of potential environmental degradation. Positive statements about the project are also received in the hearing. Members of the public frequently attempt to comment on other projects in the area. The hearing officer must keep the comments directed to the potential agency action in the application under discussion.

Specific environmental impacts identified in the comments of the public during a hearing are taken into consideration by the agency personnel in the decision-making process prior to approval or denial of a permit application. Negative impacts are often presented to the applicant for response prior to the final decision. It is rare, however, that the public hearing identifies areas not previously considered by the agencies and/or the applicant. More often, local concerns are incorporated into the conditions of a permit approval. These concerns include historic site preservation, vegetation restoration, work-hour limitation, creation of preserve location, or other mitigating factors, should the permit be approved.

The key to a successful public hearing, from the applicant's point of view, is adequate planning. An example of an applicant's checklist is given in figure 7-1.

7-1. Checklist for a public hearing.

PREPARATION

1. Call hearing officer
   a. hearing procedures (order of presentations, rebuttals, etc.)
   b. location of hearing
   c. time of hearing
   d. sequence (if other hearings will be held the same day)
   e. presentation schedule and time limits
2. Assemble applicant team to be present
   a. define representative to make presentation, if necessary
   b. define backup experts, if necessary (particularly useful if rebuttals are allowed)
3. Prepare graphics.

PRESENTATION

1. Arrive early and prepare
   a. inform hearing officer of your presence
   b. inquire about any scheduling changes
   c. set up graphics, microphones, etc.
2. Sit quietly and do not discuss the project
3. Make presentation in orderly fashion, keep to time limits, and sit down.

DURING THE HEARING

1. *Listen*
2. Take notes
3. Do not make light conversation about the project
4. If anything comes up that is worth discussing, make appointment to do so later.

REBUTTALS (if permitted)

1. Use notes from hearing and respond to specific points
2. Concentrate on specific points, not generalities
3. Follow hearing officer's instructions about time.

Presentations

Presentations by the applicant and public will be discussed in more detail here, since they are the most important preplanned presentations at a public hearing.

Applicant presentations, when made, form the basis of most of what the public knows about the project. The project description in the public hearing notice is typically very short and describes the project quite minimally, often using legal jargon. It generally consists of a one-paragraph description of the physical alterations to be performed (e.g., construct one three-story parking garage, dredge channel to ten feet, clear two acres for future development) and simple drawings. If the project design includes environmental considerations, this is often unknown to everyone but the applicant. The applicant, therefore, must first overcome the public image of the "big bad developer" and "environmental rapist." This image, of course, varies with the nature and size of the project.

The more difficult factor to handle is the "rumor factor": one resident calls another who calls another to come to the hearing. By the time the last caller hears of the project, the rumor factor has blossomed. The last individual called often comes to the hearing with multiple misconceptions that are not easily allayed by the applicant's presentation.

The applicant's presentation must be short, as the time is often limited by the hearing officer. The time limit should be obtained in advance, from the hearing officer, for proper preparation. The presentation should be accompanied by graphics, usually an aerial photograph and a project conceptual drawing, large enough to be seen by the audience at the hearing. (Again, prior contact with the hearing officer concerning the facilities to be used for the hearing and the time available for the presentation will aid in preparation.) The presentation itself should consist of a definition of the project and the identification of environmental considerations made in the planning stages (e.g., orientation of buildings for air circulation, habitat preservation, historic site preservation, or biological field-survey results).

The public attending such hearings are also subject to time requirements. However, public comments are usually presented with less preparation and fewer visual aids (unless the project is a major one). As a result, the hearing officer must often review a great deal of rambling commentary and/or scraps of supporting evidence (maps, photos, etc.). For this reason, the hearing officer will often announce a period after the hearing when the agency will be open to receipt of written comments from the public, to be included with the hearing documentation. Hearing officers will often request a speaker to submit his or her comments in writing if they are technical or lengthy.

Comments in the form of rambling recollections or questions to the agency personnel or applicant are ineffectual and do not usually influence the agency decisions. Influential comments are generally specific; e.g., "The applicant proposes to place a road through our drainage area without providing for our land to drain after the heavy September rains"; or "The applicant proposes to build a solid concrete fence that will deny access to the wooded area and pond by local wildlife." Such specific comments may be integrated directly into the review process. Agency personnel will include these comments in the final evaluation process. Applicants may decide to alter portions of a proposed project if the objections can be eliminated by minor alterations.

If a project poses major concerns, the public may attend with numerous aides, including technical experts.

It cannot be overemphasized that a public hearing is an opportunity to obtain public comment; it is *not* a forum for public debate. Agency policy will not be controlled by a public hearing. The agency decisions will never be based on results of shouting matches. Agencies exist to protect the rights of citizens. They are organized, empowered, and funded by the people. Public hearings are an important component of the permit process and should be approached by all parties in a businesslike manner.

The applicant, agency personnel, and public alike can profit by *listening* at public hearings. As simple as it might sound, this important factor is often overlooked by all parties, whose emotional states may reduce the public hearing to a shouting match. All parties attending a public hearing are presumably concerned about the environment, although their concerns are based on individual needs, desires, and standards. By listening carefully at public hearings, the applicant can learn where modifications to a project would lead to public support. The public can learn how to work with the applicant to achieve such modifications or might even find that the applicant has indeed considered particular factors and worked with that knowledge in the design of a project. Hearings tend to inspire adversarial behavior in participants. This is unfortunate since adversaries seldom develop open communication.

After the hearing, the hearing officer returns to his or her office to summarize (in a specified amount of time determined by the agency) the information in the transcripts and any written comments received after the hearing. This synthesis is then presented to the agency decision maker (usually the agency head), along with the technical reviews of the agency staff, for final decision.

TECHNICAL HEARINGS

Technical hearings, also called administrative hearings, are generated for various reasons. They may be requested by an applicant when the agency issues a statement of Intent to Deny or by other agencies or impacted individuals if approval of the application is imminent. The applicant may request a hearing after an Intent to Deny has been issued, to clarify the agency's technical basis for denial (and present technical arguments). Any additional affected parties (e.g., development offices of cities) may request a hearing after an Intent to Deny has been issued, to argue the effects of the proposed action on their well-being. In both cases, agency personnel and the affected party argue their cases and present their technical material before a hearing officer, who renders a judgment.

Technical hearings may also be generated by an agency issuing a Notice of Violation of a granted permit. The hearing provides an opportunity for the party who has been served notice to defend his or her actions or contest the penalties of the violation.

Although it is not designed to be part of the permit process, some agencies have used the technical hearing process to shift the burden of deciding whether or not to grant a permit to the hearing process. This desire to shift the responsibility for such decisions has been aggravated by the growing tendency to solve disputes by lawsuits.

Another misuse of the technical hearing is the frequent attempt to establish guidelines or rules through the hearing process. In the hearing process the decisions should be made on the basis of technical information and existing law. The decisions of a technical hearing should pertain only to the project being presented.

Technical (or administrative) hearings are generally quite formal. They are presided over by a hearing officer, who also acts as judge. This officer usually comes from a governmental legal office and not from the environmental agency. Standard courtroom procedures are followed throughout, and both parties—agency and applicant (or other party)—are represented by lawyers. Witnesses are sworn in, and technical witnesses must present credentials that satisfy all parties present. Technical witnesses may present their material and opinions either through direct questioning or through presentation with cross-examination by the opposing attorney. Supporting documents submitted as evidence may include surveys, reports, photographs, and newspaper articles.

Following the technical hearing the hearing officer submits his or her findings in the form of a recommended order. This order contains the hearing officer's determination of the facts and the law in the case presented. The agency may either accept or reject the recommendation(s). The final decision is made by the head of the agency.

8

Litigation

The law is a process for resolving disputes, rather than a set of absolutes. It is a process of human invention and involves the personalities of all relevant parties as well as the history of human decision making—its successes and failures. When it comes to environmental issues, the legal process is further hampered by the fact that these issues, as mentioned earlier, are seldom black and white. There is seldom a single correct "solution" to any given problem, but rather a set of alternatives, each with positive and negative aspects. Expert testimony does not necessarily provide the "right" answers, since there are often as many different expert opinions about a subject as there are experts. A great deal is based on prediction of the future, an inexact science. Juries can also be swayed by manner, rather than content, of presentation.

It should be noted, too, that the law is not stagnant; it is in constant flux. The judicial system is continuously altering the common law to adapt to the circumstances of modern society. The legislative system is continuously altering the law to meet state and national needs as knowledge becomes available.

A common misconception exists that all environmental conflicts pit environmentalists against businesses. In a survey of participation in environmental-disputes negotiation, the following groups were involved by the percentages indicated:

Environmental groups	33 percent
Businesses	33 percent
Local citizen groups	44 percent
Government agencies	81 percent

Only 18 percent of the cases involved businesses and environmentalists negotiating with each other (Hickerson 1984).

THE LAWS

The laws governing environmental affairs are derived from both common law and statute law. Common-law actions may be civil or criminal. Civil actions originate when someone can demonstrate that the right he or she is entitled to enjoy has been infringed upon by another. This infringement can be categorized as nuisance, negligence, trespass, or ultra-hazardous activity:

Nuisance can be defined as unreasonable and substantial interference with the use or enjoyment of one's property without actual physical trespass. Such interference might be caused by water, smoke, vibrations, germs, or noise. If a nuisance affects the general public, rather than a particular person, a suit can be brought only by the state, usually by the attorney general (Onyx Group 1974).

Negligence refers to unreasonable or imprudent conduct that causes damages. The plaintiff bears the burden of proving that the defendant was careless in causing the environmental degradation or injury.

Trespass is defined as any invasion of property. It implies an unwarranted, unlawful, or offensive intrusion. It is not necessary to prove carelessness to win a trespass action, but there is a requirement that intent be proved.

An ultra-hazardous activity is any activity not customarily performed because of inherent dangers. Anyone engaged in such an activity should be held liable without any additional evidence of wrongdoing.

Other environmental actions are based on a hodgepodge of miscellaneous theories such as:

Inverse condemnation—an unusual situation whereby activity results in deprivation of one's property or its use without prior compensation.

Antitrust action—an action based on a section of the Clayton Antitrust Act. The most famous action of this type was an antitrust suit brought because of a conspiracy by automobile manufacturers to delay the use of pollution-control devices on their products.

Maritime suits—relating to special acts and decisions governing navigable water and the high seas. Special cases that involve the polluting of water (e.g., oil spills) fall under this special body of law.

Shareholder suit—a situation whereby the stockholders bring suit against a corporation for an activity such as pollution. Minority stockholders have always had the right to sue the management of their company when company actions or activities were injurious to the value of the stock.

A suit on the grounds of nuisance is presently the most common way for a citizen to "obtain relief from noise pollution and from odors resulting from improper solid waste management" (Heer and Hagerty 1977). Negligence actions will be increasing as environmental legislation makes it easier to prove the negligence of polluting individuals and facilities (i.e., standards or minimum levels that are violated).

The only *criminal* offense committed by reason of pollution is that of public nuisance. (The distinction between public and private nuisance rests on a determination of whether the nuisance affects the rights of the public or the rights of an individual exclusively. Nearly all states have adopted broad criminal statutes covering public nuisance.) Statute law seeks to prevent or reduce pollution by creating permit procedures and instituting penalties of fine or imprisonment (McLoughlin 1972).

One suit may be brought on a number of legal grounds, even when the grounds are inconsistent or the evidence necessary to prove one "theory" undermines another. "As a rule, lawyers plead the whole world and then prove just what is going to work best at trial" (Landau and Reingold 1971).

PROCEDURES

The process involved in a legal action can be summarized as follows:

1. The injured party seeks the advice of a lawyer. The lawyer examines the complaint to establish if the law provides a remedy.
2. A period of negotiation begins. During this period a settlement is sought. If a settlement is reached, the process stops here.
3. If a settlement is not reached, a complaint is prepared. The complaint is a document that sets forth the facts. Copies of the complaint and supporting documents are served on the opposition. (The injured party is called the *plaintiff*, and the opposition is called the *defendant*.) The defendant then files a response.
4. The complaint, answer, and accompanying documents are called the *pleadings*. If the pleadings indicate that no questions of fact are at issue, a summary proceeding may decide the law and render judgment.
5. If facts are disputed, a pretrial conference is held between the attorneys and the judge to frame the issues for a trial.
6. A trial—an orderly presentation of evidence and arguments—takes place. The judge decides on issues of law (procedures, admissibility of evidence, jury instructions, etc.), while the jury decides on issues of fact. This is why technical experts are often called to give explanations and judgments that are beyond the ken of the "average laypeople" who serve as jury.
7. The decision of the jury, the *verdict*, is reached.
8. Dissatisfied parties may appeal the verdict. The factual conclusions of a jury cannot be appealed, so appeals are based on alleged errors in instructions, admission of evidence, or other rulings on points of law.

WHO MAY SUE?

The court system was designed to resolve individual issues, while the legislative system was designed to solve the problems of the general public (community to national levels). Legal actions are seldom brought by individuals, since the degree of personal rights affected must be great to justify the expense of such an action. Unfortunately, in environmental issues it is seldom an individual who is harmed but rather a segment of the population. Now even "public interest" groups, no individual members of which have been injured, are seeking court actions. The very existence of a court action depends on who is entitled to bring the action. Thus, the decision of *standing* has become a complicated issue.

Standing

Standing is defined as an individual's right to a judicial decision of a complaint. Standing must be established before a case can be decided by the courts. The existence of standing does not determine the ultimate outcome of the case. The standing is usually self-evident: an individual always has standing when he or she is injured, physically or otherwise, or instructed by law (or the government) to act or refrain from acting in a particular manner.

The determination of standing becomes complicated in issues involving the public at large or groups representing the environment. The decision to grant or deny standing to environmental groups has gone both ways in the courts. It is an area in which great flexibility is common because of the unknowns surrounding environmental issues and the law itself; indeed, some believe that "legal creativity is at its zenith here" (Landau and Rheingold 1971).

Class Action

For the most part, environmental law is concerned with the total environment and the rights and liabilities of large numbers of people who may affect, or be affected by, the environment. These laws are largely federal laws. Lawsuits are occasionally initiated on behalf of a large number of people, however. If so many people are involved in a suit that they cannot be named individually, the legal device is called a *class action.* For standing to be granted in a class-action suit, it must be proved that this large number was "injured" by the defendant's action. This type of action, representing many individuals and their causes, can be effective in resolving issues, since the leverage involved can be tremendous.

FACTORS IN ENVIRONMENTAL LAWSUITS

Contemporary law stresses "wrongful conduct." Several factors are considered as part of each environmental action which may affect its outcome:

Balancing of Equities—Whenever there is an award to compensate a plaintiff the usefulness and social need of an activity must be weighed against the potential harm it inflicts. It is always a question of degree. Factors such as reasonableness of use, degree of injury, and the needs of society will play a part in any such court decision.

State-of-the-Art—Perfect solutions are not possible. The use of state-of-the-art methods and techniques implies use of the best solutions available at any given time and place.

Compliance with Requirements—Any time a set of measurable requirements exists, such as an allowable concentration of pollutants or installation of specific equipment, compliance or failure to comply is a factor in a lawsuit.

Additional factors such as contributory negligence, assumption of risk, and statutes of limitations are less often considered but not uncommon in environmental cases.

REMEDIES

The judicial system has five basic forms of solution:

1. *Injunction*—a court order to stop an activity. (*Note:* Owing to a historical quirk, actions for injunctions are decided by the judge alone.)
2. *Compensatory Damages*—a monetary award to compensate the plaintiff for damages or injury.
3. *Punitive Damages*—a monetary award to compensate the plaintiff for the defendant's wanton disregard of the plaintiff's rights.
4. *Treble Damages*—a monetary award whereby the court, as a punishment for socially undesirable conduct, triples the damages once it has been established that the defendant was responsible.
5. *Mandamus*—an order from a superior court to a governmental official to perform a (nondiscretionary) duty.

Note: Injunctions are sometimes referred to as "equitable remedies" not because they are fairer than damage awards such as punitive or compensatory monetary awards, but because the solution originated in the "equity" court system.

Bibliography

Ackerman, B.A., S.R. Ackerman, J.W. Sawyer, Jr., and D.W. Henderson. 1974. *The uncertain search for environmental quality.* New York: Free Press.

Adkins, W.G., and D. Burke, Jr. 1974. *Social, economic and environmental factors in highway decision making.* Res. Report 148-4. College Station: Texas Transportation Institute, Texas A&M University.

Advisory Commission on Intergovernmental Relations. 1981. *The federal role in the federal system: The dynamics of growth, protecting the environment: Politics, pollution, and federal policy.* Washington, D.C.

Anderson, F.R., and R.H. Daniels. 1973. *NEPA in the courts: A legal analysis of the National Environmental Policy Act.* Baltimore: The Johns Hopkins University Press for Resources for the Future.

Anderson, F.R., A.V. Kneese, P.D. Reed, R.B. Stevenson, and S. Taylor. 1977. *Environmental improvement through economic incentives.* Baltimore: The Johns Hopkins University Press for Resources for the Future.

Andrews, R.N.L. 1976. *Environmental policy and administrative change.* Lexington, Mass.: D. C. Heath & Co.

Arbuckle, J.G., M.A. Brown, N.S. Bryson, G.W. Frick, R.M. Hall, Jr., J.G. Miller, M.L. Miller, T.F.P. Sullivan, T.A. Vanderver, Jr., and L.N. Wegman. 1985. *Environmental law handbook.* 8th ed. Rockville, Md.: Government Institutes Inc.

Arbuckle, J.G., S.W. Schroeder, and T.F.P. Sullivan. 1974. *Environmental law for non-lawyers.* Bethesda: Governmental Institutes, Inc.

Barbaro, R., and F.L. Cross, Jr. 1973. *Primer on environmental impact statements.* Lancaster, Pa.: Technomic Publishing Co.

Barnett, H.J., and C. Morse. 1963. *Scarcity and growth: The economics of natural resource availability.* Baltimore: Johns Hopkins Press.

Bateson, G. 1972. *Steps to an ecology of mind.* New York: Chandler Publishing Co.

Baumol, W.J., and W.E. Oates. 1979. *Economics, environmental policy and the quality of life.* Englewood Cliffs, N.J.: Prentice-Hall.

Bennett, G.F., and J.C. Bennett. 1973. *Environmental literature—a bibliography.* Park Ridge, N.J.: Noyes Data Corporation.

Bernarde, M.A. 1970. *Our precarious habitat: An integrated approach to understanding man's effect on his environment.* New York: W.W. Norton & Co.

Black, P.E. 1981. *Environmental impact analysis.* New York: Praeger Publishing.

Bogen, K.T. 1981. *Coordination of regulatory risk analysis: Current framework and legislative proposals,* Report No. 81-209SPR. Washington, D.C.: Congressional Research Service.

Bogert, Clinton E., ed. 1949. *Glossary of water and sewage control engineering.* New York: American Society of Civil Engineers.

Brockrath, J.T. 1977. *Environmental law for engineers, scientists, and managers.* New York: McGraw-Hill.

Burchell, R.W., and D. Listokin, eds. 1975. *Future land use: Energy, environmental and legal constraints.* New Brunswick, N.J.: Center for Urban Policy Research, Rutgers University.

———. 1975. *The environmental impact handbook.* New Brunswick, N.J.: Center for Urban Policy Research, Rutgers University.

Butler, C.H. 1984. *Cogeneration: Engineering, design, financing and regulatory compliance.* New York: McGraw-Hill.

Caldwell, L.K. 1970. *Environment: A challenge to modern society.* Garden City, N.Y.: Natural History Press.

———. 1982. *Science and the National Environmental Policy Act.* University, Ala.: University of Alabama Press.

Caldwell, L.K., L.R. Hayes, and I.M. MacWhirter. 1976. *Citizens and the environment: Case studies in popular action.* Bloomington: Indiana University Press.

California Office of Planning and Research. 1984. *California permit handbook.* Sacramento: Governor's Office.

Camougis, G. 1981. *Environmental biology for engineers.* New York: McGraw-Hill.

Campbell, R.R., and J.L. Wade. 1972. *Society and environment: The coming collision.* Newton, Mass.: Allyn and Bacon, Inc.

Canter, Larry W. 1977. *Environmental impact assessment.* New York: McGraw-Hill.

———. 1979. *Water resources assessment: Methodology and technology sourcebook.* Woburn, Mass.: Ann Arbor Science.

Canter, Larry W., and L.G. Hill. 1979. *Handbook of variables for environmental impact assessment.* Woburn, Mass.: Ann Arbor Science.

Chase, G.B. 1973. Matrix analysis in environmental impact assessment. Paper presented at the Engineering Foundation Conference on Preparing Environmental Impact Statements, 29 July-3 August, Henniker, New Hampshire.

Cheremisinoff, P.N., and A.C. Morresi. 1977. *Environmental assessment and impact statement handbook.* Woburn, Mass.: Ann Arbor Science.

Ciaccio, L.L., ed. 1971. *Water and water pollution handbook.* Vols. 1-4. New York: Marcel Dekker Inc.

Clapham, W.B., Jr. 1981. *Human ecosystems.* New York: Macmillan Publishing Co.

Clark, B.D., R. Bisset, and P. Wathern. 1980. *Environmental impact assessment: A bibliography with abstracts.* London: Mansell.

Cohon, J.L. 1978. *Multiobjective programming and planning.* New York: Academic Press.

Colinvaux, P. 1973. *Introduction to ecology.* New York: John Wiley & Sons.

Congressional Research Service. *Summaries of federal environmental laws administered by the Environmental Protection Agency.* March 1, 1984. TP 450 U.S.B.2. Washington, D.C.

The Conservation Foundation. 1984. *State of the environment: An assessment at mid-decade.* Washington, D.C.: The Conservation Foundation.

Considine, D.M., ed. 1983. *Van Nostrand's scientific encyclopedia.* 6th ed. New York: Van Nostrand Reinhold.

Corps of Engineers. 1974. *Preparation and coordination of environmental statements, app. C, Regulation 1105-2-507.* Washington, D.C.: Department of the Army.

Corwin, Ruthann, and P. H. Heffernan, eds. 1975. *Environmental impact assessments.* San Francisco: Freeman, Cooper & Company.

Council on Environmental Quality. 1980. *Environmental quality 1980, 11th annual report of the Council on Environmental Quality.* Washington, D.C.

Council on Environmental Quality. 1981. *Environmental quality 1981, 12th annual report of the Council on Environmental Quality.* Washington, D.C.

Council on Environmental Quality. 1983. *Environmental quality 1983, 14th annual report of the Council on Environmental Quality.* Washington, D.C.

Cowart, R., ed. 1976. *Land use planning, politics and policy.* Berkeley: University Extension Publications, University of California.

Craik, K.H., and E.H. Zube, eds. 1976. *Perceiving environmental quality: Research and applications.* New York: Plenum.

Creighton, J.L. 1981. *The public involvement manual.* Cambridge, Mass.: Abt Books.

Crocker, T.D., and A.J. Rogers III. 1971. *Environmental economics.* Hinsdale, Ill.: Dryden Press.

Cross, F.L., Jr., and S.M. Hennigan. 1974. *National directory of environmental experts, consultants, regulatory agencies.* Westport, Conn.: Technomic Publishing Co.

Davis, P.N. 1971. Theories of water pollution litigation. In *Wisconsin law review.* Madison: University of Wisconsin Law School.

Dean, B.V., and M.J. Nishry. 1965. Scoring and profitability models for evaluating and selecting engineering projects. *Journal of Operational Research Society of America* 13 (4).

Dee, Norbert, et al. 1972. *Environmental evaluation system for water resources planning.* Report to the U.S. Bureau of Reclamation. Columbus: Battelle Memorial Institute.

———, et al. 1973. *Planning methodology for water quality management: Environmental evaluation system.* Columbus: Battelle-Columbus.

Decker, L., and T. Price. 1980. *Directory of environmental organizations and contacts in the Washington D.C. metropolitan area.* Report No. 83-32C. Washington, D.C.: Congressional Research Service.

Dickert, G., and K.R. Domeny. 1974. *Environmental impact assessment: Guidelines and commentary.* Berkeley: University Extension, University of California.

Dickson, Kenneth L., A.W. Maki, and J. Cairns, Jr., eds. 1978. *Analyzing the hazard evaluation process; Proceedings of workshop on the application of hazard evaluation programs for chemicals in the aquatic environment.* Waterville Valley, New Hampshire, August.

Dolgin, E., and T.G.P. Guilvert, eds. 1974. *Federal environmental law.* St. Paul: West Publishing Co.

Drobny, N.L., and M.A. Smith. 1973. *Review of environmental impact assessment methodologies.* Columbus: Battelle-Columbus.

Dunne, T., and L.B. Leopold. 1978. *Water in environmental planning.* San Francisco: W.H. Freeman and Company.

Ehrlich, P.R., A.H. Ehrlich, and J.P. Holdren. 1977. *Ecoscience: Population, resources, environment.* San Francisco: W.H. Freeman and Company.

Environmental Protection Agency. 1973. *Methods for identifying and evaluating the nature and extent of non-point sources of pollutants.* Publ. EPA-430/9-73-014. Washington, D.C.: EPA.

Environmental Protection Section of the Environment and Natural Resources Policy Division. 1983. *Environmental protection: A historical review of legislation and programs of the Environmental Protection Agency.* Congressional Research Service, Report No. 83-34 ENR, Washington, D.C.

Environmental Protection Section of the Environment and Natural Resources Policy Division. 1984. *Summary of federal environmental laws administered by the Environmental Protection Agency.* Report No. 84-44 ENR, Washington, D.C.

Feenberg, D., and E.S. Mills. 1980. *Measuring the benefits of water pollution abatement.* New York: Academic Press.

Forber, R.J. 1969. *The conquest of nature: Technology and its consequences.* New York: Mentor.

Freeman, A.M. III. 1979. *The benefits of environmental improvement: Theory and practice.* Baltimore: Johns Hopkins University Press for Resources for the Future.

Freeman, A.M., III, R.H. Haveman, and A.V. Kneese. 1973. *The economics of environmental policy.* New York: John Wiley & Sons.

Frieden, B.J. 1969. *The environmental protection hustle.* Cambridge: MIT Press.

Gilpin, A. 1976. *Dictionary of environmental terms.* Queensland, Australia: University of Queensland Press.

Goldberg, E.D. 1972. *A guide to marine pollution.* New York: Gordon & Beach Science Publishers.

Golden, J., R.P. Ovellette, S. Saari, and P.N. Cheremisinoff. 1979. *Environmental impact data book.* Woburn, Mass.: Ann Arbor Science.

Goldman, M.I., ed. 1972. *Ecology and economics.* Englewood Cliffs, N.J.: Prentice Hall, Inc.

Grad, F.P. 1978. *Environmental law.* Vols. 1 & 2, 2nd ed. New York: Matthew Bender.
Gray, P. 1967. *The dictionary of the biological sciences.* New York: Reinhold Publishing Corp.
Gress, L.M., ed. 1972. *Man: An endangered species.* Woodridge, Conn.: Apollo Books.
Grossman, K. 1983. *The poison conspiracy.* New York: The Permanent Press.
Gunningham, Neil. 1974. *Pollution, social interest and the law.* Martin Robertson Law in Society Series. Bath, Great Britain: The Pittman Press.
Hagmen, D.G., ed. 1982. *Land use and environmental law review.* New York: Clark Boardman Co., Ltd.
Hall, C.A.S., and J.W. Day, Jr. 1977. *Ecosystem modeling in theory and practice: An introduction with case histories.* New York: John Wiley & Sons.
Hammond, K.A., G. Macinko, and W.B. Fairchild. 1978. *Sourcebook on the environment.* Chicago: University of Chicago Press.
Hanks, E., A.D. Tarlock, and J. Hanks, eds. 1974, 1976 (supplement). *Environmental law and policy.* St. Paul: West Publishing Co.
Hargrave, J.L., ed. *Law, institutions, and the global environment: A joint conference of the American Society of International Law and the Carnegie Endowment.* September 1971. Dobbs Ferry, N.Y.: Oceana Publications.
Harris, M. 1975. The institutional-legal face of the environmental coin. *Nebraska Law Review* 54.
Harrison, L.L. 1984. *Environmental auditing handbook.* New York: McGraw-Hill.
Hawley, G.G. 1981. *Condensed chemical dictionary.* 10th ed. New York: Van Nostrand Reinhold.
Heer, John E., Jr., and D. Joseph Hagerty. 1977. *Environmental assessments and statements.* New York: Van Nostrand Reinhold.
Henderson, I.F., and W.D. Henderson. 1960. *Dictionary of scientific terms. 7th ed.* Edinburgh: Oliver & Boyd.
Hendricks, D.W., E.C. Vlachos, L.S. Tucker, and J.C. Kellog, eds. 1975. *Environmental design for public projects.* Colorado: Water Resources Publications.
Henning, D.H. 1974. *Environmental policy and administration.* New York: American Elsevier Publishing Co.
Hickerson, K.W. 1984. Environmental disputes find solutions outside courtroom. *Engineering Times,* November.
Hirshleifer, J., J.C. DeHaven, and J.W. Milliman. 1969. *Water supply: Economics, technology, and policy.* Chicago: University of Chicago Press.
Holling, C.S. 1978. *Adaptive environmental assessment and management.* New York: John Wiley & Sons.
Hothersall, D.C., and R.J. Salter. 1977. *Transportation and the environment.* Hertfordshire, Great Britain: Granada.
Jackson, W., ed. 1971. *Man and the environment.* Dubuque: William C. Brown Pub. Co.
Jaffe, L.L., and L.H. Tribe. 1971. *Environmental Protection.* Chicago: The Bracton Press.
Jain, R.K. 1973. *Environmental impact assessment study for army military programs.* Technical Report D-13. Champaign: Construction Engineering Research Laboratory, U.S. Army.
Jain, R.K., L.V. Urban, and G.S. Stacey. 1977. *Environmental impact analysis: A new dimension in decision making.* New York: Van Nostrand Reinhold.
James, D.E., H.M.A. Jansen, and J.B. Opschoor. 1978. *Economic approaches to environmental problems.* New York: Elsevier/North Holland.
Joeres, E.F., and M.H. David, eds. 1983. *Buying a better environment.* Madison: University of Wisconsin Press.
Kneese, A.V. 1962. *Water pollution: Economic aspects and research needs.* Washington, D.C.: Resources for the Future.
Kneese, A.V., and B.T. Bower. 1968. *Managing water quality: Economics, technology*

and institutions. Baltimore: The Johns Hopkins University Press for Resources for the Future.

———. 1979. *Environmental quality and residuals management.* Baltimore: The Johns Hopkins University Press for Resources for the Future.

Krane, Paul. 1984. Help with permits in one stop. *Civil Engineering* 54(6):67–69.

Lake, L.M., ed. 1980. *Environmental mediation: The search for consensus,* Boulder: Westview Press.

———. 1982. *Environmental regulation.* New York: Praeger Publishers.

Landau, N.J., and P.D. Rheingold. 1971. *The environmental law handbook.* New York: Ballantine.

Lapodes, D.N., ed. 1974. *McGraw-Hill Encyclopedia of environmental science.* New York: McGraw-Hill.

Lee, E.Y.S. 1974. *Environmental impact computer system.* Tech. Report E-37. Champaign, Ill.: Construction Engineering Research Laboratory, U.S. Army.

Lee, K., and J. Moberg. 1975. *Environmental design evaluation: A matrix method.* Foster City, Calif.: Environmental Design and Research Center.

Leopold, L.B., et al. 1971. *A procedure for evaluating environmental impact.* Geological Survey Circular 645. Washington, D.C.: Government Printing Office.

Lichfield, N., P. Kettle, and M. Whitbread. 1975. *Evaluation in the planning process.* Oxford: Pergamon.

Liroff, R.A. 1976. *The national policy for the environment: NEPA and its aftermath.* Bloomington: Indiana University Press.

Magrab, E.B. 1975. *Environmental noise control.* New York: John Wiley & Sons.

Mandelker, D.M., ed. 1974. *New developments in land and environmental controls.* Indianapolis: Bobbs-Merrill Publishing Co.

———. 1981. *Environment and equity: A regulatory challenge.* New York: McGraw-Hill.

Masters, G.M. 1974. *Introduction to environmental science and technology.* New York: John Wiley & Sons.

McAllister, C.M. 1980. *Evaluation in environmental planning.* Cambridge: MIT Press.

McEvoy, J., III, and T. Dietz, eds. 1977. *Handbook for environmental planning: The social consequences of environmental change.* New York: John Wiley & Sons.

McHale, J. 1970. *The ecological context.* New York: George Braziller.

McHarg, Ian. 1969. *Design with nature.* Garden City, N.Y.: Natural History Press.

McKnight, A.D., P.K. Marstrand, and T.C. Sinclair, eds. *Environmental pollution control: Technical, economic and legal aspects.* London: George Allen & Unwin Ltd.

McLaughlin, J. 1972. *The law relating to pollution.* London: Manchester University Press.

Meyers, C.J., and A.D. Tarlock, eds. 1971. *Selected legal and economic aspects of environmental protection.* Mineola, N.Y.: Foundation Press.

Moore, J.L. 1973. A methodology for evaluating manufacturing environmental impact statements for Delaware's coastal zone, app. D. Report prepared by Battelle-Columbus, for the State of Delaware, June.

Munn, R.E. 1975. *Environmental impact assessment: Principles and procedures.* Paris: United Nations.

———. 1979. *Environmental impact assessment: Principles and procedures.* 2nd ed., Chichester, England: Wiley.

Murphy, E.F. 1967. *Governing nature.* Chicago: Quadrangle Books.

———. 1971. *Man and his environment: Law.* New York: Harper & Row.

———. *Nature, bureaucracy and the rules of property.* New York: North-Holland Pub. Co.

Nemerow, N.L. 1974. *Scientific stream pollution analysis.* Washington, D.C.: Scripta Book Co.

Nicholson, M. 1970. *The environmental revolution: A guide for the new masters of the Earth.* New York: McGraw-Hill.

Nijkamp, P. 1977. *Theory and application of environmental economics.* New York: Elsevier/North Holland.

Nowak, J., ed. 1976. *Environmental law: International and comparative aspects.* Dobbs Ferry, N.Y.: Oceana Publications.

Odum, E.P., 1971. *Fundamentals of ecology.* 3rd ed. Philadelphia: Saunders.

———. 1971. *Optimum pathway matrix analysis approach to the environmental decision-making process.* Athens: Institute of Ecology, University of Georgia.

The Onyx Group, Inc., eds. 1974. *Environment U.S.A.: A guide to agencies, people and resources.* New York: R.R. Bowker Co.

O'Riordan, T., and R.D. Hey, eds. 1976. *Environmental impact assessment.* Great Britain: Saxon House.

O'Riordan, T., and W.R.D. Sewell, eds. *Project appraisal and policy review.* New York: John Wiley & Sons.

Ortolano, Leonard. 1984. *Environmental planning and decision making.* New York: John Wiley & Sons.

Ott, W.R. 1978. *Environmental indices; Theory and practice.* Woburn, Mass.: Ann Arbor Science.

Overcash, M.R., and J.M. Davidson. 1980. *Environmental impact of nonpoint source pollution.* Woburn, Mass.: Ann Arbor Science.

Painter, D.E. 1974. *Air pollution technology.* Reston, Va.: Reston Publishing Co.

Pickering, W.F. 1977. *Pollution evaluation: The quantitative aspects.* New York: Marcel Dekker Inc.

Porter, A.L., F.A. Rossini, S.R. Carpenter, and A.T. Roper. 1980. *A guidebook for technology assessment and impact analysis.* New York: Elsevier/North Holland.

Pugh, R.E. 1977. *Evaluation of policy simulation models: A conceptual approach and case study.* Washington, D.C.: Information Resources Press.

Purdom, P.W., and S.H. Anderson. 1980. *Environmental science, managing the environment.* Columbus: Charles G. Merrill.

Quarles, J. 1976. *Cleaning up America: An insider's view of the Environmental Protection Agency.* Boston: Houghton-Mifflin.

Rau, J.G., and D.C. Wooten, eds. 1980. *Environmental impact analysis handbook.* New York: McGraw-Hill.

Reilley, W.K., ed. 1973. *The use of land: A citizens' policy guide to urban growth.* New York: Thomas Y. Crowell & Co.

Reitz, Arnold W., Jr. 1974. *Environmental planning: law of land and resources.* Washington, D.C.: North American International Co.

Rodgers, J.L., Jr. 1976. *Environmental impact assessment, growth management and the comprehensive plan.* Cambridge, Mass.: Ballinger.

Rohlick, G.A., ed. 1974. *Environmental impact statements: Effect on program implementation.* Austin: L.B. Johnson School of Public Affairs, University of Texas.

Rona, D.C. 1984. How to strengthen client-consultant relationships. *Civil Engineering* 54(3).

Rose, J.G. 1983. *Legal foundations of environmental planning.* New Brunswick, N.J.: Center for Urban Policy Research, Rutgers University.

Sarnoff, P. 1971. *Encyclopedic dictionary of the environment.* New York: Quadrangle Books.

Schneider, D.M., D.R. Godschalk, and N. Axler. 1978. *The carrying capacity concept as a planning tool.* Report No. 338, Planning Advisory Service. Chicago: American Planning Association.

Schneider, T., H.W. deKoning, and L.J. Brasser, 1978. *Air pollution reference measurement methods and systems.* New York: Elsevier Scientific Publishers.

Schroth, P.W. 1975. *A NEPA primer.* Dallas: Center for Urban and Environmental Studies.

Schwarz, W., ed. 1969. *Voices for the wilderness: Sierra Club wilderness conferences 1960-1966.* New York: Ballantine Books.

Scott, R.W., D.J. Bower, and D.D. Miner, eds. 1975. *Management and control of growth: Issues, techniques, problems, trends.* Washington, D.C.: Urban Land Institute.

Shahane, A.N. 1976. Interdisciplinary models of water systems. *Ecological Model* 2:117-45.

Shaw, B. 1976, *Environmental law: People, pollution, and land use.* St. Paul: West Publishing Co.

Sive, M.R. 1976. *Environmental legislation: A source book.* New York: Praeger Publishers.

Smith, F.E. 1971. *The politics of conservation.* New York: Harper Colophon.

Smith, M.A. 1974. *Field test of an environmental impact assessment methodology.* Report ERC-1574, Environmental Resources Center, Georgia Institute of Technology, Atlanta.

Soil Conservation Service. 1974. *Environmental assessment procedure.* Washington, D.C.: U.S. Department of Agriculture.

Soma, John T., and Andrea R. Stern. 1983. A survey of computerized information for lawyers: LEXIS, JURIS, WESTLAW, and FLITE. *Rutgers Computer and Technology Law Journal* 9.

Stewart, R.B., and J.E. Krier. 1978. *Environmental law and policy.* Indianapolis: Bobbs-Merrill.

Study Committee on Environmental Aspects of National Materials Policy, National Academy of Sciences, and National Academy of Engineering. 1973. *Man, materials and environment.* Cambridge: MIT Press.

Sumek, L. 1973. *Environmental management and politics: A selected bibliography.* De Kalb, Ill.: Center for Governmental Studies, Northern Illinois University.

Swainson, N.A., ed. 1976. *Managing the water environment.* Vancouver: University of British Columbia Press with the Wastewater Research Center.

Talbot, A.R. 1983. *Settling things: Six case studies in environmental mediation.* Washington, D.C.: The Conservation Foundation.

Tans, W. 1974. Priority ranking of biotic natural areas. *The Michigan Botanist* 13.

Thewlis, J. 1973. *Concise dictionary of physics.* New York: Pergamon Press.

Thomann, R.V. 1972. *Systems analysis and water quality management.* New York: Environmental Science Services Division (reissued by McGraw-Hill, New York).

Thomas, William A., ed. 1972. Indicators of environmental quality. In *Proceedings of AAAS symposium.* New York: Plenum Press.

Tomioka, S., and E.M. Tomioka. 1984. *Planned unit developments, design and regional impact.* New York: John Wiley & Sons.

Trelease, F., ed. 1974. *Cases and materials on water law.* St. Paul: West Publishing Co.

Turk, J. 1980. *Introduction to environmental studies.* Philadelphia: Saunders.

U.S. Council On Environmental Quality. 1977. *Environmental quality 1977: The eighth annual report of the council on environmental quality.* Washington, D.C.: U.S. Government Printing Office.

U.S. Department of the Interior, Geological Survey. *Directory to federal, state, and local OCS-related activities and contacts.* U.S. Geological Survey Open File Report 79-1481. Washington, D.C.

van Weringh, J., J. Patzer, R. Welsh, and R. Webster. 1978. *Computer-aided environmental legislative data system (CELDS) user manual.* United States Army Corps of Engineers, Technical Report N-56.

Vesilind, P.A. 1977. *Environmental pollution and control.* Ann Arbor: Ann Arbor Science Publishers.

Viohl, R.C., Jr., and K.G.M. Mason. *Environmental impact assessment methodologies: An annotated bibliography.* Council Planning Library Exchange Bibliography 691.

Wagner, C.E., and A. Townsend. 1980. *Federal rules of evidence manual:* Annotated. Washington, D.C.: EPA Office of Enforcement.

Walton, L.E., Jr., and J.E. Lewis. 1971. *A Manual for conducting environmental impact studies.* Norfolk: Virginia Highway Research Council.

Ward, D.V. 1978. *Biological environmental impact studies: Theory and methods.* New York: Academic Press.

Warden, R.E., and W.T. Dagodag. 1976. *Environmental impact reports.* Los Angeles: Security World Publishing Co.

Warner, M.L., and D.W. Bromley. 1974. *Environmental impact analysis: A review of three methodologies.* Technical Report, Wisconsin Water Resources Center. Madison: University of Wisconsin.

Warner, M.L., and E.H. Preston. 1974. *A review of environmental protection impact assessment methodologies.* Washington, D.C.: Environmental Protection Agency.

Warner, M.L., et al. 1973. *A review of environmental impact assessment methodologies.* PB-236-609. Columbus, Ohio: Battelle-Columbus.

Watt, K.E.F. 1968. *Ecology and resource management.* New York: McGraw-Hill.

Wenner, L.M. 1972. Enforcement of water pollution control law. *Law and Society Review* 6.

———. 1976. *One environment under law: A public-policy dilemma.* Pacific Palisades: Goodyear Publishing Co.

Whipple, W., Jr. 1977. *Planning of water quality systems.* Lexington, Mass.: Lexington Books-Heath.

Wilson, B.R., ed. 1968. *Environmental problems: Pesticides, thermal pollution and synergisms.* Philadelphia: J.B. Lippincott Co.

Young, R.A., ed. 1975. *Environmental law handbook.* Bethesda: Government Institutes Inc.

Zeleny, M. 1982. *Multiple criteria decision making.* New York: McGraw-Hill.

Zener, R.V. 1981. *Guide to federal environmental law.* New York: Practicing Law Institute.

APPENDIX A
Glossary of Environmental Terms

Abatement. A reduction in intensity or degree of pollution.
Abiotic. Referring to the absence of life.
Absorption. A process by which one material is dissolved or chemically "captured."
Acclimation. The adjustment of an organism to changes in the environment.
Acclimatization. The adaptation of a species over several generations to changes in the environment.
Activated Sludge. Raw sewage from which organic matter has been removed by saturation with air and biologically active compounds.
Acute. Severe and short-lived.
Adaptation. Change in an organism's physical structure or habitat resulting from adjustment to changes in the environment.
Adjustment. The establishment of a satisfactory relationship between individual needs and desires and the requirements of the environment.
Adsorption. The adhesion of a liquid or gaseous substance to the surface of a liquid or solid.
Aeration. The circulation of air (carrying oxygen) through a substance.
Aeroallergen. An airborne substance that causes allergy.
Aerobic. Referring to an organism or process that depends on oxygen; also, the existence of free oxygen.
Aerosol. A suspension of liquid or solid particles (generally less than 1 micron in diameter) in the air.
Air Pollution. The presence of contaminant substances in the air that directly or indirectly interfere with health, safety, or comfort or with the use and enjoyment of property.
Air Quality Criteria. Standard pollutant levels and lengths of exposure known to cause specific adverse effects to health.
Airshed. The air overlying any arbitrary geographical region; frequently, the common air supply demarcated by an urban area.
Albedo. Whiteness; also, the ratio of reflected light to total light falling upon a material.
Allelopathic. Referring to materials within certain plants that suppress the growth of other plant species.
Allogenic. Having diverse or different origins.
Alluvium. Clay, silt, gravel, sand, or detrital material deposited by running water.
Ambient Air. Unconfined portion of the atmosphere, open air.
Amensalism. An interaction between two species in which one species is inhibited and the other is not affected.
Anadromous. Ascending rivers from the sea to spawn.
Anaerobic. Referring to an organism or process that occurs in the absence of oxygen.
AQCR. Air **Q**uality **C**ontrol **R**egion.
Aquatic. Growing, living in, or frequenting water.
Aquifer. A water-bearing stratum of permeable rock, gravel, or sand.
Artificial. Produced or affected by man, with resultant characteristics that are not necessarily indicative of natural relationships.
Ash. The noncombustible material that remains after burning.
Assimilation. Absorption into the system; also, the process by which a body of water purifies itself of pollutants.
Association. A major unit in an ecological community, characterized by essential uniformity and usually by two or more dominant species.
Atmosphere. The layer of air surrounding the earth.
Attrition. The wearing down by friction.
Autotrophic. Designating plant organs with a tendency to grow in a straight line.
Avulsion. The sudden cutting of land by flood, currents, or other changes in the course of a water body.
Background Level. The level of pollutants present in the ambient environment from natural sources.

BACT. Best Available Control Technology.
Bacteria. Single-celled microorganisms that lack chlorophyll; some cause disease, others are necessary to life.
Baling. The compaction of solid waste to reduce volume.
BAT. Best Available Technology.
Benthos. Plants and animals that inhabit the bottom of a body of water.
Bioassay. The use of living organisms to measure the effect of a condition or substance.
Biochemical Oxygen Demand (BOD). The dissolved oxygen required to decompose organic material in water. Used as a measure of pollution, since heavy waste loads have a high demand for oxygen.
Biocide. A substance that can kill living organisms. A pesticide is a biocide.
Biodegradable. Capable of decomposing quickly through the action of microorganisms.
Biomass. The amount of living matter in an arbitrary unit of the environment.
Biome. A unit of plant and animal life.
Biomonitoring. The use of living organisms to test ambient water or air quality at a specific site.
Biosphere. That part of the earth in which life can exist.
Biota. All living organisms in a given area.
Biotic. Of or relating to life.
BOD. *See* Biochemical Oxygen Demand.
Bubbler. A device for measuring gaseous air pollutants.
Building Code. Government regulations establishing standards of construction in such areas as materials, minimum design forces, and hazards.
Calorie. The unit of measure of energy equivalent to the heat required to raise the temperature of one gram of water one degree centigrade.
Carcinogen. A cancer-producing substance.
Carrying Capacity. The maximum number of individuals (plants, animals, and/or humans) that can be sustained in a given area for a given amount of time.
Catchment. An area or structure that contains water, usually from runoff.
CEQ. Council on Environmental Quality.
Chemical Oxygen Demand (COD). The amount of oxygen required to oxidize all compounds, both organic and inorganic, in a given body of water.
Chlorination. The addition of chlorine to water to disinfect or oxidize undesirable compounds.
Chlorosis. Disease condition that results in loss of green color.
Chronic. Of long duration or frequent recurrence.
Clarifier. A tank for liquids in which the solids settle to the bottom.
Climate. Average weather or environmental conditions characterizing a group, period, or region.
Coagulation. The clumping of particles; often induced by chemicals.
COD. *See* Chemical Oxygen Demand.
Coefficient of Haze (COH). A measurement of visibility interference in the atmosphere.
COH. *See* Coefficient of Haze.
Conservation. The planned management of a natural resource to prevent exploitation, destruction, or neglect.
Contour. A line on a map connecting points of equal elevation.
Conurbation. A continuous network of urban communities.
Convection. Movement of fluids that transmits energy.
Criteria. Standards on which a judgment or decision may be based.
Cytology. The study of cell formation, structure, and function.
Decibel (dB). A unit of measure of the intensity of sound.
Decomposition. A change in the chemical composition and physical appearance of materials, caused by the action of bacteria.
Deme. A local population of closely related organisms.

Demography. The statistical study of population characteristics, including growth, migrations, density, and distribution.

DEP. **D**epartment of **E**nvironmental **P**rotection.

Diffusion. A process by which particles of gas, liquids, or solids intermingle as the result of their spontaneous movement.

Dilution. The process of making thinner or less concentrated.

Disintegration. Decomposition into component parts, elements, or small particles.

Dissolved Oxygen. Oxygen suspended in water in the form of microscopic bubbles. (This is not the oxygen that combines with hydrogen to form the water itself.)

Diurnal. Having a daily cycle.

Diversity Index. The ratio between the number of species and the number of individuals.

DO. **D**issolved **O**xygen.

DOE. **D**epartment **o**f **E**nergy.

Dominance. A controlling influence exerted on the environment by one or more organisms in an ecological association that largely determines what other kinds of organisms share in the association.

Dune. A hill of sand piled up by the wind.

Dust. Fine grain particles, light enough to be suspended in air.

EAW. **E**nvironmental **A**ssessment **W**orksheet. Any working outline of an **EIS** used to evaluate the potential impact of a proposed project; used to determine if an **EIS** is necessary.

Ecology. The interrelationships of living organisms to one another and to their environment; also, the study of such relationships.

Ecosystem. An integrated unit in nature, sufficient unto itself, with a balanced assortment of organisms.

Ecotone. The often indefinite boundary between two ecological communities.

Edaphic. Resulting from or influenced by the soil rather than the climate.

Eddy. A contrary or circular current.

Edge Effect. A change in species diversity and density in the transition zone between communities.

Effluent. A material discharged into an environment, usually from a relatively self-contained unit such as an industrial smokestack or sewage-treatment facility.

EIS. *See* Environmental Impact Statement.

Emission. *See* Effluent.

Endangered Species. A species in peril of becoming extinct.

Endemic. Native to a particular region.

Environment. The combination of all external influences and conditions affecting the life, development, and ultimate survival of an organism.

Environmental Impact Statement (EIS). A comprehensive statement of the potential environmental impact of a proposed project, and alternatives to the proposed project.

Environmental Index. A single number derived from two or more indicators as a measure of the status of an environment (e.g., pollutant loading).

EPA. **E**nvironmental **P**rotection **A**gency.

Epidemiology. The study of the incidence and distribution of disease in a population.

Epilimnion. The uppermost layer of a lake.

EROS. **E**arth **R**esources **O**bservation **S**ystem.

Erosion. The wearing away of land surface by wind or water.

ESECA. **E**nergy **S**upply and **E**nvironmental **C**oordination **A**ct.

Estuary. An area where fresh water meets salt water, such as a tidal river, marsh, flat, lagoon, or bay.

Etiology. The study of causes, origins, or reasons.

Euphotic Zone. The upper layers of a body of water into which sufficient light penetrates to permit the growth of green plants.

Eutrophic. Designating a well-nourished zone within a body of water, usually with seasonal oxygen deficiency.

Evapotranspiration. The loss of water from the soil by both evaporation and transpiration from the plants growing in the soil.

Fauna. Animals or animal life.

FDA. Food and Drug Administration.

Fen. Low-lying land partly covered with water.

FIFRA. Federal Insecticide, Fungicide, and Rodenticide Act.

Flocculation. The separation of suspended solids during waste-water treatment by the chemical creation of coagulated clumps.

Flora. Plant life.

Food Chain. The process by which one species becomes food for larger species, beginning with one-celled organisms and progressing up through carnivorous animals.

FRES. Forest Range Environmental Study.

Fumes. Tiny particles trapped by vapor in a gas stream.

FWQA. Federal Water Quality Act.

Garbage. Food waste; refuse.

GEMS. Global Environmental Monitoring System.

Genotype. A group of organisms sharing a specified genetic makeup.

Genus. A category of biological classification ranking between the family and species, comprising structurally or phylogenetically related species.

Geomorphology. Land and submarine relief and the study thereof.

Gradient. The change of any physical characteristic (e.g., temperature, pressure, elevation) per unit length.

Green Belt. An area in which development is restricted; usually a buffer zone between pollution sources and urban development.

Greenhouse Effect. The warming of the Earth resulting from radiant heat trapped within the atmosphere.

Grit. Hard sharp granules, usually sand or gravel.

Gross Productivity. In reference to plants, total energy production by photosynthesis, much of which will be oxidized by plant respiration.

Ground Cover. Plants on the surface of the ground that protect it from erosion.

Groundwater. Fresh water beneath the surface of the earth.

Habitat. All environmental conditions in the specific area occupied by any organism or community.

Half-Life. The time required by certain materials, such as pesticides or radioactive isotopes, to lose half their value.

Hazardous Waste. Waste materials that are inherently dangerous to handle or dispose of.

Heavy Metal. A metallic element with a high molecular weight (e.g., mercury, lead, arsenic).

Herbicide. A chemical that controls or destroys undesirable plants.

Hertz. A unit of frequency equal to one cycle per second.

Heterotrophic. Requiring complex organic compounds of nitrogen and carbon for metabolism.

Humus. Decomposed organic matter.

Hypolimnion. The bottom layer of water in a lake.

Impedance. The ratio of the pressure to the volume displacement at a given surface in a sound-transmitting medium.

Indigenous. Native to a particular region or environment.

Infiltration. A process by which a liquid or gas flows into a substance through pores or small openings.

Influent. An input of a fluid (e.g., water into a reservoir).

Interfluve. The area between two streams flowing in the same direction.

Isohaline. A line drawn on a map or chart connecting points of equal salinity.
Isohyet. A line drawn on a map or chart connecting points receiving equal rainfall.
Isopleth. A graph showing the frequency of any phenomenon as a function of two variables.
Isotope. A variation of an element having the same atomic number as the element itself but a different atomic weight, because it has a different number of neutrons. Isotopes of an element may have different radioactive behavior.
Lagoon. A shallow sound, channel, or pond, usually communicating with a larger body of water. In waste treatment, a shallow pond where bacterial action and oxygen interaction restore water purity.
Landfill. Disposal of refuse on land.
Leaching. The process of dissolution and transport of materials such as nutrients, chemicals, or contaminants to lower layers of soil or into runoff water.
Limnology. The study of physical, meteorological, chemical, and biological aspects of fresh water.
Littoral. Of or existing on a coastal region.
Marine. Of or relating to the sea or salt water.
Marsh. A periodically wet area or a very shallow flooded area.
Masking. An overlapping of factors like sound, odor, or chemicals that serves to disguise the original.
MCL. Maximum Contaminant Level.
Mercaptan. An organic compound containing sulfur.
Metabolism. The chemical changes in living cells that serve to assimilate materials and create energy.
Meteorology. The study of the atmosphere and its phenomena, especially weather.
Microclimate. The environmental conditions of a small area or of a particular organism.
Micronutrient. A substance essential to the health of an organsim but required in minute quantity.
Microorganism. An organism of microscopic size.
Mitigation. The effort to deal with the undesirable results of a proposed project through minimizing or alleviating the project's effects. Often includes rectifying impacts through repairing, rehabilitating, restoring, or providing substitute resources or environments.
Monitoring. Periodic or continuous measurement of physical or chemical conditions (e.g., pollutant levels).
Morbid. Affected with or induced by disease.
Morphology. The branch of biology that deals with the form and structure of plants and animals.
Mortality. The death rate in a given time or place.
Multiple Source. Classification of origin of group of pollutant substances when a source-by-source evaluation is not possible (e.g., industrial process, municipal incinerator).
NAAQS. National Ambient Air Quality Standard.
NAS. National Academy of Science.
NASN. National Air Sampling Network.
NEPA. National Environmental Policy Act.
Net Productivity. Gross energy productivity less losses within the plant due to respiration.
Niche. A habitat supplying the factors necessary for the existence of an organism or species.
NOI. Notice of Intention.
Noise. Any undesirable audible signal.
NPDES. National Pollution Discharge Elimination System.
NPL. Noise Pollution Level.

NRI. National Resource Inventory.
Nutrient. An element or compound essential to the growth and development of living things.
Oligotrophic. Designating lakes having abundant oxygen in the lower stratum in the summer.
Organism. Any living thing.
OSHA. Occupational Safety and Health Act.
Pathogen. An agent capable of causing disease.
Pesticide. Any substance used to control plant or animal pests.
pH. A numerical measure of acidity—the negative logarithm of the hydrogen ion concentration.
Photoperiod. The number of hours of light or darkness in a day.
Photosynthesis. The process by which sunlight in the presence of chlorophyll converts carbon dioxide, water, and inorganic salts into living tissue.
Phylogeny. The evolutionary development of a species.
Phylum. One of the primary divisions of the animal kingdom, next above a class in size.
Plankton. Microscopic floating plant and animal organisms.
Plat. A map showing property ownership.
Pollutant Variable. Any physical, chemical, or biological quantity intended as a measure of environmental pollution.
Pollution. The presence of matter or energy whose nature, location, or quantity produces undesired environmental effects.
Population. The organisms inhabiting a particular area or habitat.
Potable Water. Water suitable for drinking.
PPM. Parts per million.
Precipitate. A solid that separates from a solution as a result of chemical or physical changes.
Preservation. Any action that protects animals, trees, or other natural resources against injury or destruction.
Producers. Green plants on which all other life ultimately depends.
Productivity. Primary—the amount of organic matter made in a given time by the autotrophic organisms in an ecosystem; net—the amount of organic matter produced in excess of that used by organisms in respiration, representing potential food for consumers in an ecosystem.
PSI. Pollution Standards Index.
PUD. Planned Unit Development.
Putrefaction. The decomposition of organic matter.
Pyrolysis. Chemical change brought about by heat.
Qualitative Analysis. Chemical analysis to identify the components of a substance or mixture.
Quantitative Analysis. Chemical analysis to determine the amounts or proportions of the components of a substance or mixture.
Radioactivity. The emission of atomic particles or rays by the nucleus of an atom.
Rainout. A process in which atmospheric contaminants unite with cloud droplets, which may later grow into precipitation; part of a natural, atmospheric cleansing process.
R&D. Research and Development.
Reducer. An agent that causes chemical reduction (e.g., oxidation).
Refuse. Solid waste.
Renewable Resource. A biological resource that renews itself by growth and reproduction.
Residence Time. The average time required for a substance to flow through a system or basin.
Resource. A reserve source of supply or support. Natural resources include land and soil, water, life forms, rocks and minerals, solar and other forms of energy.

Riparian Rights. The use and access rights of a property owner whose property abuts water.

Runoff. Water from rain, melting snow, or irrigation that flows over the surface of the ground and returns to rivers, bays, and other bodies of water.

Salinity. The measure of salt in the water.

SAROAD. **S**torage **a**nd **R**etrieval **o**f **A**ir-Quality **D**ata.

Scrubber. A device that uses a liquid spray to remove aerosol and gaseous pollutants from an air stream.

Sedimentation. The settling of solids out of suspension.

Seepage. Water that flows through the soil.

Sewage. The organic solid and liquid waste generated by commercial and residential establishments.

Sewerage. An entire system of sewage collection, treatment, and disposal.

Silt. Particles of soil, less than 1/20 millimeter in diameter, which can be picked up by air or water.

SL. **S**ound **L**evel.

Sludge. Solids removed from sewage during waste-water treatment.

Smog. Polluted air; the term originated as mixture of the words *smoke* and *fog.*

Soil. The upper layer of the earth, in which plants grow.

Solid Waste. Unwanted or discarded material with insufficient liquid content to be free-flowing.

Sound Level. The amount of sound perceived by human ears, measured in **decibels.**

Sound Pressure Level. Pressure change due to sound, measured in **decibels**.

Species. A class of individuals having common attributes and designated by a common name. Biological classification ranking immediately below the genus or subgenus.

SPL. **S**ound **P**ressure **L**evel.

Succession. The process by which the population of an area changes as the available competing organisms respond to the environment.

Sullage. Refuse, sewage, or mud deposited by water.

Sump. A pit or reservoir serving as a drain or receptacle for liquids.

Surfactant. An agent used in detergents to cause lathering, usually made up of phosphates.

Suspended Solids. Small particles of solid pollutants in sewage that require special treatment to remove.

SWIRS. **S**olid **W**aste **I**nformation **R**etrieval **S**ervice.

Symbiosis. The association of two or more organisms for mutual benefit.

Synergism. The cooperative action of two substances that results in an effect greater than the sum of the effects of the substances acting independently.

Systematics. The science of classification.

Systems Analysis. A systematic approach to assessing any problem, arriving at alternatives, and projecting the probable consequences of employing the alternatives in order to facilitate decision making.

TDG. *See* Total Dissolved Gases.

TDS. *See* Total Dissolved Solids.

Territory. An area sufficient in size to meet the food and social requirements of an animal or group of animals.

Thermocline. Temperature gradient.

Threshold. The point above which an effect is measurable.

Total Dissolved Gases (TDG). A measure of concentration of gases in solution in water that affects the metabolism of aquatic life forms.

Total Dissolved Solids (TDS). The aggregate of salts of calcium, magnesium, sodium, potassium, carbonates, chlorides, sulfates, phosphates, nitrates, and other substances in water.

Toxicant. A chemical that controls pests by killing rather than repelling.

Toxic Substance. A chemical or mixture of chemicals that may present an unreasonable risk of injury to health or the environment.
Transect. A cross section of an area used as a sample for study.
Transpiration. A process by which water vapor is excreted through a living membrane (plant or animal tissue).
Trophic. Of or relating to nutrition.
Trophic Level. The energy level at which an organism sustains itself.
Turbidity. Reduction in light transmission in air or water due to particles in suspension.
UNEP. United Nations Environment Program.
USDA. United States Department of Agriculture.
USGS. United States Geological Survey.
Washout. The removal of air pollutants by rain.
Waste. Unwanted materials left over from manufacturing processes and from places of human or animal habitation.
Water Quality Criteria. Levels of pollutants that affect the use of water.
Watershed. The land area that drains into a stream or river.
Wetland. Land that is often wet or flooded because of water-table rises, tidal fluctuations, or flooding by storms.
WHO. World Health Organization.
Wildlife. Living things that are neither human nor domesticated.
Wind Rose. A graphic presentation of wind speed and direction over time.
WRC. Water Resources Council.
Zooplankton. Planktonic animals that supply food for fish.

APPENDIX B
Federal Government Environmental Departments

Department of Agriculture
14th Street and Jefferson Drive, S.W.
Washington, DC 20250
Soil Conservation Service, Forest Service

Department of the Air Force
Washington, DC 20330

Department of the Army
The Pentagon
Washington, DC 20310

Department of Commerce
Commerce Building
14th Street
Washington, DC 20230
National Oceanic and Atmospheric Administration, National Marine Fisheries, National Technical Information Service

Department of Defense
The Pentagon
Washington, DC 20301

Department of Energy
Forrestal Building
Independence Avenue
Washington, DC 20314
Federal Energy Regulatory Commission

Department of Health and Human Services
200 Independence Avenue, S.W.
Washington, DC 20201

Department of the Interior
Interior Building
C Street
Washington, DC 20240
U.S. Geological Survey, Office of Surface Reclamation and Enforcement, Bureau of Land Management, Bureau of Reclamation, Office of Water Research and Technology, National Park Service, U.S. Fish and Wildlife Service

Department of Justice
Constitution Avenue and 10th Street, N.W.
Washington, DC 20530

Department of Labor
Third Street and Constitution Avenue, N.W.
Washington, DC 20210

Department of the Navy
200 Stovall Street
Alexandria, VA 22332

Department of State
Washington, DC 20520

Department of Transportation
400 7th Street, S.W.
Washington, DC 20590
Federal Highway Administration, Federal Railroad Administration, Saint Lawrence Seaway Corporation, U.S. Coast Guard

APPENDIX C
Federal Environmental Agencies

Advisory Council on Historic
Preservation
1100 Pennsylvania Avenue, N.W.
Washington, DC 20004

Agriculture Department
Natural Resources and Environment
14th Street and Independence Avenue
Washington, DC 20250

Appalachian Regional Commission
1666 Connecticut Avenue, N.W.
Washington, DC 20235

Council on Environmental Quality
722 Jackson Place, N.W.
Washington, DC 20503

Delaware River Basin Commission
1100 L Street, N.W.
Washington, DC 20240

Endangered Species Scientific Authority
Eighteenth and C Streets, N.W.
Washington, DC 20240

Energy Department
1000 Independence Avenue, S.W.
Washington, DC 20585

Environmental Protection Agency
401 M Street, S.W.
Washington, DC 20460

Federal Coordinating Council for
Science, Engineering, and Technology
New Executive Office Building
Washington, DC 20506

Federal Power Commission
Room 4306
825 North Capitol Street, N.E.
Washington, DC 20426

Great Lakes Basin Commission
3475 Plymouth Road
Ann Arbor, MI 48106

Housing and Urban Development
Department
HUD Building
Washington, DC 20410

Interstate Commerce Commission
12th Street and Constitution
Avenue, N.W.
Washington, DC 20423

Justice Department
Land and Natural Resources
1200 Pennsylvania Avenue, N.W.
Washington, DC 20026

Missouri River Basin Commission
10050 Regency Circle
Omaha, NE 68114

National Science Foundation
1880 G Street, N.W.
Washington, DC 20006

National Transportation Safety Board
800 Independence Avenue, S.W.
Washington, DC 20594

New England River Basin Commission
55 Court Street
Boston, MA 02108

Nuclear Regulatory Commission
1717 H Street, N.W.
Washington, DC 20555

Ohio River Basin Commission
36 East 4th Street
Suite 208
Cincinnati, OH 45202

Pacific Northwest River Basin
Commission
P.O. Box 908
Vancouver, WA 98660

Souris-Red-Rainy River Basins
Commission
Suite 6
Professional Building
Holiday Mall
Moorhead, MN 56560

Susquehanna River Basin Commission
1100 L Street, N.W.
Washington, DC 20240

Tennessee Valley Authority
412 1st Street, S.E.
Washington, DC 20444

U.S. Water Resources Council
Suite 800
2120 L Street, N.W.
Washington, DC 20037

APPENDIX D
State Environmental Agencies

ALABAMA

Department of Environmental Management
1751 Federal Drive
Montgomery 36130
(clearinghouse agency)

ALASKA

Department of Environmental Conservation
Pouch O
Juneau 99811

Department of Fish and Game
Subport Building
Juneau 99801

Department of Natural Resources
Pouch M
Juneau 99811

Department of Public Safety
450 Whittier Street
Juneau 99801

ARIZONA

Board of Pesticide Control
1624 W. Adams, Suite 103
Phoenix 85007

Department of Health Services
Division of Environmental Health Services
1740 W. Adams Street
Phoenix 85007

Department of Water Resources
99 W. Virginia
Phoenix 85004

Game and Fish Department
2222 W. Greenway Road
Phoenix 85023

Land Development
1624 W. Adams Street
Phoenix 85007

Oil and Gas Conservation Commission
1645 W. Jefferson, Suite 420
Phoenix 85007

ARKANSAS

Department of Parks and Tourism
One Capitol Mall
Little Rock 72201

Department of Pollution Control and Ecology
8001 National Drive
Little Rock 72219

Forestry Commission
3821 W. Roosevelt Road
Little Rock 72214

Game and Fish Commission
#2 Natural Resources Drive
Little Rock 72205

Natural Heritage Commission
Suite 500, Continental Bldg.
Main & Markham
Little Rock 72201

State Plant Board
#1 Natural Resources Drive
Little Rock 72205

CALIFORNIA

Air Resources Board
1102 Q Street
Sacramento 95814

California Coastal Commission
613 Howard Street, 4th floor
San Francisco 94105

California Energy Commission
1516 Ninth Street, Room 200
Sacramento 95814

Caltrans—Planning
1120 N Street
Sacramento 95814

Department of Boating and Waterways
1629 S Street
Sacramento 95814

Department of Conservation
1416 Ninth Street, Room 1354
Sacramento 95814

Department of Fish and Game
1416 Ninth Street
Sacramento 95814

Department of Food and Agriculture
1220 N Street
Sacramento 95814

Department of Forestry
1416 Ninth Street, Room 1506-17
Sacramento 95814

Department of General Services
1125 Tenth Street
Sacramento 95814

Department of Health
714 P Street, Room 430
Sacramento 95814

Department of Housing and
Community Development
921 10th Street, 5th floor
Sacramento 95814

Department of Parks and Recreation
P.O. Box 2390
Sacramento 95811

Department of Water Resources
1416 Ninth Street
Sacramento 95814

Native American Heritage Commission
915 Capitol Mall, Room 288
Sacramento 95814

Office of Historic Preservation
1050 20th Street
Sacramento 95814

Public Utilities Commission
350 McAllister Street
San Francisco 94102

Public Works Board
1025 P Street, 4th floor
Sacramento 95814

Reclamation Board
1416 Ninth Street
Sacramento 95814

San Francisco Bay Conservation and
Development Commission
30 Van Ness Avenue, Room 2011
San Francisco 94102

Solid Waste Management Board
1020 Ninth Street, Room 300
Sacramento 95814

State Lands Commission
1807 13th Street
Sacramento 95814

State Water Resources Control Board
P.O. Box 100
Sacramento 95801

COLORADO

Department of Agriculture
1525 Sherman Street
Denver 80203

Department of Health
4210 East 11th Avenue
Denver 80220

Department of Natural Resources
1313 Sherman Street
Denver 80203

Office of Energy Conservation
112 East 14th Avenue
Denver 80203

CONNECTICUT

Connecticut Siting Council
1 Central Park Plaza
New Britain 06051

Council on Environmental Quality
165 Capitol Avenue
Hartford 06115

Department of Agriculture
165 Capitol Avenue
Hartford 06115

Department of Environmental
Protection
165 Capitol Avenue
Hartford 06115

DELAWARE

Department of Agriculture
Drawer D
Dover 19901

Department of Natural Resources and
Environmental Control
The Edward Tatnall Building
Dover 19901

Geological Survey
University of Delaware
Newark 19711

DISTRICT OF COLUMBIA

Department of Environmental Services
415 12th Street, N.W.
Washington, DC 20004

FLORIDA

Department of Agriculture and
Consumer Services
State Capitol
Tallahassee 32301

Department of Community Affairs
2571 Executive Center Circle, E.
Tallahassee 32304

Department of Environmental
Regulation
2600 Blairstone Road
Tallahassee 32301

Department of Natural Resources
Commonwealth Building
Tallahassee 32303

Department of State
Division of Archives, History and
Records Management
The Capitol
Tallahassee 32301

Game and Fresh Water Fish
Commission
620 S. Meridian Street
Tallahassee 32304

NW Florida Water Management
District
Route 1, Box 3100
Havana 32333

St. Johns River Water Management
District
Post Office Box 1429
Palatka 32077

South Florida Water Management
District
3301 Gun Club Road
West Palm Beach 33402

State Department of Health and
Rehabilitative Services
1323 Winewood Building
Tallahassee 32301

Suwannee River Water Management
District
Route 3, Box 64
Live Oak 32060

SW Florida Water Management District
2379 Broad Street
Brooksville 33512

GEORGIA

Department of Agriculture
Agriculture Bldg.
Capitol Square
Atlanta 30334

Department of Natural Resources
270 Washington Street, S.W.
Atlanta 30334

HAWAII

Department of Agriculture
P.O. Box 22159
Honolulu 96822

Department of Health
P.O. Box 3378
Honolulu 96801

Department of Land and Natural
Resources
P.O. Box 621
Honolulu 96809

Department of Planning and Economic
Development
250 King Street
Honolulu 96813

Office of Environmental Quality
Control
Department of Health
550 Halekauurla Street, Room 301
Honolulu 96809

IDAHO

Department of Agriculture
P.O. Box 790
Boise 83701

Department of Health and Welfare
Statehouse
Boise 83720

Department of Lands
State Capitol Building
Boise 83720

Department of Parks and Recreation
Statehouse
Boise 83720

Department of Water Resources
Statehouse
Boise 83720

Fish and Game Department
600 S. Walnut
Boise 83707

ILLINOIS

Department of Agriculture
State Fairgrounds
Springfield 62706

Department of Conservation
524 S. 2nd
Springfield 62706

Department of Energy and Natural
Resources
325 West Adams
Springfield 62706

DENR
State Geological Survey Division
615 E. Peabody
Urbana 61820

DENR
State Natural History Survey Division
607 E. Peabody
Urbana 61820

DENR
State Water Survey
605 E. Springfield Street
Champaign 61820

Department of Mines and Minerals
704 Stratton Building
Springfield 62706

Department of Nuclear Safety
1035 Outer Park Drive
5th Floor
Springfield 62704

Department of Transportation
Division of Water Resources
2300 S. Dirksen Parkway
Springfield 62764

Illinois Energy Resource Commission
Stratton Building
3rd Floor South
Springfield 62706

Illinois Environmental Protection Agency
2200 Churchill Road
Springfield 62706

Illinois Nature Preserves Commission
600 North Grand West
Springfield 62706

Pollution Control Board
309 W. Washington
Chicago 60606

INDIANA

Department of Natural Resources
608 State Office Building
Indianapolis 46204

State Board of Health
1330 W. Michigan Street
Indianapolis 46106

State Geological Survey
611 N. Walnut Grove
Bloomington 47401

State Soil and Water Conservation Committee
AGAD Building
Purdue University
West Lafayette 47907

Stream Pollution Control Board
1330 W. Michigan Street
Indianapolis 46106

IOWA

Department of Agriculture
Wallace State Office Bldg.
Des Moines 50319

Department of Soil Conservation
State Capitol Complex
Des Moines 50319

Department of Water, Air, and Waste Management
Henry A. Wallace Bldg.
900 East Grand
Des Moines 50319

Geological Survey
123 N. Capitol Street
Iowa City 52242

State Conservation Commission
Wallace State Office Bldg.
Des Moines 50319

Water Quality Commission
Henry A. Wallace Bldg.
900 East Grand
Des Moines 50319

KANSAS

Fish and Game Commission
Box 54A—R.R. 2
Pratt 67124

Forestry Extension
Kansas State University
2610 Claflin Road
Manhattan 66506

Geological Survey
University of Kansas
303 Moore
Lawrence 66045

State Biological Survey
University of Kansas
210 Snow
Lawrence 66045

State Board of Agriculture
109 West 9th
Mezzanine
Topeka 66612

State Department of Health and Environment
Forbes Field
Building 740
Topeka 66612

State Park and Resources Authority
503 Kansas Avenue
Topeka 66603

Water Resources Board
503 Kansas Avenue, Suite 303
Topeka 66603

KENTUCKY

Department of Agriculture
Capital Plaza Tower
Frankfort 40601

Department of Fish and Wildlife Resources
Arnold Mitchell Building
#1 Game Farm Road
Frankfort 40601

Department of Mines and Minerals
P.O. Box 680
Lexington 40586

Department of Parks
Capital Plaza Tower
Frankfort 40601

Geological Survey
311 Breckinridge Hall
University of Kentucky
Lexington 40506

Kentucky Heritage Council and State Historic Preservation Officer
Capital Plaza Tower
Frankfort 40601

Natural Resources and Environmental Protection Cabinet
Capital Plaza Tower
Frankfort 40601

LOUISIANA

Department of Commerce and Industry
P.O. Box 44185, Capitol Station
Baton Rouge 70804
(clearinghouse agency)

Department of Environmental Quality
P.O. Box 44066, Capitol Station
Baton Rouge 70804

Department of Wildlife and Fisheries
P.O. Box 15570
Baton Rouge 70895

MAINE

Bureau of Forestry
State House
Augusta 04333

Bureau of Parks and Recreation
State House
Augusta 04333

Bureau of Public Lands
Augusta 04333

Department of Agriculture
State Office Bldg.
Augusta 04333

Department of Conservation
State Office Bldg.
Augusta 04333

Department of Environmental Protection
Ray Building
Hospital Street
Augusta 04333

Department of Inland Fisheries and Wildlife
284 State Street
Augusta 04333

Department of Marine Resources
State House
Augusta 04333

Land-Use Regulation Commission
State House
Augusta 04333

Maine Geological Survey
State House
Augusta 04333

State Soil and Water Conservation Commission
State House
Augusta 04333

MARYLAND

Department of Agriculture
50 Harry S. Truman Parkway
Annapolis 21401

Department of Natural Resources
Tawes State Office Building
Annapolis 21401

State Department of Health and Mental Hygiene
201 W. Preston Street
Baltimore 21203

MASSACHUSETTS

Coastal Zone Management Office
Executive Office of Environmental Affairs
100 Cambridge Street
Boston 02202

Department of Environmental Management
100 Cambridge Street
Boston 02202

Department of Environmental Quality Engineering
100 Cambridge Street
Boston 02202

Department of Fisheries, Wildlife, and Recreation Vehicles
100 Cambridge Street
Boston 02202

Department of Food and Agriculture
100 Cambridge Street
Boston 02202

Department of Metropolitan District Commission
100 Cambridge Street
Boston 02202

Department of Public Health
600 Washington Street
Boston 02111

Department of Public Works
100 Nashua Street
Boston 02114

Department of Wetlands
100 Cambridge Street
Boston 02202

Executive Office of Environmental Affairs
Leverett Saltonstall Building
100 Cambridge Street
Boston 02202

Massachusetts Environmental Policy Act (MEPA) Unit
Executive Office of Environmental Affairs
100 Cambridge Street
Boston 02202

MICHIGAN

Department of Commerce
6545 Mercantile Way
Lansing 48909

Department of Natural Resources
P.O. Box 30028
Lansing 48909

Department of Public Health
3500 North Logan
P.O. Box 30035
Lansing 48909

Environmental Review Coordinator
History Division
Department of State
208 North Capitol Avenue
Lansing 48918

MINNESOTA

Bureau of Business Licenses
Minnesota Department of Energy and Economic Development
900 American Center
150 East Kellogg Boulevard
St. Paul 55101
(clearinghouse agency)

MISSISSIPPI

Bureau of Land and Water Resources
P.O. Box 10631
Jackson 39209

Department of Agriculture and Commerce
P.O. Box 1609
Jackson 39205

Department of Natural Resources
Bureau of Pollution Control
P.O. Box 10385
Jackson 39209

Department of Natural Resources
Bureau of Recreation and Parks
P.O. Box 10600
Jackson 39209

Department of Wildlife Conservation
Bureau of Marine Resources
P.O. Box 959
Long Beach 39560

Forestry Commission
908 Robert E. Lee Building
Jackson 39201

Game and Fish Commission
239 N. Lamar Street
Jackson 39205

State Board of Health
P.O. Box 1700
Jackson 39205

MISSOURI

Department of Agriculture
P.O. Box 630
Jefferson City 65102

Department of Conservation
P.O. Box 180
Jefferson City 65102

Department of Natural Resources
P.O. Box 1368
Jefferson City 65102

MONTANA
Department of Agriculture
Scott Hart Building
303 Roberts Street
Helena 59620

Department of Commerce
1424 9th Avenue
Helena 59620

Department of Fish, Wildlife, and Parks
1420 East Sixth Avenue
Helena 59620

Department of Health and Environmental Sciences
Cogswell Building
Helena 59620

Department of Highways
2701 Prospect Avenue
Helena 59620

Department of Natural Resources and Conservation
32 South Ewing
Helena 59620

Department of State Lands
Capitol Station
Helena 59620

Environmental Quality Council
Capitol Station
Helena 59620

Montana Historical Society
225 North Roberts
Helena 59620

NEBRASKA
Department of Agriculture
301 Centennial Mall South
Lincoln 68509

Department of Environmental Control
State House Station
Box 94877
Lincoln 68509

Department of Water Resources
State House Station
Box 94676
Lincoln 68509

Game and Parks Commission
2200 N. 33rd Street
Lincoln 68503

Natural Resources Commission
301 Centennial Mall South
Lincoln 68509

Oil and Gas Conservation Commission
P.O. Box 399
Sidney 69162

NEVADA
Bureau of Mines and Geology
University of Nevada
Reno 89557

Department of Agriculture
350 Capitol Hill Avenue
Reno 89510

Department of Conservation and Natural Resources
Division of Environmental Protection
201 South Fall Street, Room 221
Carson City 89710

Department of Human Resources
Division of Health
Kinkead Building
505 E. King Street
Carson City 89710

Department of Wildlife
P.O. Box 10678
Reno 89520

NEW HAMPSHIRE
Council of Resources and Development
State House Annex
Concord 03301

Department of Agriculture
85 Manchester Street
Concord 03301

Department of Resource and Economic Development
Christian Mutual Building
Concord 03301

Fish and Game Department
34 Bridge Street
Concord 03301

State Conservation Committee
85 Manchester Street
Concord 03301

Water Resources Board
37 Pleasant Street
Concord 03301

Water Supply and Pollution Control Commission
Hazen Drive
Concord 03301

NEW JERSEY
Department of Agriculture
CN 330
Trenton 08625

Department of Environmental
Protection
CN 402
Trenton 08625

NEW MEXICO
Department of Agriculture
P.O. Box 3189
Las Cruces 88003

Health and Environmental Department
P.O. Box 968
Santa Fe 87503

Natural Resources Department
Villagra Building
Santa Fe 87503

State Bureau of Mines and Mineral
Resources
Campus Station
Socorro 87801

NEW YORK
Department of Environmental
Conservation
Regional Permit Administrator

Building 40, Room 219
SUNY at Stony Brook
Stony Brook 11794

2 World Trade Center
Room 6126
New York 10047

21 South Putt Corners Road
New Paltz 12561

3176 Guilderland Avenue
Schenectady 12306

Route 86
Ray Brook 12167

State Office Building
317 Washington Street
Watertown 13601

P.O. Box 5170, Fisher Avenue
Cortland 13045

6274 East Avon-Lima Road
Avon 14414

600 Delaware Avenue
Buffalo 14202

Office of Business Permits
New York State Executive Department
Gov. Alfred E. Smith Office Building
Albany 12225

NORTH CAROLINA
Department of Natural Resources and
Community Development
14th floor, Archdale Building
Raleigh 27611
(information clearinghouse)

Division of Environmental Management
9th floor, Archdale Building
Raleigh 27611

Division of Forest Resources
10th floor, Archdale Building
Raleigh 27611

Division of Land Resources
5th floor, Archdale Building
Raleigh 27611

Division of Marine Fisheries
P.O. Box 769
Morehead City 28557

Division of Parks and Recreation
7th floor, Archdale Building
Raleigh 27611

Division of Soil and Water Conservation
5th floor, Archdale Building
Raleigh 27611

Natural Resources Planning and
Assessment
8th floor, Archdale Building
Raleigh 27611

Office of Coastal Management
13th floor, Archdale Building
Raleigh 27611

Office of Legal Affairs
12th floor, Archdale Building
Raleigh 27611

Office of Water Resources
11th floor, Archdale Building
Raleigh 27611

Wildlife Resources Commission
3rd floor, Archdale Building
Raleigh 27611

NORTH DAKOTA
Department of Agriculture
State Capitol
Bismarck 58505

Department of Health
Bismarck 58505

Geological Survey
University Station
Grand Forks 58202

Natural Resources Council
Governor's Office
State Capitol
Bismarck 58505

Parks and Recreation Department
R.D. 2, Box 139
Mandan 58554

Public Service Commission
State Capitol Building
Bismarck 58505

State Game and Fish Department
2121 Lovett Avenue
Bismarck 58505

State Soil and Conservation Committee
State Capitol Building
Bismarck 58505

Water Commission
Bismarck 58505

OHIO

Department of Agriculture
65 S. Front Street
Columbus 43215

Department of Natural Resources
Fountain Square
Columbus 43224

Environmental Protection Agency
361 East Broad Street
Columbus 43216

Hazardous Waste Facility Approval Board
361 East Broad Street
Columbus 43216

Ohio Power Siting Board
580 South High Street
Columbus 43215

Public Utilities Commission of Ohio
180 East Broad Street
Columbus 43215

OKLAHOMA

Conservation Commission
2800 North Lincoln, Room 160
Oklahoma City 73105

Department of Mines
4040 N. Lincoln Boulevard, Suite 107
Oklahoma City 73105

Environmental Health Services
State Department of Health
1000 N.E. Tenth
Oklahoma City 73152

Environmental Section
Department of Wildlife Conservation
1801 North Lincoln
Oklahoma City 73105

Oklahoma Corporation Commission
Jim Thorpe Building
Oklahoma City 73105

State Department of Agriculture
2800 North Lincoln
Oklahoma City 73105

Water Quality Division
Water Resources Board
1000 N.E. Tenth
Oklahoma City 73152

OREGON

Department of Agriculture
Salem 97310

Department of Environmental Quality
522 S.W. Fifth Avenue
Portland 97207

Department of Fish and Wildlife
P.O. Box 3503
Portland 97208

Department of Geology and Mineral Industries
1069 State Office Building
Portland 97201

Department of Transportation
Transportation Building
Salem 97310

State Department of Forestry
2600 State Street
Salem 97310

PENNSYLVANIA

Department of Agriculture
2301 N. Cameron Street
Harrisburg 17120

Department of Environmental Resources
Fulton Bldg.
Harrisburg 17120

Fish Commission
P.O. Box 1673
Harrisburg 17120

Game Commission
P.O. Box 1567
Harrisburg 17120

Governor's Energy Council
P.O. Box 8010
Harrisburg 17102

State Conservation Commission
P.O. Box 2357
Executive House
Harrisburg 17120

RHODE ISLAND

Department of Environmental Management
83 Park Street
Providence 02903

Department of Transportation
210 State Office Bldg.
Providence 02903

State Water Resources Board
265 Melrose Street
Providence 02907

SOUTH CAROLINA

Department of Agriculture
P.O. Box 11280
Columbia 29211

Department of Health and Environmental Control
J. Marion Sims Bldg.
2600 Bull Street
Columbia 29201

Department of Parks, Recreation, and Tourism
Edgar Brown Bldg.
1205 Pendleton Street
Columbia 29201

Division of Natural Resources
P.O. Box 1145
Columbia 29211

Geological Survey
Harbison Forest Road
Columbia 29210

State Commission of Forestry
P.O. Box 21707
Columbia 29221

State Land Resources Conservation Commission
2221 Devine Street
Columbia 29205

Water Resources Commission
3830 Forest Drive
Columbia 29240

Wildlife and Marine Resources Department
Building D
Dutch Plaza
Columbia 29202

SOUTH DAKOTA

Department of Agriculture
Sigurd Anderson Bldg.
Pierre 57501

Department of Water and Natural Resources
Joe Foss Bldg.
Pierre 57501

Game, Fish, and Parks Department
Sigurd Anderson Bldg.
Pierre 57501

TENNESSEE

Department of Agriculture
P.O. Box 40627
Melrose Branch
Nashville 37204

Department of Conservation
2611 W. End Avenue
Nashville 37203

Energy, Environment, and Resources Center
University of Tennessee
Knoxville 37916

State Soil Conservation Committee
P.O. Box 1071
Knoxville 37901

Water Quality Control Board
621 Condell Hall Building
Nashville 37219

Wildlife Resources Agency
Ellington Agricultural Center
Nashville 37204

TEXAS

Bureau of Economic Geology
University Station
Box X
Austin 78712

Department of Agriculture
P.O. Box 12847
Capitol Station
Austin 78711

Department of Health
1100 W. 49th Street
Austin 78756

Department of Water Resources
1700 N. Congress
Austin 78701

Forest Service
College Station 77843

General Land Office
Stephen Austin State Office Building
Austin 78701

Guadalupe-Blanco River Authority
P.O. Box 271
Seguin 78155

Parks and Wildlife Department
4200 Smith School Road
Austin 78744

State Soil and Water Conservation Board
1002 First National Bldg.
Temple 76501

UTAH

Department of Agriculture
350 North Redwood Road
Salt Lake City 84116

Department of Health
150 West North Temple
P.O. Box 2500
Salt Lake City 84110

Department of Natural Resources
1636 West North Temple
Salt Lake City 84116

State Soil Conservation Commission
147 North 200 West
Salt Lake City 84103

VERMONT

Agency of Environmental Conservation
Montpelier 05602

Department of Agriculture
116 State Street
Montpelier 05602

Department of Health
60 Main Street
Burlington 05401

Office of the State Geologist
Montpelier 05602

State Natural Resources Conservation Council
State Office Building
Montpelier 05602

VIRGINIA

Commission of Game and Inland Fisheries
4010 W. Broad Street
Richmond 23230

Council on the Environment
903 Ninth Street Office Building
Richmond 23219

Department of Agriculture and Consumer Services
P.O. Box 1163
Richmond 23209

Department of Conservation and Economic Development
1100 Washington Bldg.
Richmond 23219

Department of Health
James Madison Building
109 Governor Street
Richmond 23219

Division of Parks
1201 Washington Building
Richmond 23219

Marine Resources Commission
P.O. Box 1248
Newport News 23601

Northern Virginia Regional Park Authority
11001 Popes Head Road
Fairfax 22030

Soil and Water Conservation Commission
203 Governor Street
Richmond 23219

State Water Control Board
2111 N. Hamilton Street
Richmond 23230

WASHINGTON

Environmental Permit Information Center
Washington Department of Ecology
Headquarters Office, PV-11
St. Martin's College Campus—Lacey
Olympia 98504
(information clearinghouse)

WEST VIRGINIA

Air Pollution Control Commission
1558 Washington Street, East
Charleston 25311

Department of Agriculture
State Capitol Building
Charleston 25305

Department of Health
Building 3, Room 206
State Capitol Building
Charleston 25305

Department of Natural Resources
1800 Washington Street, East
Charleston 25305

Geological and Economic Survey
P.O. Box 879
Morgantown 26505

WISCONSIN

Department of Agriculture, Trade, and Consumer Protection
801 W. Badger Road
Madison 53708

Department of Natural Resources
P.O. Box 7921
Madison 53707

Geological and Natural History Survey
1815 University Avenue
Madison 53706

State Board of Soil and Water Conservation Districts
1815 University Avenue
Madison 53706

WYOMING

Department of Agriculture
2219 Carey Avenue
Cheyenne 82002

Department of Economic Planning and Development
Herschler Building
Cheyenne 82002

Department of Health and Social Services
Hathaway Building
Cheyenne 82002

Environmental Quality Department
Herschler Building
Cheyenne 82002

Game and Fish Department
5400 Bishop
Cheyenne 82002

Geological Survey of Wyoming
Box 3008
University Station
Laramie 82071

Office of Industrial Siting Administration
Boyd Building
Cheyenne 82002

Recreation Commission
Herschler Building
Cheyenne 82002

State Board of Land Commissioners
Herschler Building
Cheyenne 82002

State Conservation Commission
22119 Carey Avenue
Cheyenne 82002

State Forestry Division
1100 West 22nd Street
Cheyenne 82002

APPENDIX E
Environmental Organizations

Air Pollution Control Association
4400 Fifth Avenue
Pittsburgh, PA 15213

Alliance for Environmental Education
3421 M Street, N.W.
Washington, DC 20007

American Conservation Association, Inc.
30 Rockefeller Plaza
Room 5510
New York, NY 10020

American Forestry Association
1319 18th Street, N.W.
Washington, DC 20036

American Rivers Conservation Council
323 Pennsylvania Avenue, S.E.
Washington, DC 20003

Americans for the Environment
322 4th Street, N.W.
Washington, DC 20002

Center for Action on Endangered Species
175 West Main Street
Ayer, MA 01432

Center for Environmental Education, Inc.
1925 K Street
Suite 206
Washington, DC 20006

Center for Law and Social Policy
1751 N Street, N.W.
Washington, DC 20036

Center for Renewable Resources
1001 Connecticut Avenue, N.W., Room 510
Washington, DC 20036

Clean Water Action Project
1341 G Street, N.W., Suite 204
Washington, DC 20001

Coastal Society
5410 Grosvenor Lane, Suite 150
Bethesda, MD 20814

Common Cause
2030 M Street, N.W.
Washington, DC 20036

Concern, Inc.
1794 Columbia Road, N.W.
Washington, DC 20009

Conservation Associates
1500 Mills Tower
220 Bush Street
San Francisco, CA 94104

Conservation Foundation
1255 23rd Street, N.W.
Washington, DC 20037

Council of Pollution Control Financing Agencies
477 H Street, N.W.
Washington, DC 20031

Defenders of Wildlife
1244 19th Street, N.W.
Washington, DC 20036

Environmental Action, Inc.
1525 New Hampshire Avenue, N.W.
Washington, DC 20036

Environmental Action Foundation, Inc.
1525 New Hampshire Avenue, N.W.
Washington, DC 20036

Environmental Defense Fund, Inc.
475 Park Avenue South
New York, NY 10016

Environmental Fund
1302 18th Street, N.W.
Washington, DC 20036

Environmental Law Institute
Dupont Circle Building, Suite 600
1346 Connecticut Avenue, N.W.
Washington, DC 20036

Environmental Policy Center
317 Pennsylvania Avenue, S.E.
Washington, DC 20003

Environmental Task Force
1012 14th Street, N.W.
Washington, DC 20005

Friends of the Earth
124 Spear Street
San Francisco, CA 94105

GREENPEACE USA
1611 Connecticut Avenue, N.W.
Washington, DC 20009

International Association of Fish and Wildlife Agencies
1412 16th Street, N.W.
Washington, DC 20036

International Institute for Environment and Development
1717 Massachusetts Avenue, N.W.
Washington, DC 20036

INTERNET: The International Environmental Resources Network
P.O. Box 417
Concord, MA 01742

Izaak Walton League of America, Inc.
1701 N. Fort Myer Drive
Arlington, VA 22209

League of Conservation Voters
317 Pennsylvania Avenue, S.E.
Washington, DC 20003

National Audubon Society
950 Third Avenue
New York, NY 10022

National Clean Air Coalition
530 Seventh Street, S.E.
Washington, DC 20003

National Parks and Conservation Association
1701 18th Street, N.W.
Washington, DC 20009

National Resource Defense Council, Inc.
122 East 42nd Street
New York, NY 10017

National Waterways Conference, Inc.
1130 17th Street, N.W., Suite 200
Washington, DC 20036

National Wetlands Technical Council
Suite 300
1717 Massachusetts Avenue, N.W.
Washington, DC 20036

National Wildlife Federation
1325 Massachusetts Avenue, N.W.
Washington, DC 20036

Nature Conservancy
1800 North Kent Street, Suite 800
Arlington, VA 22209

North American Wildlife Foundation
709 Wire Building
1000 Vermont Avenue, N.W.
Washington, DC 20005

Rachel Carson Council, Inc.
8940 Jones Mill Road
Chevy Chase, MD 20815

Resources for the Future
1616 P Street, N.W.
Washington, DC 20036

ROMCOE, Center for Environmental Problem Solving
1115 Grant Street
Denver, CO 80203

Sierra Club
530 Bush Street
San Francisco, CA 94108

Union of Concerned Scientists
1346 Connecticut Avenue, N.W., Suite 1101
Washington, DC 20036

Urban Environmental Conference, Inc.
1314 14th Street, N.W., 3rd Floor
Washington, DC 20005

Urban Wildlife Research Center, Inc.
10921 Trotting Ridge Way
Columbia, MD 21044

Water Pollution Control Federation
601 Rythe Street
Alexandria, VA 22314

Water Resources Congress
3800 N. Fairfax Drive
Arlington, VA 22203

Wildfowl Foundation
709 Wire Building
1000 Vermont Avenue, N.W.
Washington, DC 20005

Wilderness Society
1901 Pennsylvania Ave., N.W.
Washington, DC 20006

Wildlife Legislative Fund of America
1612 K Street, N.W., Suite 1101
Washington, DC 20006

Wildlife Management Institute
709 Wire Building
1000 Vermont Avenue, N.W.
Washington, DC 20005

Wildlife Society
7101 Wisconsin Avenue, N.W., Suite 611
Washington, DC 20014

Worldwatch Institute
1776 Massachusetts Avenue, N.W.
Washington, DC 20036

World Wildlife Fund
1601 Connecticut Avenue, N.W., Suite 800
Washington, DC 20009

APPENDIX F
Regulations for Implementing the National Environmental Policy Act

CONTENTS

For further information, contact:

Nicholas C. Yost, General Counsel
Council on Environmental Quality
Executive Office of the President
722 Jackson Pl. N.W.
Washington, D.C. 20006
(202) 395-5750

For sale by the Superintendent of Documents, U.S. Government Printing Office
Washington, D.C. 20402

Stock Number 041-011-00042-4

TABLE OF CONTENTS

PART 1500—PURPOSE, POLICY, AND MANDATE

AUTHORITY: NEPA, the Environmental Quality Improvement Act of 1970, as amended (42 U.S.C. 4371 et seq.), section 309 of the Clean Air Act, as amended (42 U.S.C. 7609) and Executive Order 11514, Protection and Enhancement of Environmental Quality (March 5, 1970 as amended by Executive Order 11991, May 24, 1977).

§ 1500.1 Purpose.

(a) The National Environmental Policy Act (NEPA) is our basic national charter for protection of the environment. It establishes policy, sets goals (section 101), and provides means (section 102) for carrying out the policy. Section 102(2) contains "action-forcing" provisions to make sure that federal agencies act according to the letter and spirit of the Act. The regulations that follow implement Section 102(2). Their purpose is to tell federal agencies what they must do to comply with the procedures and achieve the goals of the Act. The President, the federal agencies, and the courts share responsibility for enforcing the Act so as to achieve the substantive requirements of section 101.

(b) NEPA procedures must insure that environmental information is available to public officials and citizens before decisions are made and before actions are taken. The information must be of high quality. Accurate scientific analysis, expert agency comments, and public scrutiny are essential to implementing NEPA. Most important, NEPA documents must concentrate on the issues that are truly significant to the action in question, rather than amassing needless detail.

(c) Ultimately, of course, it is not better documents but better decisions that count. NEPA's purpose is not to generate paperwork—even excellent paperwork—but to foster excellent action. The NEPA process is intended to help public officials make decisions that are based on understanding of environmental consequences, and take actions that protect, restore, and enhance the environment. These regulations provide the direction to achieve this purpose.

§ 1500.2 Policy.

Federal agencies shall to the fullest extent possible:

(a) Interpret and administer the policies, regulations, and public laws of the United States in accordance with the policies set forth in the Act and in these regulations.

(b) Implement procedures to make the NEPA process more useful to decisionmakers and the public; to reduce paperwork and the accumulation of extraneous background data; and to emphasize real environmental issues and alternatives. Environmental impact statements shall be concise, clear, and to the point, and shall be supported by evidence that agencies have made the necessary environmental analyses.

(c) Integrate the requirements of NEPA with other planning and environmental review procedures required by law or by agency practice so that all such procedures run concurrently rather than consecutively.

(d) Encourage and facilitate public involvement in decisions which affect the quality of the human environment.

(e) Use the NEPA process to identify and assess the reasonable alternatives to proposed actions that will avoid or minimize adverse effects of these actions upon the quality of the human environment.

(f) Use all practicable means, consistent with the requirements of the Act and other essential considerations of national policy, to restore and enhance the quality of the human environment and avoid or minimize any possible adverse effects of their actions upon the quality of the human environment.

§ 1500.3 Mandate.

Parts 1500–1508 of this Title provide regulations applicable to and binding on all Federal agencies for implementing the procedural provisions of the National Environmental Policy Act of 1969, as amended (Pub. L. 91-190, 42 U.S.C. 4321 et seq.) (NEPA or the Act) except where compliance would be inconsistent with other statutory requirements. These regulations are issued pursuant to NEPA, the Environmental Quality Improvement Act of 1970, as amended (42 U.S.C. 4371 et seq.) Section 309 of the Clean Air Act, as amended (42 U.S.C. 7609) and Executive Order 11514, Protection and Enhancement of Environmental Quality (March 5, 1970, as amended by Executive Order 11991, May 24, 1977). These regulations, unlike the predecessor guidelines, are not confined to Sec. 102(2)(C) (environmental impact statements). The regulations apply to the whole of section 102(2). The provisions of the Act and of these regulations must be read together as a whole in order to comply with the spirit and letter of the law. It is the Council's intention that judicial review of agency compliance with these regulations not occur before an agency has filed the final environmental impact statement, or has made a final finding of no significant impact (when such a finding will result in action affecting the environment), or takes action that will result in irreparable injury. Furthermore, it is the Council's intention that any trivial violation of these regulations not give rise to any independent cause of action.

§ 1500.4 Reducing paperwork.

Agencies shall reduce excessive paperwork by:

(a) Reducing the length of environmental impact statements (§ 1502.2(c)), by means such as setting appropriate page limits (§§ 1501.7(b)(1) and 1502.7).

(b) Preparing analytic rather than encyclopedic environmental impact statements (§ 1502.2(a)).

(c) Discussing only briefly issues other than significant ones (§ 1502.2(b)).

(d) Writing environmental impact statements in plain language (§ 1502.8).

(e) Following a clear format for environmental impact statements (§ 1502.10).

(f) Emphasizing the portions of the environmental impact statement that are useful to decisionmakers and the public (§§ 1502.14 and 1502.15) and reducing emphasis on background material (§ 1502.16).

(g) Using the scoping process, not only to identify significant environmental issues deserving of study, but also to deemphasize insignificant issues, narrowing the scope of the environmental impact statement process accordingly (§ 1501.7).

(h) Summarizing the environmental impact statement (§ 1502.12) and circulating the summary instead of the entire environmental impact statement if the latter is unusually long (§ 1502.19).

(i) Using program, policy, or plan environmental impact statements and tiering from statements of broad scope to those of narrower scope, to eliminate repetitive discussions of the same issues (§§ 1502.4 and 1502.20).

(j) Incorporating by reference (§ 1502.21).

(k) Integrating NEPA requirements with other environmental review and consultation requirements (§ 1502.25).

(l) Requiring comments to be as specific as possible (§ 1503.3).

(m) Attaching and circulating only changes to the draft environmental impact statement, rather than rewriting and circulating the entire statement when changes are minor (§ 1503.4(c)).

(n) Eliminating duplication with State and local procedures, by providing for joint preparation (§ 1506.2), and with other Federal procedures, by providing that an agency may adopt appropriate environmental documents prepared by another agency (§ 1506.3).

(o) Combining environmental documents with other documents (§ 1506.4).

(p) Using categorical exclusions to define categories of actions which do not individually or cumulatively have a significant effect on the human environment and which are therefore exempt from requirements to prepare an environmental impact statement (§ 1508.4).

(q) Using a finding of no significant impact when an action not otherwise excluded will not have a significant effect on the human environment and is therefore exempt from requirements to prepare an environmental impact statement (§ 1508.13).

§ 1500.5 Reducing delay.

Agencies shall reduce delay by:

(a) Integrating the NEPA process into early planning (§ 1501.2).

(b) Emphasizing interagency cooperation before the environmental impact statement is prepared, rather than submission of adversary comments on a completed document (§ 1501.6).

(c) Insuring the swift and fair resolution of lead agency disputes (§ 1501.5).

(d) Using the scoping process for an early identification of what are and what are not the real issues (§ 1501.7).

(e) Establishing appropriate time limits for the environmental impact statement process (§§ 1501.7(b)(2) and 1501.8).

(f) Preparing environmental impact statements early in the process (§ 1502.5).

(g) Integrating NEPA requirements with other environmental review and consultation requirements (§ 1502.25).

(h) Eliminating duplication with State and local procedures by providing for joint preparation (§ 1506.2) and with other Federal procedures by providing that an agency may adopt appropriate environmental documents prepared by another agency (§ 1506.3).

(i) Combining environmental documents with other documents (§ 1506.4).

(j) Using accelerated procedures for proposals for legislation (§ 1506.8).

(k) Using categorical exclusions to define categories of actions which do not individually or cumulatively have a significant effect on the human environment (§ 1508.4) and which are therefore exempt from requirements to prepare an environmental impact statement.

(l) Using a finding of no significant impact when an action not otherwise excluded will not have a significant effect on the human environment (§ 1508.13) and is therefore exempt from requirements to prepare an environmental impact statement.

§ 1500.6 Agency authority.

Each agency shall interpret the provisions of the Act as a supplement to its existing authority and as a mandate to view traditional policies and missions in the light of the Act's national environmental objectives. Agencies shall review their policies, procedures, and regulations accordingly and revise them as necessary to insure full compliance with the purposes and provisions of the Act. The phrase "to the fullest extent possible" in section 102 means that each agency of the Federal Government shall comply with that section unless existing law applicable to the agency's operations expressly prohibits or makes compliance impossible.

PART 1501—NEPA AND AGENCY PLANNING

AUTHORITY: NEPA, the Environmental Quality Improvement Act of 1970, as amended (42 U.S.C. 4371 et seq.), Section 309 of the Clean Air Act, as amended (42 U.S.C. 7609), and Executive Order 11514, Protection and Enhancement of Environmental Quality (March 5, 1970, as amended by Executive Order 11991, May 24 1977).

§ 1501.1 Purpose.

The purposes of this part include:

(a) Integrating the NEPA process into early planning to insure appropriate consideration of NEPA's policies and to eliminate delay.

(b) Emphasizing cooperative consultation among agencies before the environmental impact statement is prepared rather than submission of adversary comments on a completed document.

(c) Providing for the swift and fair resolution of lead agency disputes.

(d) Identifying at an early stage the significant environmental issues deserving of study and deemphasizing insignificant issues, narrowing the scope of the environmental impact statement accordingly.

(e) Providing a mechanism for putting appropriate time limits on the environmental impact statement process.

§ 1501.2 Apply NEPA early in the process.

Agencies shall integrate the NEPA process with other planning at the earliest possible time to insure that planning and decisions reflect environmental values, to avoid delays later in the process, and to head off potential conflicts. Each agency shall:

(a) Comply with the mandate of section 102(2)(A) to "utilize a systematic, interdisciplinary approach which will insure the integrated use of the natural and social sciences and the environmental design arts in planning and in decisionmaking which may have an impact on man's environment," as specified by § 1507.2.

(b) Identify environmental effects and values in adequate detail so they can be compared to economic and technical analyses. Environmental documents and appropriate analyses shall be circulated and reviewed at the same time as other planning documents.

(c) Study, develop, and describe appropriate alternatives to recommended courses of action in any proposal which involves unresolved conflicts concerning alternative uses of available resources as provided by section 102(2)(E) of the Act.

(d) Provide for cases where actions are planned by private applicants or other non-Federal entities before Federal involvement so that:

(1) Policies or designated staff are available to advise potential applicants of studies or other information foreseeably required for later Federal action.

(2) The Federal agency consults early with appropriate State and local agencies and Indian tribes and with interested private persons and organizations when its own involvement is reasonably foreseeable.

(3) The Federal agency commences its NEPA process at the earliest possible time.

§ 1501.3 When to prepare an environmental assessment.

(a) Agencies shall prepare an environmental assessment (§ 1508.9) when necessary under the procedures adopted by individual agencies to supplement these regulations as described in § 1507.3. An assessment is not necessary if the agency has decided to prepare an environmental impact statement.

(b) Agencies may prepare an environmental assessment on any action at any time in order to assist agency planning and decisionmaking.

§ 1501.4 Whether to prepare an environmental impact statement.

In determining whether to prepare an environmental impact statement the Federal agency shall:

(a) Determine under its procedures supplementing these regulations (described in § 1507.3) whether the proposal is one which:

(1) Normally requires an environmental impact statement, or

(2) Normally does not require either an environmental impact statement or an environmental assessment (categorical exclusion).

(b) If the proposed action is not covered by paragraph (a) of this section, prepare an environmental assessment (§ 1508.9). The agency shall involve environmental agencies, applicants, and the public, to the extent practicable, in preparing assessments required by § 1508.9(a)(1).

(c) Based on the environmental assessment make its determination whether to prepare an environmental impact statement.

(d) Commence the scoping process (§ 1501.7), if the agency will prepare an environmental impact statement.

(e) Prepare a finding of no significant impact (§ 1508.13), if the agency determines on the basis of the environmental assessment not to prepare a statement.

(1) The agency shall make the finding of no significant impact available to the affected public as specified in § 1506.6.

(2) In certain limited circumstances, which the agency may cover in its procedures under § 1507.3, the agency shall make the finding of no significant impact available for public review (including State and areawide clearinghouses) for 30 days before the agency makes its final determination whether to prepare an environmental impact statement and before the action may begin. The circumstances are:

(i) The proposed action is, or is closely similar to, one which normally requires the preparation of an environmental impact statement under the procedures adopted by the agency pursuant to § 1507.3, or

(ii) The nature of the proposed action is one without precedent.

§ 1501.5 Lead agencies.

(a) A lead agency shall supervise the preparation of an environmental impact statement if more than one Federal agency either:

(1) Proposes or is involved in the same action; or

(2) Is involved in a group of actions directly related to each other because of their functional interdependence or geographical proximity.

(b) Federal, State, or local agencies, including at least one Federal agency, may act as joint lead agencies to prepare an environmental impact statement (§ 1506.2).

(c) If an action falls within the provisions of paragraph (a) of this section the potential lead agencies shall determine by letter or memorandum which agency shall be the lead agency and which shall be cooperating agencies. The agencies shall resolve the lead agency question so as not to cause delay. If there is disagreement among the agencies, the following factors (which are listed in order of descending importance) shall determine lead agency designation:

(1) Magnitude of agency's involvement.

(2) Project approval/disapproval authority.

(3) Expertise concerning the action's environmental effects.

(4) Duration of agency's involvement.

(5) Sequence of agency's involvement.

(d) Any Federal agency, or any State or local agency or private person substantially affected by the absence of lead agency designation, may make a written request to the potential lead agencies that a lead agency be designated.

(e) If Federal agencies are unable to agree on which agency will be the lead agency or if the procedure described in paragraph (c) of this section has not resulted within 45 days in a lead agency designation, any of the agencies or persons concerned may file a request with the Council asking it to determine which Federal agency shall be the lead agency.

A copy of the request shall be transmitted to each potential lead agency. The request shall consist of:

(1) A precise description of the nature and extent of the proposed action.

(2) A detailed statement of why each potential lead agency should or should not be the lead agency under the criteria specified in paragraph (c) of this section.

(f) A response may be filed by any potential lead agency concerned within 20 days after a request is filed with the Council. The Council shall determine as soon as possible but not later than 20 days after receiving the request and all responses to it which Federal agency shall be the lead agency and which other Federal agencies shall be cooperating agencies.

§ 1501.6 Cooperating agencies.

The purpose of this section is to emphasize agency cooperation early in the NEPA process. Upon request of the lead agency, any other Federal agency which has jurisdiction by law shall be a cooperating agency. In addition any other Federal agency which has special expertise with respect to any environmental issue, which should be addressed in the statement may be a cooperating agency upon request of the lead agency. An agency may request the lead agency to designate it a cooperating agency.

(a) The lead agency shall:

(1) Request the participation of each cooperating agency in the NEPA process at the earliest possible time.

(2) Use the environmental analysis and proposals of cooperating agencies with jurisdiction by law or special expertise, to the maximum extent possible consistent with its responsibility as lead agency.

(3) Meet with a cooperating agency at the latter's request.

(b) Each cooperating agency shall:

(1) Participate in the NEPA process at the earliest possible time.

(2) Participate in the scoping process (described below in § 1501.7).

(3) Assume on request of the lead agency responsibility for developing information and preparing environmental analyses including portions of the environmental impact statement concerning which the cooperating agency has special expertise.

(4) Make available staff support at the lead agency's request to enhance the latter's interdisciplinary capability.

(5) Normally use its own funds. The lead agency shall, to the extent available funds permit, fund those major activities or analyses it requests from cooperating agencies. Potential lead agencies shall include such funding requirements in their budget requests.

(c) A cooperating agency may in response to a lead agency's request for assistance in preparing the environmental impact statement (described in paragraph (b) (3), (4), or (5) of this section) reply that other program commitments preclude any involvement or the degree of involvement requested in the action that is the subject of the environmental impact statement. A copy of this reply shall be submitted to the Council.

§ 1501.7 Scoping.

There shall be an early and open process for determining the scope of issues to be addressed and for identifying the significant issues related to a proposed action. This process shall be termed scoping. As soon as practicable after its decision to prepare an environmental impact statement and before the scoping process the lead agency shall publish a notice of intent (§ 1508.22) in the FEDERAL REGISTER except as provided in § 1507.3(e).

(a) As part of the scoping process the lead agency shall:

(1) Invite the participation of affected Federal, State, and local agencies, any affected Indian tribe, the proponent of the action, and other interested persons (including those who might not be in accord with the action on environmental grounds), unless there is a limited exception under § 1507.3(c). An agency may give notice in accordance with § 1506.6.

(2) Determine the scope (§ 1508.25) and the significant issues to be analyzed in depth in the environmental impact statement.

(3) Identify and eliminate from detailed study the issues which are not significant or which have been covered by prior environmental review

(§ 1506.3), narrowing the discussion of these issues in the statement to a brief presentation of why they will not have a significant effect on the human environment or providing a reference to their coverage elsewhere.

(4) Allocate assignments for preparation of the environmental impact statement among the lead and cooperating agencies, with the lead agency retaining responsibility for the statement.

(5) Indicate any public environmental assessments and other environmental impact statements which are being or will be prepared that are related to but are not part of the scope of the impact statement under consideration.

(6) Identify other environmental review and consultation requirements so the lead and cooperating agencies may prepare other required analyses and studies concurrently with, and integrated with, the environmental impact statement as provided in § 1502.25.

(7) Indicate the relationship between the timing of the preparation of environmental analyses and the agency's tentative planning and decisionmaking schedule.

(b) As part of the scoping process the lead agency may:

(1) Set page limits on environmental documents (§ 1502.7).

(2) Set time limits (§ 1501.8).

(3) Adopt procedures under § 1507.3 to combine its environmental assessment process with its scoping process.

(4) Hold an early scoping meeting or meetings which may be integrated with any other early planning meeting the agency has. Such a scoping meeting will often be appropriate when the impacts of a particular action are confined to specific sites.

(c) An agency shall revise the determinations made under paragraphs (a) and (b) of this section if substantial changes are made later in the proposed action, or if significant new circumstances or information arise which bear on the proposal or its impacts.

§ 1501.8 Time limits.

Although the Council has decided that prescribed universal time limits for the entire NEPA process are too inflexible, Federal agencies are encouraged to set time limits appropriate to individual actions (consistent with the time intervals required by § 1506.10). When multiple agencies are involved the reference to agency below means lead agency.

(a) The agency shall set time limits if an applicant for the proposed action requests them: *Provided,* That the limits are consistent with the purposes of NEPA and other essential considerations of national policy.

(b) The agency may:

(1) Consider the following factors in determining time limits:

(i) Potential for environmental harm.

(ii) Size of the proposed action.

(iii) State of the art of analytic techniques.

(iv) Degree of public need for the proposed action, including the consequences of delay.

(v) Number of persons and agencies affected.

(vi) Degree to which relevant information is known and if not known the time required for obtaining it.

(vii) Degree to which the action is controversial.

(viii) Other time limits imposed on the agency by law, regulations, or executive order.

(2) Set overall time limits or limits for each constituent part of the NEPA process, which may include:

(i) Decision on whether to prepare an environmental impact statement (if not already decided).

(ii) Determination of the scope of the environmental impact statement.

(iii) Preparation of the draft environmental impact statement.

(iv) Review of any comments on the draft environmental impact statement from the public and agencies.

(v) Preparation of the final environmental impact statement.

(vi) Review of any comments on the final environmental impact statement.

(vii) Decision on the action based in part on the environmental impact statement.

(3) Designate a person (such as the project manager or a person in the agency's office with NEPA responsibilities) to expedite the NEPA process.

(c) State or local agencies or members of the public may request a Federal Agency to set time limits.

PART 1502—ENVIRONMENTAL IMPACT STATEMENT

AUTHORITY: NEPA, the Environmental Quality Improvement Act of 1970, as amended (42 U.S.C. 4371 et seq.), Section 309 of the Clean Air Act, as amended (42 U.S.C. 7609), and Executive Order 11514, Protection and Enhancement of Environmental Quality (March 5, 1970, as amended by Executive Order 11991, May 24, 1977).

§ 1502.1 Purpose.

The primary purpose of an environmental impact statement is to serve as an action-forcing device to insure that the policies and goals defined in the Act are infused into the ongoing programs and actions of the Federal Government. It shall provide full and fair discussion of significant environmental impacts and shall inform decisionmakers and the public of the reasonable alternatives which would avoid or minimize adverse impacts or enhance the quality of the human environment. Agencies shall focus on significant environmental issues and alternatives and shall reduce paperwork and the accumulation of extraneous background data. Statements shall be concise, clear, and to the point, and shall be supported by evidence that the agency has made the necessary environmental analyses. An environmental impact statement is more than a disclosure document. It shall be used by Federal officials in conjunction with other relevant material to plan actions and make decisions.

§ 1502.2 Implementation.

To achieve the purposes set forth in § 1502.1 agencies shall prepare environmental impact statements in the following manner:

(a) Environmental impact statements shall be analytic rather than encyclopedic.

(b) Impacts shall be discussed in proportion to their significance. There shall be only brief discussion of other than significant issues. As in a finding of no significant impact, there should be only enough discussion to show why more study is not warranted.

(c) Environmental impact statements shall be kept concise and shall be no longer than absolutely necessary to comply with NEPA and with these regulations. Length should vary first with potential environmental problems and then with project size.

(d) Environmental impact statements shall state how alternatives considered in it and decisions based on it will or will not achieve the requirements of sections 101 and 102(1) of the Act and other environmental laws and policies.

(e) The range of alternatives discussed in environmental impact statements shall encompass those to be considered by the ultimate agency decisionmaker.

(f) Agencies shall not commit resources prejudicing selection of alternatives before making a final decision (§ 1506.1).

(g) Environmental impact statements shall serve as the means of assessing the environmental impact of proposed agency actions, rather than justifying decisions already made.

§ 1502.3 Statutory requirements for statements.

As required by sec. 102(2)(C) of NEPA environmental impact statements (§ 1508.11) are to be included in every recommendation or report

On proposals (§ 1508.23)

For legislation and (§ 1508.17)

Other major Federal actions (§ 1508.18)

Significantly (§ 1508.27)

Affecting (§§ 1508.3, 1508.8)

The quality of the human environment (§ 1508.14).

§ 1502.4 Major Federal actions requiring the preparation of environmental impact statements.

(a) Agencies shall make sure the proposal which is the subject of an environmental impact statement is properly defined. Agencies shall use the criteria for scope (§ 1508.25) to determine which proposal(s) shall be the subject of a particular statement. Proposals or parts of proposals which are related to each other closely enough to be, in effect, a single course of action shall be evaluated in a single impact statement.

(b) Environmental impact statements may be prepared, and are sometimes required, for broad Federal actions such as the adoption of new agency programs or regulations (§ 1508.18). Agencies shall prepare statements on broad actions so that they are relevant to policy and are timed to coincide with meaningful points in agency planning and decisionmaking.

(c) When preparing statements on broad actions (including proposals by more than one agency), agencies may find it useful to evaluate the proposal(s) in one of the following ways:

(1) Geographically, including actions occurring in the same general location, such as body of water, region, or metropolitan area.

(2) Generically, including actions which have relevant similarities, such as common timing, impacts, alternatives, methods of implementation, media, or subject matter.

(3) By stage of technological development including federal or federally assisted research, development or demonstration programs for new technologies which, if applied, could significantly affect the quality of the human environment. Statements shall be prepared on such programs and shall be available before the program has reached a stage of investment or commitment to implementation likely to determine subsequent development or restrict later alternatives.

(d) Agencies shall as appropriate employ scoping (§ 1501.7), tiering (§ 1502.20), and other methods listed in §§ 1500.4 and 1500.5 to relate broad and narrow actions and to avoid duplication and delay.

§ 1502.5 Timing.

An agency shall commence preparation of an environmental impact statement as close as possible to the time the agency is developing or is presented with a proposal (§ 1508.23) so that preparation can be completed in time for the final statement to be included in any recommendation or report on the proposal. The statement shall be prepared early enough so that it can serve practically as an important contribution to the decisionmaking process and will not be used to rationalize or justify decisions already made (§§ 1500.2(c), 1501.2, and 1502.2). For instance:

(a) For projects directly undertaken by Federal agencies the environmental impact statement shall be prepared at the feasibility analysis (go-no go) stage and may be supplemented at a later stage if necessary.

(b) For applications to the agency appropriate environmental assessments or statements shall be commenced no later than immediately after the application is received. Federal agencies are encouraged to begin preparation of such assessments or statements earlier, preferably jointly with applicable State or local agencies.

(c) For adjudication, the final environmental impact statement shall normally precede the final staff recommendation and that portion of the public hearing related to the impact study. In appropriate circumstances the statement may follow preliminary hearings designed to gather information for use in the statements.

(d) For informal rulemaking the draft environmental impact statement shall normally accompany the proposed rule.

§ 1502.6 Interdisciplinary preparation.

Environmental impact statements shall be prepared using an inter-disciplinary approach which will insure the integrated use of the natural and social sciences and the environmental design arts (section 102(2)(A) of the Act). The disciplines of the preparers shall be appropriate to the scope and issues identified in the scoping process (§ 1501.7).

§ 1502.7 Page limits.

The text of final environmental impact statements (e.g., paragraphs (d) through (g) of § 1502.10) shall normally be less than 150 pages and for proposals of unusual scope or complexity shall normally be less than 300 pages.

§ 1502.8 Writing.

Environmental impact statements shall be written in plain language and may use appropriate graphics so that decisionmakers and the public can readily understand them. Agencies should employ writers of clear prose or editors to write, review, or edit statements, which will be based upon the analysis and supporting data from the natural and social sciences and the environmental design arts.

§ 1502.9 Draft, final, and supplemental statements.

Except for proposals for legislation as provided in § 1506.8 environmental impact statements shall be prepared in two stages and may be supplemented.

(a) Draft environmental impact statements shall be prepared in accordance with the scope decided upon in the scoping process. The lead agency shall work with the cooperating agencies and shall obtain comments as required in Part 1503 of this chapter. The draft statement must fulfill and satisfy to the fullest extent possible the requirements established for final statements in section 102(2)(C) of the Act. If a draft statement is so inadequate as to preclude meaningful analysis, the agency shall prepare and circulate a revised draft of the appropriate portion. The agency shall make every effort to disclose and discuss at appropriate points in the draft statement all major points of view on the environmental impacts of the alternatives including the proposed action.

(b) Final environmental impact statements shall respond to comments as required in Part 1503 of this chapter. The agency shall discuss at appropriate points in the final statement any responsible opposing view which was not adequately discussed in the draft statement and shall indicate the agency's response to the issues raised.

(c) Agencies:

(1) Shall prepare supplements to either draft or final environmental impact statements if:

(i) The agency makes substantial changes in the proposed action that are relevant to environmental concerns; or

(ii) There are significant new circumstances or information relevant to environmental concerns and bearing on the proposed action or its impacts.

(2) May also prepare supplements when the agency determines that the purposes of the Act will be furthered by doing so.

(3) Shall adopt procedures for introducing a supplement into its formal administrative record, if such a record exists.

(4) Shall prepare, circulate, and file a supplement to a statement in

the same fashion (exclusive of scoping) as a draft and final statement unless alternative procedures are approved by the Council.

§ 1502.10 Recommended format.

Agencies shall use a format for environmental impact statements which will encourage good analysis and clear presentation of the alternatives including the proposed action. The following standard format for environmental impact statements should be followed unless the agency determines that there is a compelling reason to do otherwise:

(a) Cover sheet.

(b) Summary.

(c) Table of Contents.

(d) Purpose of and Need for Action.

(e) Alternatives Including Proposed Action (secs. 102(2)(C)(iii) and 102(2)(E) of the Act).

(f) Affected Environment.

(g) Environmental Consequences (especially sections 102(2)(C) (i), (ii), (iv), and (v) of the Act).

(h) List of Preparers.

(i) List of Agencies, Organizations, and Persons to Whom Copies of the Statement Are Sent.

(j) Index.

(k) Appendices (if any).

If a different format is used, it shall include paragraphs (a), (b), (c), (h), (i), and (j), of this section and shall include the substance of paragraphs (d), (e), (f), (g), and (k) of this section, as further described in §§ 1502.11-1502.18, in any appropriate format.

§ 1502.11 Cover sheet.

The cover sheet shall not exceed one page. It shall include:

(a) A list of the responsible agencies including the lead agency and any cooperating agencies.

(b) The title of the proposed action that is the subject of the statement (and if appropriate the titles of related cooperating agency actions), together with the State(s) and county(ies) (or other jurisdiction if applicable) where the action is located.

(c) The name, address, and telephone number of the person at the agency who can supply further information.

(d) A designation of the statement as a draft, final, or draft or final supplement.

(e) A one paragraph abstract of the statement.

(f) The date by which comments must be received (computed in cooperation with EPA under § 1506.10).

The information required by this section may be entered on Standard Form 424 (in items 4, 6, 7, 10, and 18).

§ 1502.12 Summary.

Each environmental impact statement shall contain a summary which adequately and accurately summarizes the statement. The summary shall stress the major conclusions, areas of controversy (including issues raised by agencies and the public), and the issues to be resolved (including the choice among alternatives). The summary will normally not exceed 15 pages.

§ 1502.13 Purpose and need.

The statement shall briefly specify the underlying purpose and need to which the agency is responding in proposing the alternatives including the proposed action.

§ 1502.14 Alternatives including the proposed action.

This section is the heart of the environmental impact statement. Based on the information and analysis presented in the sections on the Affected Environment (§ 1502.15) and the Environmental Consequences (§ 1502.16), it should present the environmental impacts of the proposal and the alternatives in comparative form, thus sharply defining the issues and providing a clear basis for choice among options by the decisionmaker and the public. In this section agencies shall:

(a) Rigorously explore and objectively evaluate all reasonable alternatives, and for alternatives which were eliminated from detailed study, briefly discuss the reasons for their having been eliminated.

(b) Devote substantial treatment to each alternative considered in detail including the proposed action so that reviewers may evaluate their comparative merits.

(c) Include reasonable alternatives not within the jurisdiction of the lead agency.

(d) Include the alternative of no action.

(e) Identify the agency's preferred alternative or alternatives, if one or more exists, in the draft statement and identify such alternative in the final statement unless another law prohibits the expression of such a preference.

(f) Include appropriate mitigation measures not already included in the proposed action or alternatives.

§ 1502.15 Affected environment.

The environmental impact statement shall succinctly describe the environment of the area(s) to be affected or created by the alternatives under consideration. The descriptions shall be no longer than is necessary to understand the effects of the alternatives. Data and analyses in a statement shall be commensurate with the importance of the impact, with less important material summarized, consolidated, or simply referenced. Agencies shall avoid useless bulk in statements and shall concentrate effort and attention on important issues. Verbose descriptions of the affected environment are themselves no measure of the adequacy of an environmental impact statement.

§ 1502.16 Environmental consequences.

This section forms the scientific and analytic basis for the comparisons under § 1502.14. It shall consolidate the discussions of those elements required by secs. 102(2)(C) (i), (ii), (iv), and (v) of NEPA which are within the scope of the statement and as much of sec. 102(2)(C)(iii) as is necessary to support the comparisons. The discussion will include the environmental impacts of the alternatives including the proposed action, any adverse environmental effects which cannot be avoided should the proposal be implemented, the relationship between short-term uses of man's environment and the maintenance and enhancement of long-term productivity, and any irreversible or irretrievable commitments of resources which would be involved in the proposal should it be implemented. This section should not duplicate discussions in § 1502.14. It shall include discussions of:

(a) Direct effects and their significance (§ 1508.8).

(b) Indirect effects and their significance (§ 1508.8).

(c) Possible conflicts between the proposed action and the objectives of Federal, regional, State, and local (and in the case of a reservation, Indian tribe) land use plans, policies and controls for the area concerned. (See § 1506.2(d).)

(d) The environmental effects of alternatives including the proposed action. The comparisons under § 1502.14 will be based on this discussion.

(e) Energy requirements and conservation potential of various alternatives and mitigation measures.

(f) Natural or depletable resource requirements and conservation potential of various alternatives and mitigation measures.

(g) Urban quality, historic and cultural resources, and the design of the built environment, including the reuse and conservation potential of various alternatives and mitigation measures.

(h) Means to mitigate adverse environmental impacts (if not fully covered under § 1502.14(f)).

§ 1502.17 List of preparers.

The environmental impact statement shall list the names, together with their qualifications (expertise, experience, professional disciplines), of the persons who were primarily responsible for preparing the environmental impact statement or significant background papers, including basic components of the statement (§§ 1502.6 and 1502.8). Where possible the persons who are responsible for a particular analysis, including analyses in background papers, shall be identified. Normally the list will not exceed two pages.

§ 1502.18 Appendix.

If an agency prepares an appendix to an environmental impact statement the appendix shall:

(a) Consist of material prepared in connection with an environmental impact statement (as distinct from material which is not so prepared and which is incorporated by reference (§ 1502.21)).

(b) Normally consist of material which substantiates any analysis fundamental to the impact statement.

(c) Normally be analytic and relevant to the decision to be made.

(d) Be circulated with the environmental impact statement or be readily available on request.

§ 1502.19 Circulation of the environmental impact statement.

Agencies shall circulate the entire draft and final environmental impact statements except for certain appendices as provided in § 1502.18(d) and unchanged statements as provided in § 1503.4(c). However, if the statement is unusually long, the agency may circulate the summary instead, except that the entire statement shall be furnished to:

(a) Any Federal agency which has jurisdiction by law or special expertise with respect to any environmental impact involved and any appropriate Federal, State or local agency authorized to develop and enforce environmental standards.

(b) The applicant, if any.

(c) Any person, organization, or agency requesting the entire environmental impact statement.

(d) In the case of a final environmental impact statement any person, organization, or agency which submitted substantive comments on the draft.

If the agency circulates the summary and thereafter receives a timely request for the entire statement and for additional time to comment, the time for that requestor only shall be extended by at least 15 days beyond the minimum period.

§ 1502.20 Tiering.

Agencies are encouraged to tier their environmental impact statements to eliminate repetitive discussions of the same issues and to focus on the actual issues ripe for decision at each level of environmental review (§ 1508.28). Whenever a broad environmental impact statement has been prepared (such as a program or policy statement) and a subsequent statement or environmental assessment is then prepared on an action included within the entire program or policy (such as a site specific action) the subsequent statement or environmental assessment need only summarize the issues discussed in the broader statement and incorporate discussions from the broader statement by reference and shall concentrate on the issues specific to the subsequent action. The subsequent document shall state where the earlier document is available. Tiering may also be appropriate for different stages of actions. (Sec. 1508.28).

§ 1502.21 Incorporation by reference.

Agencies shall incorporate material into an environmental impact statement by reference when the effect will be to cut down on bulk without impeding agency and public review of the action. The incorporated material shall be cited in the statement and its content briefly described. No material may be incorporated by reference unless it is reasonably available for inspection by potentially interested persons within the time allowed for comment. Material based on proprietary data which is itself not available for review and comment shall not be incorporated by reference.

§ 1502.22 Incomplete or unavailable information.

When an agency is evaluating significant adverse effects on the human environment in an environmental impact statement and there are gaps in relevant information or scientific uncertainty, the agency shall always make clear that such information is lacking or that uncertainty exists.

(a) If the information relevant to adverse impacts is essential to a reasoned choice among alternatives and is not known and the overall costs of obtaining it are not exorbitant, the agency shall include the information in the environmental impact statement.

(b) If (1) the information relevant to adverse impacts is essential to a reasoned choice among alternatives and is not known and the overall costs of obtaining it are exorbitant or (2) the information relevant to adverse impacts is important to the decision and the means to obtain it are not known (e.g., the means for obtaining it are beyond the state of the art) the agency shall weigh the need for the action against the risk and severity of possible adverse impacts were the action to proceed in the face of uncertainty. If the agency proceeds, it shall include a worst case analysis and an indication of the probability or improbability of its occurrence.

§ 1502.23 Cost-benefit analysis.

If a cost-benefit analysis relevant to the choice among environmentally different alternatives is being considered for the proposed action, it shall be incorporated by reference or appended to the statement as an aid in evaluating the environmental consequences. To assess the adequacy of compliance with sec. 102(2)(B) of the Act the statement shall, when a cost-benefit analysis is prepared, discuss the relationship between that analysis and any analyses of unquantified environmental impacts, values, and amenities. For purposes of complying with the Act, the weighing of the merits and drawbacks of the various alternatives need not be displayed in a monetary cost-benefit analysis and should not be when there are important qualitative considerations. In any event, an environmental impact statement should at least indicate those considerations, including factors not related to environmental quality, which are likely to be relevant and important to a decision.

§ 1502.24 Methodology and scientific accuracy.

Agencies shall insure the professional integrity, including scientific integrity, of the discussions and analyses in environmental impact statements. They shall identify any methodologies used and shall make explicit reference by footnote to the scientific and other sources relied upon for conclusions in the statement. An agency may place discussion of methodology in an appendix.

§ 1502.25 Environmental review and consultation requirements.

(a) To the fullest extent possible, agencies shall prepare draft environmental impact statements concurrently with and integrated with environmental impact analyses and related surveys and studies required by the Fish and Wildlife Coordination Act (16 U.S.C. Sec. 661 et seq.), the National Historic Preservation Act of 1966 (16 U.S.C. Sec. 470 et seq.), the Endangered Species Act of 1973 (16 U.S.C. Sec. 1531 et seq.), and other environmental review laws and executive orders.

(b) The draft environmental impact statement shall list all Federal permits, licenses, and other entitlements which must be obtained in implementing the proposal. If it is uncertain whether a Federal permit, license, or other entitlement is necessary, the draft environmental impact statement shall so indicate.

PART 1503—COMMENTING

Sec.

AUTHORITY: NEPA, the Environmental Quality Improvement Act of 1970, as amended (42 U.S.C. 4371 et seq.), Section 309 of the Clean Air Act, as amended (42 U.S.C. 7609), and Executive Order 11514, Protection and Enhancement of Environmental Quality (March 5, 1970, as amended by Executive Order 11991, May 24, 1977).

§ 1503.1 Inviting comments.

(a) After preparing a draft environmental impact statement and before preparing a final environmental impact statement the agency shall:

(1) Obtain the comments of any Federal agency which has jurisdic-

tion by law or special expertise with respect to any environmental impact involved or which is authorized to develop and enforce environmental standards.

(2) Request the comments of:

(i) Appropriate State and local agencies which are authorized to develop and enforce environmental standards;

(ii) Indian tribes, when the effects may be on a reservation; and

(iii) Any agency which has requested that it receive statements on actions of the kind proposed.

Office of Management and Budget Circular A-95 (Revised), through its system of clearinghouses, provides a means of securing the views of State and local environmental agencies. The clearinghouses may be used, by mutual agreement of the lead agency and the clearinghouse, for securing State and local reviews of the draft environmental impact statements.

(3) Request comments from the applicant, if any.

(4) Request comments from the public, affirmatively soliciting comments from those persons or organizations who may be interested or affected.

(b) An agency may request comments on a final environmental impact statement before the decision is finally made. In any case other agencies or persons may make comments before the final decision unless a different time is provided under § 1506.10.

§ 1503.2 Duty to comment.

Federal agencies with jurisdiction by law or special expertise with respect to any environmental impact involved and agencies which are authorized to develop and enforce environmental standards shall comment on statements within their jurisdiction, expertise, or authority. Agencies shall comment within the time period specified for comment in § 1506.10. A Federal agency may reply that it has no comment. If a cooperating agency is satisfied that its views are adequately reflected in the environmental impact statement, it should reply that it has no comment.

§ 1503.3 Specificity of comments.

(a) Comments on an environmental impact statement or on a proposed action shall be as specific as possible and may address either the adequacy of the statement or the merits of the alternatives discussed or both.

(b) When a commenting agency criticizes a lead agency's predictive methodology, the commenting agency should describe the alternative methodology which it prefers and why.

(c) A cooperating agency shall specify in its comments whether it needs additional information to fulfill other applicable environmental reviews or consultation requirements and what information it needs. In particular, it shall specify any additional information it needs to comment adequately on the draft statement's analysis of significant site-specific effects associated with the granting or approving by that cooperating agency of necessary Federal permits, licenses, or entitlements.

(d) When a cooperating agency with jurisdiction by law objects to or expresses reservations about the proposal on grounds of environmental impacts, the agency expressing the objection or reservation shall specify the mitigation measures it considers necessary to allow the agency to grant or approve applicable permit, license, or related requirements or concurrences.

§ 1503.4 Response to comments.

(a) An agency preparing a final environmental impact statement shall assess and consider comments both individually and collectively, and shall respond by one or more of the means listed below, stating its response in the final statement. Possible responses are to:

(1) Modify alternatives including the proposed action.

(2) Develop and evaluate alternatives not previously given serious consideration by the agency.

(3) Supplement, improve, or modify its analyses.

(4) Make factual corrections.

(5) Explain why the comments do

not warrant further agency response, citing the sources, authorities, or reasons which support the agency's position and, if appropriate, indicate those circumstances which would trigger agency reappraisal or further response.

(b) All substantive comments received on the draft statement (or summaries thereof where the response has been exceptionally voluminous), should be attached to the final statement whether or not the comment is thought to merit individual discussion by the agency in the text of the statement.

(c) If changes in response to comments are minor and are confined to the responses described in paragraphs (a) (4) and (5) of this section, agencies may write them on errata sheets and attach them to the statement instead of rewriting the draft statement. In such cases only the comments, the responses, and the changes and not the final statement need be circulated (§ 1502.19). The entire document with a new cover sheet shall be filed as the final statement (§ 1506.9).

PART 1504—PREDECISION REFERRALS TO THE COUNCIL OF PROPOSED FEDERAL ACTIONS DETERMINED TO BE ENVIRONMENTALLY UNSATISFACTORY

AUTHORITY: NEPA, the Environmental Quality Improvement Act of 1970, as amended (42 U.S.C. 4371 et seq.), Section 309 of the Clean Air Act, as amended (42 U.S.C. 7609), and Executive Order 11514, Protection and Enhancement of Environmental Quality (March 5, 1970, as amended by Executive Order 11991, May 24, 1977).

§ 1504.1 Purpose.

(a) This part establishes procedures for referring to the Council Federal interagency disagreements concerning proposed major Federal actions that might cause unsatisfactory environmental effects. It provides means for early resolution of such disagreements.

(b) Under section 309 of the Clean Air Act (42 U.S.C. 7609), the Administrator of the Environmental Protection Agency is directed to review and comment publicly on the environmental impacts of Federal activities, including actions for which environmental impact statements are prepared. If after this review the Administrator determines that the matter is "unsatisfactory from the standpoint of public health or welfare or environmental quality," section 309 directs that the matter be referred to the Council (hereafter "environmental referrals").

(c) Under section 102(2)(C) of the Act other Federal agencies may make similar reviews of environmental impact statements, including judgments on the acceptability of anticipated environmental impacts. These reviews must be made available to the President, the Council and the public.

§ 1504.2 Criteria for referral.

Environmental referrals should be made to the Council only after concerted, timely (as early as possible in the process), but unsuccessful attempts to resolve differences with the lead agency. In determining what environmental objections to the matter are appropriate to refer to the Council, an agency should weigh potential adverse environmental impacts, considering:

(a) Possible violation of national environmental standards or policies.

(b) Severity.

(c) Geographical scope.

(d) Duration.

(e) Importance as precedents.

(f) Availability of environmentally preferable alternatives.

§ 1504.3 Procedure for referrals and response.

(a) A Federal agency making the referral to the Council shall:

(1) Advise the lead agency at the earliest possible time that it intends to refer a matter to the Council unless a satisfactory agreement is reached.

(2) Include such advice in the referring agency's comments on the draft environmental impact statement, except when the statement does not contain adequate information to permit an assessment of the matter's environmental acceptability.

(3) Identify any essential information that is lacking and request that it be made available at the earliest possible time.

(4) Send copies of such advice to the Council.

(b) The referring agency shall deliver its referral to the Council not later than twenty-five (25) days after the final environmental impact statement has been made available to the Environmental Protection Agency, commenting agencies, and the public. Except when an extension of this period has been granted by the lead agency, the Council will not accept a referral after that date.

(c) The referral shall consist of:

(1) A copy of the letter signed by the head of the referring agency and delivered to the lead agency informing the lead agency of the referral and the reasons for it, and requesting that no action be taken to implement the matter until the Council acts upon the referral. The letter shall include a copy of the statement referred to in (c)(2) below.

(2) A statement supported by factual evidence leading to the conclusion that the matter is unsatisfactory from the standpoint of public health or welfare or environmental quality. The statement shall:

(i) Identify any material facts in controversy and incorporate (by reference if appropriate) agreed upon facts,

(ii) Identify any existing environmental requirements or policies which would be violated by the matter,

(iii) Present the reasons why the referring agency believes the matter is environmentally unsatisfactory,

(iv) Contain a finding by the agency whether the issue raised is of national importance because of the threat to national environmental resources or policies or for some other reason,

(v) Review the steps taken by the referring agency to bring its concerns to the attention of the lead agency at the earliest possible time, and

(vi) Give the referring agency's recommendations as to what mitigation alternative, further study, or other course of action (including abandonment of the matter) are necessary to remedy the situation.

(d) Not later than twenty-five (25) days after the referral to the Council the lead agency may deliver a response to the Council and the referring agency. If the lead agency requests more time and gives assurance that the matter will not go forward in the interim, the Council may grant an extension. The response shall:

(1) Address fully the issues raised in the referral.

(2) Be supported by evidence.

(3) Give the lead agency's response to the referring agency's recommendations.

(e) Interested persons (including the applicant) may deliver their views in writing to the Council. Views in support of the referral should be delivered not later than the referral. Views in support of the response shall be delivered not later than the response.

(f) Not later than twenty-five (25) days after receipt of both the referral and any response or upon being informed that there will be no response (unless the lead agency agrees to a longer time), the Council may take one or more of the following actions:

(1) Conclude that the process of referral and response has successfully resolved the problem.

(2) Initiate discussions with the agencies with the objective of mediation with referring and lead agencies.

(3) Hold public meetings or hearings to obtain additional views and information.

(4) Determine that the issue is not one of national importance and request the referring and lead agencies to pursue their decision process.

(5) Determine that the issue should be further negotiated by the referring and lead agencies and is not appropriate for Council consideration until one or more heads of

agencies report to the Council that the agencies' disagreements are irreconcilable.

(6) Publish its findings and recommendations (including where appropriate a finding that the submitted evidence does not support the position of an agency).

(7) When appropriate, submit the referral and the response together with the Council's recommendation to the President for action.

(g) The Council shall take no longer than 60 days to complete the actions specified in paragraph (f) (2), (3), or (5) of this section.

(h) When the referral involves an action required by statute to be determined on the record after opportunity for agency hearing, the referral shall be conducted in a manner consistent with 5 U.S.C. 557(d) (Administrative Procedure Act).

PART 1505—NEPA AND AGENCY DECISIONMAKING

AUTHORITY: NEPA, the Environmental Quality Improvement Act of 1970, as amended (42 U.S.C. 4371 et seq.), Section 309 of the Clean Air Act, as amended (42 U.S.C. 7609), and Executive Order 11514, Protection and Enhancement of Environmental Quality (March 5, 1970, as amended by Executive Order 11991, May 24, 1977).

§ 1505.1 Agency decisionmaking procedures.

Agencies shall adopt procedures (§ 1507.3) to ensure that decisions are made in accordance with the policies and purposes of the Act. Such procedures shall include but not be limited to:

(a) Implementing procedures under section 102(2) to achieve the requirements of sections 101 and 102(1).

(b) Designating the major decision points for the agency's principal programs likely to have a significant effect on the human environment and assuring that the NEPA process corresponds with them.

(c) Requiring that relevant environmental documents, comments, and responses be part of the record in formal rulemaking or adjudicatory proceedings.

(d) Requiring that relevant environmental documents, comments, and responses accompany the proposal through existing agency review processes so that agency officials use the statement in making decisions.

(e) Requiring that the alternatives considered by the decisionmaker are encompassed by the range of alternatives discussed in the relevant environmental documents and that the decisionmaker consider the alternatives described in the environmental impact statement. If another decision document accompanies the relevant environmental documents to the decisionmaker, agencies are encouraged to make available to the public before the decision is made any part of that document that relates to the comparison of alternatives.

§ 1505.2 Record of decision in cases requiring environmental impact statements.

At the time of its decision (§ 1506.10) or, if appropriate, its recommendation to Congress, each agency shall prepare a concise public record of decision. The record, which may be integrated into any other record prepared by the agency, including that required by OMB Circular A-95 (Revised), part I, sections 6 (c) and (d), and part II, section 5(b)(4), shall:

(a) State what the decision was.

(b) Identify all alternatives considered by the agency in reaching its decision, specifying the alternative or alternatives which were considered to be environmentally preferable. An agency may discuss preferences among alternatives based on relevant factors including economic and technical considerations and agency statutory missions. An agency shall identify and discuss all such factors including any essential considerations of national policy which were balanced by the agency in making its decision and state how those considerations entered into its decision.

(c) State whether all practicable means to avoid or minimize environmental harm from the alternative selected have been adopted, and if not, why they were not. A monitoring and enforcement program shall be adopted and summarized where applicable for any mitigation.

§ 1505.3 Implementing the decision.

Agencies may provide for monitoring to assure that their decisions are carried out and should do so in important cases. Mitigation (§ 1505.2(c)) and other conditions established in the environmental impact statement or during its review and committed as part of the decision shall be implemented by the lead agency or other appropriate consenting agency. The lead agency shall:

(a) Include appropriate conditions in grants, permits or other approvals.

(b) Condition funding of actions on mitigation.

(c) Upon request, inform cooperating or commenting agencies on progress in carrying out mitigation measures which they have proposed and which were adopted by the agency making the decision.

(d) Upon request, make available to the public the results of relevant monitoring.

PART 1506—OTHER REQUIREMENTS OF NEPA

Sec.

AUTHORITY: NEPA, the Environmental Quality Improvement Act of 1970, as amended (42 U.S.C. 4371 et seq.), Section 309 of the Clean Air Act, as amended (42 U.S.C. 7609), and Executive Order 11514, Protection and Enhancement of Environmental Quality (March 5, 1970, as amended by Executive Order 11991, May 24, 1977).

§ 1506.1 Limitations on actions during NEPA process.

(a) Until an agency issues a record of decision as provided in § 1505.2 (except as provided in paragraph (c) of this section), no action concerning the proposal shall be taken which would:

(1) Have an adverse environmental impact; or

(2) Limit the choice of reasonable alternatives.

(b) If any agency is considering an application from a non-Federal entity, and is aware that the applicant is about to take an action within the agency's jurisdiction that would meet either of the criteria in paragraph (a) of this section, then the agency shall promptly notify the applicant that the agency will take appropriate action to insure that the objectives and procedures of NEPA are achieved.

(c) While work on a required program environmental impact statement is in progress and the action is not covered by an existing program statement, agencies shall not undertake in the interim any major Federal action covered by the program which may significantly affect the quality of the human environment unless such action:

(1) Is justified independently of the program;

(2) Is itself accompanied by an adequate environmental impact statement; and

(3) Will not prejudice the ultimate decision on the program. Interim action prejudices the ultimate decision on the program when it tends to determine subsequent development or limit alternatives.

(d) This section does not preclude development by applicants of plans or designs or performance of other work necessary to support an application for Federal, State or local permits or assistance. Nothing in this section shall preclude Rural Electrification Administration approval of minimal expenditures not affecting the environment (*e.g.* long leadtime equipment and purchase options) made by non-governmental entities seeking loan guarantees from the Administration.

§ 1506.2 Elimination of duplication with State and local procedures.

(a) Agencies authorized by law to cooperate with State agencies of statewide jurisdiction pursuant to section 102(2)(D) of the Act may do so.

(b) Agencies shall cooperate with State and local agencies to the fullest extent possible to reduce duplication between NEPA and State and local requirements, unless the agencies are specifically barred from doing so by some other law. Except for cases covered by paragraph (a) of this section, such cooperation shall to the fullest extent possible include:

(1) Joint planning processes.

(2) Joint environmental research and studies.

(3) Joint public hearings (except where otherwise provided by statute).

(4) Joint environmental assessments.

(c) Agencies shall cooperate with State and local agencies to the fullest extent possible to reduce duplication between NEPA and comparable State and local requirements, unless the agencies are specifically barred from doing so by some other law. Except for cases covered by paragraph (a) of this section, such cooperation shall to the fullest extent possible include joint environmental impact statements. In such cases one or more Federal agencies and one or more State or local agencies shall be joint lead agencies. Where State laws or local ordinances have environmental impact statement requirements in addition to but not in conflict with those in NEPA, Federal agencies shall cooperate in fulfilling these requirements as well as those of Federal laws so that one document will comply with all applicable laws.

(d) To better integrate environmental impact statements into State or local planning processes, statements shall discuss any inconsistency of a proposed action with any approved State or local plan and laws (whether or not federally sanctioned). Where an inconsistency exists, the statement should describe the extent to which the agency would reconcile its proposed action with the plan or law.

§ 1506.3 Adoption.

(a) An agency may adopt a Federal draft or final environmental impact statement or portion thereof provided that the statement or portion thereof meets the standards for an adequate statement under these regulations.

(b) If the actions covered by the original environmental impact statement and the proposed action are substantially the same, the agency adopting another agency's statement is not required to recirculate it except as a final statement. Otherwise the adopting agency shall treat the statement as a draft and recirculate it (except as provided in paragraph (c) of this section).

(c) A cooperating agency may adopt without recirculating the environmental impact statement of a lead agency when, after an independent review of the statement, the cooperating agency concludes that its comments and suggestions have been satisfied.

(d) When an agency adopts a statement which is not final within the agency that prepared it, or when the action it assesses is the subject of a referral under part 1504, or when the statement's adequacy is the subject of a judicial action which is not final, the agency shall so specify.

§ 1506.4 Combining documents.

Any environmental document in compliance with NEPA may be combined with any other agency document to reduce duplication and paperwork.

§ 1506.5 Agency responsibility.

(a) *Information.* If an agency requires an applicant to submit environmental information for possible use by the agency in preparing an environmental impact statement, then the agency should assist the applicant by outlining the types of information required. The agency shall independently evaluate the information submitted and shall be responsible for its accuracy. If the agency chooses to use the information submitted by the applicant in the environmental impact statement, either directly or by reference, then the names of the persons responsible for the independent evaluation shall be included in the list of preparers (§ 1502.17). It is the intent of this subparagraph that acceptable work not be redone, but that it be verified by the agency.

(b) *Environmental assessments.* If an agency permits an applicant to prepare an environmental assessment, the agency, besides fulfilling the requirements of paragraph (a) of this section, shall make its own evaluation of the environmental issues and take responsibility for the scope and content of the environmental assessment.

(c) *Environmental impact statements.* Except as provided in §§ 1506.2 and 1506.3 any environmental impact statement prepared pursuant to the requirements of NEPA shall be prepared directly by or by a contractor selected by the lead agency or where appropriate under § 1501.6(b), a cooperating agency. It is the intent of these regulations that the contractor be chosen solely by the lead agency, or by the lead agency in cooperation with cooperating agencies, or where appropriate by a cooperating agency to avoid any conflict of interest. Contractors shall execute a disclosure statement prepared by the lead agency, or where appropriate the cooperating agency, specifying that they have no financial or other interest in the outcome of the project. If the document is prepared by contract, the responsible Federal official shall furnish guidance and participate in the preparation and shall independently evaluate the statement prior to its approval and take responsibility for its scope and contents. Nothing in this section is intended to prohibit any agency from requesting any person to submit information to it or to prohibit any person from submitting information to any agency.

§ 1506.6 Public involvement.

Agencies shall: (a) Make diligent efforts to involve the public in preparing and implementing their NEPA procedures.

(b) Provide public notice of NEPA-related hearings, public meetings, and the availability of environmental documents so as to inform those persons and agencies who may be interested or affected.

(1) In all cases the agency shall mail notice to those who have requested it on an individual action.

(2) In the case of an action with effects of national concern notice shall include publication in the FEDERAL REGISTER and notice by mail to national organizations reasonably expected to be interested in the matter and may include listing in the *102 Monitor.* An agency engaged in rulemaking may provide notice by mail to national organizations who have requested that notice regularly be provided. Agencies shall maintain a list of such organizations.

(3) In the case of an action with effects primarily of local concern the notice may include:

(i) Notice to State and areawide clearinghouses pursuant to OMB Circular A-95 (Revised).

(ii) Notice to Indian tribes when effects may occur on reservations.

(iii) Following the affected State's public notice procedures for comparable actions.

(iv) Publication in local newspapers (in papers of general circulation rather than legal papers).

(v) Notice through other local media.

(vi) Notice to potentially interested community organizations including small business associations.

(vii) Publication in newsletters that may be expected to reach potentially interested persons.

(viii) Direct mailing to owners and occupants of nearby or affected property.

(ix) Posting of notice on and off site in the area where the action is to be located.

(c) Hold or sponsor public hearings or public meetings whenever appropriate or in accordance with statutory requirements applicable to the agency. Criteria shall include whether there is:

(1) Substantial environmental controversy concerning the proposed action or substantial interest in holding the hearing.

(2) A request for a hearing by another agency with jurisdiction over the action supported by reasons why a hearing will be helpful. If a draft environmental impact statement is to be considered at a public hearing, the agency should make the statement available to the public at least 15 days in advance (unless the purpose of the hearing is to provide information for the draft environmental impact statement).

(d) Solicit appropriate information from the public.

(e) Explain in its procedures where interested persons can get information or status reports on environmental impact statements and other elements of the NEPA process.

(f) Make environmental impact statements, the comments received, and any underlying documents available to the public pursuant to the provisions of the Freedom of Information Act (5 U.S.C. 552), without regard to the exclusion for interagency memoranda where such memoranda transmit comments of Federal agencies on the environmental impact of the proposed action. Materials to be made available to the public shall be provided to the public without charge to the extent practicable, or at a fee which is not more than the actual costs of reproducing copies required to be sent to other Federal agencies, including the Council.

§ 1506.7 Further guidance.

The Council may provide further guidance concerning NEPA and its procedures including:

(a) A handbook which the Council may supplement from time to time, which shall in plain language provide guidance and instructions concerning the application of NEPA and these regulations.

(b) Publication of the Council's Memoranda to Heads of Agencies.

(c) In conjunction with the Environmental Protection Agency and the publication of the 102 Monitor, notice of:

(1) Research activities;

(2) Meetings and conferences related to NEPA; and

(3) Successful and innovative procedures used by agencies to implement NEPA.

§ 1506.8 Proposals for legislation.

(a) The NEPA process for proposals for legislation (§ 1508.17) significantly affecting the quality of the human environment shall be integrated with the legislative process of the Congress. A legislative environmental impact statement is the detailed statement required by law to be included in a recommendation or report on a legislative proposal to Congress. A legislative environmental impact statement shall be considered part of the formal transmittal of a legislative proposal to Congress; however, it may be transmitted to Congress up to 30 days later in order to allow time for completion of an accurate statement which can serve as the basis for public and Congressional debate. The statement must be available in time for Congressional hearings and deliberations.

(b) Preparation of a legislative environmental impact statement shall conform to the requirements of these regulations except as follows:

(1) There need not be a scoping process.

(2) The legislative statement shall be prepared in the same manner as a draft statement, but shall be considered the "detailed statement" required by statute; *Provided,* That when any of the following conditions exist both the draft and final environmental impact statement on the legislative proposal shall be prepared and circulated as provided by §§ 1503.1 and 1506.10.

(i) A Congressional Committee with jurisdiction over the proposal has a rule requiring both draft and final environmental impact statements.

(ii) The proposal results from a study process required by statute (such as those required by the Wild and Scenic Rivers Act (16 U.S.C.

1271 et seq.) and the Wilderness Act (16 U.S.C. 1131 et seq.)).

(iii) Legislative approval is sought for Federal or federally assisted construction or other projects which the agency recommends be located at specific geographic locations. For proposals requiring an environmental impact statement for the acquisition of space by the General Services Administration, a draft statement shall accompany the Prospectus or the 11(b) Report of Building Project Surveys to the Congress, and a final statement shall be completed before site acquisition.

(iv) The agency decides to prepare draft and final statements.

(c) Comments on the legislative statement shall be given to the lead agency which shall forward them along with its own responses to the Congressional committees with jurisdiction.

§ 1506.9 Filing requirements.

Environmental impact statements together with comments and responses shall be filed with the Environmental Protection Agency, attention Office of Federal Activities (A-104), 401 M Street SW., Washington, D.C. 20460. Statements shall be filed with EPA no earlier than they are also transmitted to commenting agencies and made available to the public. EPA shall deliver one copy of each statement to the Council, which shall satisfy the requirement of availability to the President. EPA may issue guidelines to agencies to implement its responsibilities under this section and § 1506.10 below.

§ 1506.10 Timing of agency action.

(a) The Environmental Protection Agency shall publish a notice in the FEDERAL REGISTER each week of the environmental impact statements filed during the preceding week. The minimum time periods set forth in this section shall be calculated from the date of publication of this notice.

(b) No decision on the proposed action shall be made or recorded under § 1505.2 by a Federal agency until the later of the following dates:

(1) Ninety (90) days after publication of the notice described above in paragraph (a) of this section for a draft environmental impact statement.

(2) Thirty (30) days after publication of the notice described above in paragraph (a) of this section for a final environmental impact statement.

An exception to the rules on timing may be made in the case of an agency decision which is subject to a formal internal appeal. Some agencies have a formally established appeal process which allows other agencies or the public to take appeals on a decision and make their views known, after publication of the final environmental impact statement. In such cases, where a real opportunity exists to alter the decision, the decision may be made and recorded at the same time the environmental impact statement is published. This means that the period for appeal of the decision and the 30-day period prescribed in paragraph (b)(2) of this section may run concurrently. In such cases the environmental impact statement shall explain the timing and the public's right of appeal. An agency engaged in rulemaking under the Administrative Procedure Act or other statute for the purpose of protecting the public health or safety, may waive the time period in paragraph (b)(2) of this section and publish a decision on the final rule simultaneously with publication of the notice of the availability of the final environmental impact statement as described in paragraph (a) of this section.

(c) If the final environmental impact statement is filed within ninety (90) days after a draft environmental impact statement is filed with the Environmental Protection Agency, the minimum thirty (30) day period and the minimum ninety (90) day period may run concurrently. However, subject to paragraph (d) of this section agencies shall allow not less than 45 days for comments on draft statements.

(d) The lead agency may extend prescribed periods. The Environmental Protection Agency may upon a showing by the lead agency of compelling reasons of national policy reduce the prescribed periods and may upon a showing by any other Federal agency of compelling reasons of national policy also extend prescribed periods, but only after consultation with the lead agency. (Also see § 1507.3(d).) Failure to file timely comments shall not be a sufficient reason for extending a period. If the lead agency does not concur with the extension of time, EPA may not extend it for more than 30 days. When the Environmental Protection Agency reduces or extends any period of time it shall notify the Council.

§ 1506.11 Emergencies.

Where emergency circumstances make it necessary to take an action with significant environmental impact without observing the provisions of these regulations, the Federal agency taking the action should consult with the Council about alternative arrangements. Agencies and the Council will limit such arrangements to actions necessary to control the immediate impacts of the emergency. Other actions remain subject to NEPA review.

§ 1506.12 Effective date.

The effective date of these regulations is July 30, 1979, except that for agencies that administer programs that qualify under sec. 102(2)(D) of the Act or under sec. 104(h) of the Housing and Community Development Act of 1974 an additional four months shall be allowed for the State or local agencies to adopt their implementing procedures.

(a) These regulations shall apply to the fullest extent practicable to ongoing activities and environmental documents begun before the effective date. These regulations do not apply to an environmental impact statement or supplement if the draft statement was filed before the effective date of these regulations. No completed environmental documents need be redone by reason of these regulations. Until these regulations are applicable, the Council's guidelines published in the FEDERAL REGISTER of August 1, 1973, shall continue to be applicable. In cases where these regulations are applicable the guidelines are superseded. However, nothing shall prevent an agency from proceeding under these regulations at an earlier time.

(b) NEPA shall continue to be applicable to actions begun before January 1, 1970, to the fullest extent possible.

PART 1507—AGENCY COMPLIANCE

AUTHORITY: NEPA, the Environmental Quality Improvement Act of 1970, as amended (42 U.S.C. 4371 et seq.), Section 309 of the Clean Air Act, as amended (42 U.S.C. 7609), and Executive Order 11514, Protection and Enhancement of Environmental Quality (March 5, 1970, as amended by Executive Order 11991, May 24, 1977).

§ 1507.1 Compliance.

All agencies of the Federal Government shall comply with these regulations. It is the intent of these regulations to allow each agency flexibility in adapting its implementing procedures authorized by § 1507.3 to the requirements of other applicable laws.

§ 1507.2 Agency capability to comply.

Each agency shall be capable (in terms of personnel and other resources) of complying with the requirements enumerated below. Such compliance may include use of other's resources, but the using agency shall itself have sufficient capability to evaluate what others do for it. Agencies shall:

(a) Fulfill the requirements of Sec. 102(2)(A) of the Act to utilize a systematic, interdisciplinary approach which will insure the integrated use of the natural and social sciences and the environmental design arts in planning and in decisionmaking which may have an impact on the human environment. Agencies shall designate a person to be responsible for overall review of agency NEPA compliance.

(b) Identify methods and procedures required by Sec. 102(2)(B) to insure that presently unquantified environmental amenities and values may be given appropriate consideration.

(c) Prepare adequate environmental impact statements pursuant to Sec. 102(2)(C) and comment on statements in the areas where the agency has jurisdiction by law or special expertise or is authorized to develop and enforce environmental standards.

(d) Study, develop, and describe alternatives to recommended courses of action in any proposal which involves unresolved conflicts concerning alternative uses of available resources. This requirement of Sec. 102(2)(E) extends to all such proposals, not just the more limited scope of Sec. 102(2)(C)(iii) where the discussion of alternatives is confined to impact statements.

(e) Comply with the requirements of Sec. 102(2)(H) that the agency initiate and utilize ecological information in the planning and development of resource-oriented projects.

(f) Fulfill the requirements of sections 102(2)(F), 102(2)(G), and 102(2)(I), of the Act and of Executive Order 11514, Protection and Enhancement of Environmental Quality, Sec. 2.

§ 1507.3 Agency procedures.

(a) Not later than eight months after publication of these regulations as finally adopted in the FEDERAL REGISTER, or five months after the establishment of an agency, whichever shall come later, each agency shall as necessary adopt procedures to supplement these regulations. When the agency is a department, major subunits are encouraged (with the consent of the department) to adopt their own procedures. Such procedures shall not paraphrase these regulations. They shall confine themselves to implementing procedures. Each agency shall consult with the Council while developing its procedures and before publishing them in the FEDERAL REGISTER for comment. Agencies with similar programs should consult with each other and the Council to coordinate their procedures, especially for programs requesting similar information from applicants. The procedures shall be adopted only after an opportunity for public review and after review by the Council for conformity with the Act and these regulations. The Council shall complete its review within 30 days. Once in effect they shall be filed with the Council and made readily available to the public. Agencies are encouraged to publish explanatory guidance for these regulations and their own procedures. Agencies shall continue to review their policies and procedures and in consultation with the Council to revise them as necessary to ensure full compliance with the purposes and provisions of the Act.

(b) Agency procedures shall comply with these regulations except where compliance would be inconsistent with statutory requirements and shall include:

(1) Those procedures required by §§ 1501.2(d), 1502.9(c)(3), 1505.1, 1506.6(e), and 1508.4.

(2) Specific criteria for and identification of those typical classes of action:

(i) Which normally do require environmental impact statements.

(ii) Which normally do not require either an environmental impact statement or an environmental assessment (categorical exclusions (§ 1508.4)).

(iii) Which normally require environmental assessments but not necessarily environmental impact statements.

(c) Agency procedures may include specific criteria for providing limited exceptions to the provisions of these regulations for classified proposals. They are proposed actions which are specifically authorized under criteria established by an Executive Order or statute to be kept secret in the interest of national defense or foreign policy and are in fact properly classified pursuant to such Executive Order or statute. Environmental assessments and environmental impact statements which address classified proposals may be safeguarded and restricted from public dissemination in accordance with agencies' own regulations applicable to classified information. These documents may be organized so that classified portions can be included as annexes, in order that the unclassified portions can be made available to the public.

(d) Agency procedures may provide for periods of time other than those presented in § 1506.10 when necessary to comply with other specific statutory requirements.

(e) Agency procedures may provide that where there is a lengthy period between the agency's decision to prepare an environmental impact statement and the time of actual preparation, the notice of intent required by § 1501.7 may be published at a reasonable time in advance of preparation of the draft statement.

PART 1508—TERMINOLOGY AND INDEX

AUTHORITY: NEPA, the Environmental Quality Improvement Act of 1970, as amended (42 U.S.C. 4371 *et seq.*), Section 309 of the Clean Air Act, as amended (42 U.S.C. 7609), and Executive Order 11514, Protection and Enhancement of Environmental Quality (March 5, 1970, as amended by Executive Order 11991, May 24, 1977).

§ 1508.1 Terminology.

The terminology of this part shall be uniform throughout the Federal Government.

§ 1508.2 Act.

"Act" means the National Environmental Policy Act, as amended (42 U.S.C. 4321, et seq.) which is also referred to as "NEPA."

§ 1508.3 Affecting.

"Affecting" means will or may have an effect on.

§ 1508.4 Categorical exclusion.

"Categorical Exclusion" means a category of actions which do not individually or cumulatively have a significant effect on the human environment and which have been found to have no such effect in procedures adopted by a Federal agency in implementation of these regulations (§ 1507.3) and for which, therefore, neither an environmental assessment nor an environmental impact statement is required. An agency may decide in its procedures or otherwise, to prepare environmental assessments for the reasons stated in § 1508.9 even though it is not required to do so. Any procedures under this section shall provide for extraordinary circumstances in which a normally excluded action may have a significant environmental effect.

§ 1508.5 Cooperating agency.

"Cooperating Agency" means any Federal agency other than a lead agency which has jurisdiction by law or special expertise with respect to any environmental impact involved in a proposal (or a reasonable alternative) for legislation or other major Federal action significantly affecting the quality of the human environment. The selection and responsibilities of a cooperating agency are described in § 1501.6. A State or local agency of similar qualifications or, when the effects are on a reservation, an Indian Tribe, may by agreement with the lead agency become a cooperating agency.

§ 1508.6 Council.

"Council" means the Council on Environmental Quality established by Title II of the Act.

§ 1508.7 Cumulative impact.

"Cumulative impact" is the impact on the environment which results from the incremental impact of the action when added to other past, present, and reasonably foreseeable future actions regardless of what agency (Federal or non-Federal) or person undertakes such other actions. Cumulative impacts can result from individually minor but collectively significant actions taking place over a period of time.

§ 1508.8 Effects.

"Effects" include:

(a) Direct effects, which are caused by the action and occur at the same time and place.

(b) Indirect effects, which are caused by the action and are later in time or farther removed in distance, but are still reasonably foreseeable. Indirect effects may include growth inducing effects and other effects related to induced changes in the pattern of land use, population density or growth rate, and related effects on air and water and other natural systems, including ecosystems.

Effects and impacts as used in these regulations are synonymous. Effects includes ecological (such as the effects on natural resources and on the components, structures, and functioning of affected ecosystems), aesthetic, historic, cultural, economic, social, or health, whether direct, indirect, or cumulative. Effects may also include those resulting from actions which may have both beneficial and detrimental effects, even if on balance the agency believes that the effect will be beneficial.

§ 1508.9 Environmental assessment.

"Environmental Assessment":

(a) Means a concise public document for which a Federal agency is responsible that serves to:

(1) Briefly provide sufficient evidence and analysis for determining whether to prepare an environmental impact statement or a finding of no significant impact.

(2) Aid an agency's compliance with the Act when no environmental impact statement is necessary.

(3) Facilitate preparation of a statement when one is necessary.

(b) Shall include brief discussions of the need for the proposal, of alternatives as required by sec. 102(2)(E), of the environmental impacts of the proposed action and alternatives, and a listing of agencies and persons consulted.

§ 1508.10 Environmental document.

"Environmental document" includes the documents specified in § 1508.9 (environmental assessment), § 1508.11 (environmental impact statement), § 1508.13 (finding of no significant impact), and § 1508.22 (notice of intent).

§ 1508.11 Environmental impact statement.

"Environmental Impact Statement" means a detailed written statement as required by Sec. 102(2)(C) of the Act.

§ 1508.12 Federal agency.

"Federal agency" means all agencies of the Federal Government. It does not mean the Congress, the Judiciary, or the President, including the performance of staff functions for the President in his Executive Office. It also includes for purposes of these regulations States and units of general local government and Indian tribes assuming NEPA responsibilities under section 104(h) of the Housing and Community Development Act of 1974.

§ 1508.13 Finding of no significant impact.

"Finding of No Significant Impact" means a document by a Federal agency briefly presenting the reasons why an action, not otherwise excluded (§ 1508.4), will not have a significant effect on the human environment and for which an environmental impact statement therefore will not be prepared. It shall include the environmental assessment or a summary of it and shall note any other environmental documents related to it (§ 1501.7(a)(5)). If the assessment is included, the finding need not repeat any of the discussion in the assessment but may incorporate it by reference.

§ 1508.14 Human Environment.

"Human Environment" shall be interpreted comprehensively to include the natural and physical environment and the relationship of people with that environment. (See the definition of "effects" (§ 1508.8).) This means that economic or social effects are not intended by themselves to require preparation of an environmental impact statement. When an environmental impact statement is prepared and economic or social and natural or physical environmental effects are interrelated, then the environmental impact statement will discuss all of these effects on the human environment.

§ 1508.15 Jurisdiction By Law.

"Jurisdiction by law" means agency authority to approve, veto, or finance all or part of the proposal.

§ 1508.16 Lead agency.

"Lead Agency" means the agency or agencies preparing or having taken primary responsibility for preparing the environmental impact statement.

§ 1508.17 Legislation.

"Legislation" includes a bill or legislative proposal to Congress developed by or with the significant cooperation and support of a Federal agency, but does not include requests for appropriations. The test for significant cooperation is whether the proposal is in fact predominantly that of the agency rather than another source. Drafting does not by itself constitute significant cooperation. Proposals for legislation include requests for ratification of treaties. Only the agency which has primary responsibility for the subject matter involved will prepare a legislative environmental impact statement.

§ 1508.18 Major Federal action.

"Major Federal action" includes actions with effects that may be major and which are potentially subject to Federal control and responsibility. Major reinforces but does not have a meaning independent of significantly (§ 1508.27). Actions include the circumstance where the responsible officials fail to act and that failure to act is reviewable by courts or administrative tribunals under the Administrative Procedure Act or other applicable law as agency action.

(a) Actions include new and continuing activities, including projects and programs entirely or partly financed, assisted, conducted, regulated, or approved by federal agencies; new or revised agency rules, regulations, plans, policies, or procedures; and legislative proposals (§§ 1506.8, 1508.17). Actions do not include funding assistance solely in the form of general revenue sharing funds, distributed under the State and Local Fiscal Assistance Act of 1972, 31 U.S.C. 1221 et seq., with no Federal agency control over the subsequent use of such funds. Actions do not include bringing judicial or administrative civil or criminal enforcement actions.

(b) Federal actions tend to fall within one of the following categories:

(1) Adoption of official policy, such as rules, regulations, and interpretations adopted pursuant to the Administrative Procedure Act, 5 U.S.C. 551 et seq.; treaties and international conventions or agreements; formal documents establishing an agency's policies which will result in or substantially alter agency programs.

(2) Adoption of formal plans, such as official documents prepared or approved by federal agencies which guide or prescribe alternative uses of federal resources, upon which future agency actions will be based.

(3) Adoption of programs, such as a group of concerted actions to implement a specific policy or plan; systematic and connected agency decisions allocating agency resources to implement a specific statutory program or executive directive.

(4) Approval of specific projects, such as construction or management activities located in a defined geographic area. Projects include actions approved by permit or other regulatory decision as well as federal and federally assisted activities.

§ 1508.19 Matter.

"Matter" includes for purposes of Part 1504:

(a) With respect to the Environmental Protection Agency, any proposed legislation, project, action or regulation as those terms are used in Section 309(a) of the Clean Air Act (42 U.S.C. 7609).

(b) With respect to all other agencies, any proposed major federal action to which section 102(2)(C) of NEPA applies.

§ 1508.20 Mitigation.

"Mitigation" includes:

(a) Avoiding the impact altogether by not taking a certain action or parts of an action.

(b) Minimizing impacts by limiting the degree or magnitude of the action and its implementation.

(c) Rectifying the impact by repairing, rehabilitating, or restoring the affected environment.

(d) Reducing or eliminating the impact over time by preservation and maintenance operations during the life of the action.

(e) Compensating for the impact by replacing or providing substitute resources or environments.

§ 1508.21 NEPA process.

"NEPA process" means all measures necessary for compliance with the requirements of Section 2 and Title I of NEPA.

§ 1508.22 Notice of intent.

"Notice of Intent" means a notice that an environmental impact statement will be prepared and considered. The notice shall briefly:

(a) Describe the proposed action and possible alternatives.

(b) Describe the agency's proposed scoping process including whether, when, and where any scoping meeting will be held.

(c) State the name and address of a person within the agency who can answer questions about the proposed action and the environmental impact statement.

§ 1508.23 Proposal.

"Proposal" exists at that stage in the development of an action when an agency subject to the Act has a goal and is actively preparing to make a decision on one or more alternative means of accomplishing that goal and the effects can be meaningfully evaluated. Preparation of an environmental impact statement on a proposal should be timed (§ 1502.5) so that the final statement may be completed in time for the statement to be included in any recommendation or report on the proposal. A proposal may exist in fact as well as by agency declaration that one exists.

§ 1508.24 Referring agency.

"Referring agency" means the federal agency which has referred any matter to the Council after a determination that the matter is unsatisfactory from the standpoint of public health or welfare or environmental quality.

§ 1508.25 Scope.

Scope consists of the range of actions, alternatives, and impacts to be considered in an environmental impact statement. The scope of an individual statement may depend on its relationships to other statements (§§1502.20 and 1508.28). To determine the scope of environmental impact statements, agencies shall consider 3 types of actions, 3 types of alternatives, and 3 types of impacts. They include:

(a) Actions (other than unconnected single actions) which may be:

(1) Connected actions, which means that they are closely related and therefore should be discussed in the same impact statement. Actions are connected if they:

(i) Automatically trigger other actions which may require environmental impact statements.

(ii) Cannot or will not proceed unless other actions are taken previously or simultaneously.

(iii) Are interdependent parts of a larger action and depend on the larger action for their justification.

(2) Cumulative actions, which when viewed with other proposed actions have cumulatively significant impacts and should therefore be discussed in the same impact statement.

(3) Similar actions, which when

viewed with other reasonably foreseeable or proposed agency actions, have similarities that provide a basis for evaluating their environmental consequencies together, such as common timing or geography. An agency may wish to analyze these actions in the same impact statement. It should do so when the best way to assess adequately the combined impacts of similar actions or reasonable alternatives to such actions is to treat them in a single impact statement.

(b) Alternatives, which include: (1) No action alternative. (2) Other reasonable courses of actions. (3) Mitigation measures (not in the proposed action).

(c) Impacts, which may be: (1) Direct. (2) Indirect. (3) Cumulative.

§ 1508.26 Special expertise.

"Special expertise" means statutory responsibility, agency mission, or related program experience.

§ 1508.27 Significantly.

"Significantly" as used in NEPA requires considerations of both context and intensity:

(a) *Context.* This means that the significance of an action must be analyzed in several contexts such as society as a whole (human, national), the affected region, the affected interests, and the locality. Significance varies with the setting of the proposed action. For instance, in the case of a site-specific action, significance would usually depend upon the effects in the locale rather than in the world as a whole. Both short- and long-term effects are relevant.

(b) *Intensity.* This refers to the severity of impact. Responsible officials must bear in mind that more than one agency may make decisions about partial aspects of a major action. The following should be considered in evaluating intensity:

(1) Impacts that may be both beneficial and adverse. A significant effect may exist even if the Federal agency believes that on balance the effect will be beneficial.

(2) The degree to which the proposed action affects public health or safety.

(3) Unique characteristics of the geographic area such as proximity to historic or cultural resources, park lands, prime farmlands, wetlands, wild and scenic rivers, or ecologically critical areas.

(4) The degree to which the effects on the quality of the human environment are likely to be highly controversial.

(5) The degree to which the possible effects on the human environment are highly uncertain or involve unique or unknown risks.

(6) The degree to which the action may establish a precedent for future actions with significant effects or represents a decision in principle about a future consideration.

(7) Whether the action is related to other actions with individually insignificant but cumulatively significant impacts. Significance exists if it is reasonable to anticipate a cumulatively significant impact on the environment. Significance cannot be avoided by terming an action temporary or by breaking it down into small component parts.

(8) The degree to which the action may adversely affect districts, sites, highways, structures, or objects listed in or eligible for listing in the National Register of Historic Places or may cause loss or destruction of significant scientific, cultural, or historical resources.

(9) The degree to which the action may adversely affect an endangered or threatened species or its habitat that has been determined to be critical under the Endangered Species Act of 1973.

(10) Whether the action threatens a violation of Federal, State, or local law or requirements imposed for the protection of the environment.

§ 1508.28 Tiering.

"Tiering" refers to the coverage of general matters in broader environmental impact statements (such as national program or policy statements) with subsequent narrower statements or environmental analyses (such as regional or basinwide program statements or ultimately site-specific statements) incorporating by reference the general discussions and concentrating solely on the issues specific to the statement subsequently prepared. Tiering is appropriate when the sequence of statements or analyses is:

(a) From a program, plan, or policy environmental impact statement to a program, plan, or policy statement or analysis of lesser scope or to a site-specific statement or analysis.

(b) From an environmental impact statement on a specific action at an early stage (such as need and site selection) to a supplement (which is preferred) or a subsequent statement or analysis at a later stage (such as environmental mitigation). Tiering in such cases is appropriate when it helps the lead agency to focus on the issues which are ripe for decision and exclude from consideration issues already decided or not yet ripe.

Index

Index

Index

THE NATIONAL ENVIRONMENTAL POLICY ACT OF 1969, AS AMENDED*

An Act to establish a national policy for the environment, to provide for the establishment of a Council on Environmental Quality, and for other purposes.

Be it enacted by the Senate and House of Representatives of the United States of America in Congress assembled, That this Act may be cited as the "National Environmental Policy Act of 1969."

PURPOSE

SEC. 2. The purposes of this Act are: To declare a national policy which will encourage productive and enjoyable harmony between man and his environment; to promote efforts which will prevent or eliminate damage to the environment and biosphere and stimulate the health and welfare of man; to enrich the understanding of the ecological systems and natural resources important to the Nation; and to establish a Council on Environmental Quality.

TITLE I

DECLARATION OF NATIONAL ENVIRONMENTAL POLICY

SEC. 101. (a) The Congress, recognizing the profound impact of man's activity on the interrelations of all components of the natural environment, particularly the profound influences of population growth, high-density urbanization, industrial expansion, resource exploitation, and new and expanding technological advances and recognizing further the critical importance of restoring and maintaining environmental quality to the overall welfare and development of man, declares that it is the continuing policy of the Federal Government, in cooperation with State and local governments, and other concerned public and private organizations, to use all practicable means and measures, including financial and technical assistance, in a manner calculated to foster and promote the general welfare, to create and maintain conditions under which man and nature can exist in productive harmony, and fulfill the social, economic, and other requirements of present and future generations of Americans.

(b) In order to carry out the policy set forth in this Act, it is the continuing responsibility of the Federal Government to use all practicable means, consistent with other essential considerations of national policy, to improve and coordinate Federal plans, functions, programs, and resources to the end that the Nation may—

(1) fulfill the responsibilities of each generation as trustee of the environment for succeeding generations;

(2) assure for all Americans safe, healthful, productive, and esthetically and culturally pleasing surroundings;

(3) attain the widest range of beneficial uses of the environment without degradation, risk to health or safety, or other undesirable and unintended consequences;

(4) preserve important historic, cultural, and natural aspects of our national heritage, and maintain, wherever possible, an environment which supports diversity, and variety of individual choice;

(5) achieve a balance between population and resource use which will permit high standards of living and a wide sharing of life's amenities; and

(6) enhance the quality of renewable resources and approach the maximum attainable recycling of depletable resources.

---

*Pub. L. 91–190, 42 U.S.C. 4321–4347, January 1, 1970, as amended by Pub. L. 94–52, July 3, 1975, and Pub. L. 94–83, August 9, 1975.

(c) The Congress recognizes that each person should enjoy a healthful environment and that each person has a responsibility to contribute to the preservation and enhancement of the environment.

SEC. 102. The Congress authorizes and directs that, to the fullest extent possible: (1) the policies, regulations, and public laws of the United States shall be interpreted and administered in accordance with the policies set forth in this Act, and (2) all agencies of the Federal Government shall—

(A) Utilize a systematic, interdisciplinary approach which will insure the integrated use of the natural and social sciences and the environmental design arts in planning and in decisionmaking which may have an impact on man's environment;

(B) Identify and develop methods and procedures, in consultation with the Council on Environmental Quality established by title II of this Act, which will insure that presently unquantified environmental amenities and values may be given appropriate consideration in decision-making along with economic and technical considerations;

(C) Include in every recommendation or report on proposals for legislation and other major Federal actions significantly affecting the quality of the human environment, a detailed statement by the responsible official on—

(i) The environmental impact of the proposed action,

(ii) Any adverse environmental effects which cannot be avoided should the proposal be implemented,

(iii) Alternatives to the proposed action,

(iv) The relationship between local short-term uses of man's environment and the maintenance and enhancement of long-term productivity, and

(v) Any irreversible and irretrievable commitments of resources which would be involved in the proposed action should it be implemented.

Prior to making any detailed statement, the responsible Federal official shall consult with and obtain the comments of any Federal agency which has jurisdiction by law or special expertise with respect to any environmental impact involved. Copies of such statement and the comments and views of the appropriate Federal, State, and local agencies, which are authorized to develop and enforce environmental standards, shall be made available to the President, the Council on Environmental Quality and to the public as provided by section 552 of title 5, United States Code, and shall accompany the proposal through the existing agency review processes;

(d) Any detailed statement required under subparagraph (c) after January 1, 1970, for any major Federal action funded under a program of grants to States shall not be deemed to be legally insufficient solely by reason of having been prepared by a State agency or official, if:

(i) the State agency or official has statewide jurisdiction and has the responsibility for such action,

(ii) the responsible Federal official furnishes guidance and participates in such preparation,

(iii) the responsible Federal official independently evaluates such statement prior to its approval and adoption, and

(iv) after January 1, 1976, the responsible Federal official provides early notification to, and solicits the views of, any other State or any Federal land management entity of any action or any alternative thereto which may have significant impacts upon such State or affected Federal land management entity and, if there is any disagreement on such impacts, prepares a written assessment of such impacts and views for incorporation into such detailed statement.

The procedures in this subparagraph shall not relieve the Federal official of his responsibilities for the scope, objectivity, and content of the entire statement or of any other responsibility under this Act; and further, this subparagraph does not affect the legal sufficiency of statements prepared by State agencies with less than statewide jurisdiction.

(e) Study, develop, and describe appropriate alternatives to recommended courses of action in any proposal which involves unresolved conflicts concerning alternative uses of available resources;

(f) Recognize the worldwide and long-range character of environmental problems and, where consistent with the foreign policy of the United States, lend appropriate support to initiatives, resolutions, and programs designed to maximize international cooperation in anticipating and preventing a decline in the quality of mankind's world environment;

(g) Make available to States, counties, municipalities, institutions, and individuals, advice and information useful in restoring, maintaining, and enhancing the quality of the environment;

(h) Initiate and utilize ecological information in the planning and development of resource-oriented projects; and

(i) Assist the Council on Environmental Quality established by title II of this Act.

SEC. 103. All agencies of the Federal Government shall review their present statutory authority, administrative regulations, and current policies and procedures for the purpose of determining whether there are any deficiencies or inconsistencies therein which prohibit full compliance with the purposes and provisions of this Act and shall propose to the President not later than July 1, 1971, such measures as may be necessary to bring their authority and policies into conformity with the intent, purposes, and procedures set forth in this Act.

SEC. 104. Nothing in section 102 or 103 shall in any way affect the specific statutory obligations of any Federal agency (1) to comply with criteria or standards of environmental quality, (2) to coordinate or consult with any other Federal or State agency, or (3) to act, or refrain from acting contingent upon the recommendations or certification of any other Federal or State agency.

SEC. 105. The policies and goals set forth in this Act are supplementary to those set forth in existing authorizations of Federal agencies.

TITLE II

COUNCIL ON ENVIRONMENTAL QUALITY

SEC. 201. The President shall transmit to the Congress annually beginning July 1, 1970, an Environmental Quality Report (hereinafter referred to as the "report") which shall set forth (1) the status and condition of the major natural, manmade, or altered environmental classes of the Nation, including, but not limited to, the air, the aquatic, including marine, estuarine, and fresh water, and the terrestrial environment, including, but not limited to, the forest, dryland, wetland, range, urban, suburban and rural environment; (2) current and foreseeable trends in the quality, management and utilization of such environments and the effects of those trends on the social, economic, and other requirements of the Nation; (3) the adequacy of available natural resources for fulfilling human and economic requirements of the Nation in the light of expected population pressures; (4) a review of the programs and activities (including regulatory activities) of the Federal Government, the State and local governments, and nongovernmental entities or individuals with particular reference to their effect on the environment and on the conservation, development and utilization of natural resources; and (5) a program for remedying the deficiencies of existing programs and activities, together with recommendations for legislation.

SEC. 202. There is created in the Executive Office of the President a Council on Environmental Quality (hereinafter referred to as the "Council"). The Council shall be composed of three members who shall be appointed by the President to serve at his pleasure, by and with the advice and consent of the Senate. The President shall designate one of the members of the Council to serve as Chairman. Each member shall be a person who, as a result of his training, experience, and attainments, is exceptionally well qualified to analyze and interpret environmental trends and information of all kinds; to appraise programs and activities of the Federal Government in the light of the policy set forth in title I of this Act; to be conscious of and responsive to the scientific, economic, social, esthetic, and cultural needs and interests of the Nation; and to formulate and recommend national policies to promote the improvement of the quality of the environment.

SEC. 203. The Council may employ such officers and employees as may be necessary to carry out its functions under this Act. In addition, the Council may employ and fix the compensation of such experts and consultants as may be necessary for the carrying out of its functions under this Act, in accordance with section 3109 of title 5, United States Code (but without regard to the last sentence thereof).

SEC. 204. It shall be the duty and function of the Council—

(1) to assist and advise the President in the preparation of the Environmental Quality Report required by section 201 of this title;

(2) to gather timely and authoritative information concerning the conditions and trends in the quality of the environment both current and prospective, to analyze and interpret such information for the purpose of determining whether such conditions and trends are interfering, or are likely to interfere, with the achievement of the policy set forth in title I of this Act, and to compile and submit to the President studies relating to such conditions and trends;

(3) to review and appraise the various programs and activities of the Federal Government in the light of the policy set forth in title I of this Act for the purpose of determining the extent to which such programs and activities are contributing to the achievement of such policy, and to make recommendations to the President with respect thereto;

(4) to develop and recommend to the President national policies to foster and promote the improvement of environmental quality to meet the conservation, social, economic, health, and other requirements and goals of the Nation;

(5) to conduct investigations, studies, surveys, research, and analyses relating to ecological systems and environmental quality;

(6) to document and define changes in the natural environment, including the plant and animal systems, and to accumulate necessary data and other information for a continuing analysis of these changes or trends and an interpretation of their underlying causes;

(7) to report at least once each year to the President on the state and condition of the environment; and

(8) to make and furnish such studies, reports thereon, and recommendations with respect to matters of policy and legislation as the President may request.

SEC. 205. In exercising its powers, functions, and duties under this Act, the Council shall—

(1) Consult with the Citizens' Advisory Committee on Environmental Quality established by Executive Order No. 11472, dated May 29, 1969, and with such representatives of science, industry, agriculture, labor, conservation organizations, State and local governments and other groups, as it deems advisable; and

(2) Utilize, to the fullest extent possible, the services, facilities and information (including statistical information) of public and private agencies and organizations, and individuals, in order that duplication of effort and expense may be avoided, thus assuring that the Council's activities will not unnecessarily overlap or conflict with similar activities authorized by law and performed by established agencies.

SEC. 206. Members of the Council shall serve full time and the Chairman of the Council shall be compensated at the rate provided for Level II of the Executive Schedule Pay Rates (5 U.S.C. 5313). The other members of the Council shall be compensated at the rate provided for Level IV of the Executive Schedule Pay Rates (5 U.S.C. 5315).

SEC. 207. The Council may accept reimbursements from any private nonprofit organization or from any department, agency, or instrumentality of the Federal Government, any State, or local government, for the reasonable travel expenses incurred by an officer or employee of the Council in connection with his attendance at any conference, seminar, or similar meeting conducted for the benefit of the Council.

SEC. 208. The Council may make expenditures in support of its international activities, including expenditures for: (1) international travel; (2) activities in implementation of international agreements; and (3) the support of international exchange programs in the United States and in foreign countries.

SEC. 209. There are authorized to be appropriated to carry out the provisions of this chapter not to exceed $300,000 for fiscal year 1970, $700,000 for fiscal year 1971, and $1,000,000 for each fiscal year thereafter.

THE ENVIRONMENTAL QUALITY IMPROVEMENT ACT OF 1970*

TITLE II—ENVIRONMENTAL QUALITY (OF THE WATER QUALITY IMPROVEMENT ACT OF 1974)

SHORT TITLE

SEC. 201. This title may be cited as the "Environmental Quality Improvement Act of 1970."

FINDINGS, DECLARATIONS, AND PURPOSES

SEC. 202. (a) The Congress finds—

(1) That man has caused changes in the environment;

(2) That many of these changes may affect the relationship between man and his environment; and

(3) That population increases and urban concentration contribute directly to pollution and the degradation of our environment.

(b)(1) The Congress declares that there is a national policy for the environment which provides for the enhancement of environmental quality. This policy is evidenced by statutes heretofore enacted relating to the prevention, abatement, and control of environmental pollution, water and land resources, transportation, and economic and regional development.

(2) The primary responsibility for implementing this policy rests with State and local governments.

(3) The Federal Government encourages and supports implementation of this policy through appropriate regional organizations established under existing law.

(c) The purposes of this title are—

(1) To assure that each Federal department and agency conducting or supporting public works activities which affect the environment shall implement the policies established under existing law; and

(2) To authorize an Office of Environmental Quality, which, notwithstanding any other provision of law, shall provide the professional and administrative staff for the Council on Environmental Quality established by Public Law 91–190.

OFFICE OF ENVIRONMENTAL QUALITY

SEC. 203. (a) There is established in the Executive Office of the President an office to be known as the Office of Environmental Quality (hereafter in this title referred to as the "Office"). The Chairman of the Council on Environmental Quality established by Public Law 91–190 shall be the Director of the Office. There shall be in the Office a Deputy Director who shall be appointed by the President, by and with the advice and consent of the Senate.

(b) The compensation of the Deputy Director shall be fixed by the President at a rate not in excess of the annual rate of compensation payable to the Deputy Director of the Bureau of the Budget.

(c) The Director is authorized to employ such officers and employees (including experts and consultants) as may be necessary to enable the Office to carry out its functions under this title and Public Law 91–190, except that he may employ no more than 10 specialists and other experts without regard to the provisions of title 5, United States Code, governing appointments in the competitive service, and pay such specialists and experts without regard to the provisions of chapter 51 and subchapter 111 of chapter 53 of such title relating to classification and General Schedule pay rates, but no such specialist or

*Pub. L. 91–224, 42 U.S.C. 4371–4374, April 3, 1970.

expert shall be paid at a rate in excess of the maximum rate for GS–18 of the General Schedule under section 5330 of title 5.

(d) In carrying out his functions the Director shall assist and advise the President on policies and programs of the Federal Government affecting environmental quality by—

(1) Providing the professional and administrative staff and support for the Council on Environmental Quality established by Public Law 91–190;

(2) Assisting the Federal agencies and departments in appraising the effectiveness of existing and proposed facilities, programs, policies, and activities of the Federal Government, and those specific major projects designated by the President which do not require individual project authorization by Congress, which affect environmental quality;

(3) Reviewing the adequacy of existing systems for monitoring and predicting environmental changes in order to achieve effective coverage and efficient use of research facilities and other resources;

(4) Promoting the advancement of scientific knowledge of the effects of actions and technology on the environment and encourage the development of the means to prevent or reduce adverse effects that endanger the health and well-being of man;

(5) Assisting in coordinating among the Federal departments and agencies those programs and activities which affect, protect, and improve environmental quality;

(6) Assisting the Federal departments and agencies in the development and interrelationship of environmental quality criteria and standards established through the Federal Government;

(7) Collecting, collating, analyzing, and interpreting data and information on environmental quality, ecological research, and evaluation.

(e) The Director is authorized to contract with public or private agencies, institutions, and organizations and with individuals without regard to sections 3618 and 3709 of the Revised Statutes (31 U.S.C. 529; 41 U.S.C. 5) in carrying out his functions.

REPORT

Sec. 204. Each Environmental Quality Report required by Public Law 91–190 shall, upon transmittal to Congress, be referred to each standing committee having jurisdiction over any part of the subject matter of the Report.

AUTHORIZATION

Sec. 205. There are hereby authorized to be appropriated not to exceed $500,000 for the fiscal year ending June 30, 1970, not to exceed $750,000 for the fiscal year ending June 30, 1971, not to exceed $1,250,000 for the fiscal year ending June 30, 1972, and not to exceed $1,500,000 for the fiscal year ending June 30, 1973. These authorizations are in addition to those contained in Public Law 91–190.

Approved April 3, 1970.

THE CLEAN AIR ACT § 309*

§ 7609. Policy review

(a) The Administrator shall review and comment in writing on the environmental impact of any matter relating to duties and responsibilities granted pursuant to this chapter or other provisions of the authority of the Administrator, contained in any (1) legislation proposed by any Federal department or agency, (2) newly authorized Federal projects for construction and any major Federal agency action (other than a project for construction) to which section 4332(2)(C) of this title applies, and (3) proposed regulations published by any department or agency of the Federal Government. Such written comment shall be made public at the conclusion of any such review.

(b) In the event the Administrator determines that any such legislation, action, or regulation is unsatisfactory from the standpoint of public health or welfare or environmental quality, he shall publish his determination and the matter shall be referred to the Council on Environmental Quality.

---

*July 14, 1955, c. 360, § 309, as added Dec. 31, 1970, Pub. L. 91-604 § 12(a), 42 U.S.C. § 7609 (1970).

Executive Order 11514. March 5, 1970

PROTECTION AND ENHANCEMENT OF ENVIRONMENTAL QUALITY

As amended by Executive Order 11991. (Secs. 2(g) and (3(h)). May 24, 1977*

By virtue of the authority vested in me as President of the United States and in furtherance of the purpose and policy of the National Environmental Policy Act of 1969 (Public Law No. 91-190, approved January 1, 1970), it is ordered as follows:

Section 1. *Policy.* The Federal Government shall provide leadership in protecting and enhancing the quality of the Nation's environment to sustain and enrich human life. Federal agencies shall initiate measures needed to direct their policies, plans and programs so as to meet national environmental goals. The Council on Environmental Quality, through the Chairman, shall advise and assist the President in leading this national effort.

Sec. 2. *Responsibilities of Federal agencies.* Consonant with Title I of the National Environmental Policy Act of 1969, hereafter referred to as the "Act", the heads of Federal agencies shall:

(a) Monitor, evaluate, and control on a continuing basis their agencies' activities so as to protect and enhance the quality of the environment. Such activities shall include those directed to controlling pollution and enhancing the environment and those designed to accomplish other program objectives which may affect the quality of the environment. Agencies shall develop programs and measures to protect and enhance environmental quality and shall assess progress in meeting the specific objectives of such activities. Heads of agencies shall consult with appropriate Federal, State and local agencies in carrying out their activities as they affect the quality of the environment.

(b) Develop procedures to ensure the fullest practicable provision of timely public information and understanding of Federal plans and programs with environmental impact in order to obtain the views of interested parties. These procedures shall include, whenever appropriate, provision for public hearings, and shall provide the public with relevant information, including information on alternative courses of action. Federal agencies shall also encourage State and local agencies to adopt similar procedures for informing the public concerning their activities affecting the quality of the environment.

(c) Insure that information regarding existing or potential environmental problems and control methods developed as part of research, development, demonstration, test, or evaluation activities is made available to Federal agencies, States, counties, municipalities, institutions, and other entities, as appropriate.

---

*The Preamble to Executive Order 11991 is as follows:

By virtue of the authority vested in me by the Constitution and statutes of the United States of America, and as President of the United States of America, in furtherance of the purpose and policy of the National Environmental Policy Act of 1969, as amended (42 U.S.C. 4321 *et seq.*), the Environmental Quality Improvement Act of 1970 (42 U.S.C. 4371 *et seq.*), and Section 309 of the Clean Air Act, as amended (42 U.S.C. 1857h-7), it is hereby ordered as follows:

(d) Review their agencies' statutory authority, administrative regulations, policies, and procedures, including those relating to loans, grants, contracts, leases, licenses, or permits, in order to identify any deficiencies or inconsistencies therein which prohibit or limit full compliance with the purposes and provisions of the Act. A report on this review and the corrective actions taken or planned, including such measures to be proposed to the President as may be necessary to bring their authority and policies into conformance with the intent, purposes, and procedures of the Act, shall be provided to the Council on Environmental Quality not later than September 1, 1970.

(e) Engage in exchange of data and research results, and cooperate with agencies of other governments to foster the purposes of the Act.

(f) Proceed, in coordination with other agencies, with actions required by section 102 of the Act.

(g) In carrying out their responsibilites under the Act and this Order, comply with the regulations issued by the Council except where such compliance would be inconsistent with statutory requirements.

Sec. 3. *Responsibilities of Council on Environmental Quality.* The Council on Environmental Quality shall:

(a) Evaluate existing and proposed policies and activities of the Federal Government directed to the control of pollution and the enhancement of the environment and to the accomplishment of other objectives which affect the quality of the environment. This shall include continuing review of procedures employed in the development and enforcement of Federal standards affecting environmental quality. Based upon such evaluations the Council shall, where appropriate, recommend to the President policies and programs to achieve more effective protection and enhancement of environmental quality and shall, where appropriate, seek resolution of significant environmental issues.

(b) Recommend to the President and to the agencies priorities among programs designed for the control of pollution and for enhancement of the environment.

(c) Determine the need for new policies and programs for dealing with environmental problems not being adequately addressed.

(d) Conduct, as it determines to be appropriate, public hearings or conferences on issues of environmental significance.

(e) Promote the development and use of indices and monitoring systems (1) to assess environmental conditions and trends, (2) to predict the environmental impact of proposed public and private actions, and (3) to determine the effectiveness of programs for protecting and enhancing environmental quality.

(f) Coordinate Federal programs related to environmental quality.

(g) Advise and assist the President and the agencies in achieving international cooperation for dealing with environmental problems, under the foreign policy guidance of the Secretary of State.

(h) Issue regulations to Federal agencies for the implementation of the procedural provisions of the Act (42 U.S.C. 4332(2)). Such regulations shall be developed after consultation with affected agencies and after such public hearings as may be appropriate. They will be designed to make the environmental impact statement process more useful to decisionmakers and the public; and to reduce paperwork and the accumulation of extraneous background data, in order to emphasize the need to focus on real environmental issues and alternatives. They will require impact statements to be concise, clear, and to the point, and supported by evidence that agencies have made the necessary environmental analyses. The Council shall include in its regulations procedures (1) for the early preparation of environmental impact statements, and (2) for the referral to the Council of conflicts between agencies concerning the implementation of the National Environmental Policy Act of 1969, as amended, and

Section 309 of the Clean Air Act, as amended, for the Council's recommendation as to their prompt resolution.

(i) Issue such other instructions to agencies, and request such reports and other information from them, as may be required to carry out the Council's responsibilities under the Act.

(j) Assist the President in preparing the annual Environmental Quality Report provided for in section 201 of the Act.

(k) Foster investigations, studies, surveys, research, and analyses relating to (i) ecological systems and environmental quality, (ii) the impact of new and changing technologies thereon, and (iii) means of preventing or reducing adverse effects from such technologies.

Sec. 4. *Amendments of E.O. 11472.* Executive Order No. 11472 of May 29, 1969, including the heading thereof, is hereby amended:

(1) By substituting for the term "the Environmental Quality Council", wherever it occurs, the following: "the Cabinet Committee on the Environment".

(2) By substituting for the term "the Council", wherever it occurs, the following: "the Cabinet Committee".

(3) By inserting in subsection (f) of section 101, after "Budget,", the following: "the Director of the Office of Science and Technology,".

(4) By substituting for subsection (g) of section 101 the following:

"(g) The Chairman of the Council on Environmental Quality (established by Public Law 91-190) shall assist the President in directing the affairs of the Cabinet Committee."

(5) by deleting subsection (c) of section 102.

(6) By substituting for "the Office of Science and Technology", in section 104, the following: "the Council on Environmental Quality (established by Public Law 91-190)".

(7) By substituting for "(hereinafter referred to as the 'Committee')", in section 201, the following: "(hereinafter referred to as the 'Citizens' Committee')".

(8) By substituting for the term "the Committee", wherever it occurs, the following: "the Citizens' Committee".

☆U.S. GOVERNMENT PRINTING OFFICE : 1980- 315-426/6358

APPENDIX G
Implementation of Procedures for the National Environmental Policy Act

ENVIRONMENTAL PROTECTION AGENCY

40 CFR PART 6
[FRL 1315-6]

Implementation of Procedures on the National Environmental Policy Act

AGENCY: Environmental Protection Agency (EPA)

ACTION: Rule.

SUMMARY: On November 29, 1978, the Council on Environmental Quality (CEQ) promulgated Regulations establishing uniform procedures for implementing the procedural provisions of the National Environmental Policy Act. CEQ required Federal agencies to adopt appropriate procedures to supplement their Regulations. As a result, EPA has amended its procedures contained in 40 CFR Part 6 to take into account this initiative.

DATES: These regulations will be effective on December 15, 1979.

ADDRESSES: Comments submitted on the regulations may be inspected at the Public Information Reference Unit, EPA Headquarters, Room 2922, Waterside Mall, 401 M Street, S.W., Washington, D.C., between 8:00 A.M. and 4:30 P.M. on business days.

FOR FURTHER INFORMATION CONTACT: Thomas Sheckells, Office of Environmental Review (A-104), EPA, 401 M Street, S.W., Washington, D.C. 20460; telephone 202/755-0790.

SUPPLEMENTARY INFORMATION:

Introduction

The National Environmental Policy Act of 1969 (NEPA), 42 U.S.C. 4321 *et seq.*, as implemented by Executive Orders 11514 and 11991, and the Council on Environmental Quality (CEQ) Regulations (40 CFR Sections 1500-1508) requires that all agencies of the Federal Government to the fullest extent possible carry out the provisions of NEPA by building into agency decision-making appropriate and careful consideration of the environmental effects of proposed actions, and avoiding or minimizing the adverse effects of these actions. The environmental impact statement (EIS) requirement under section 102(2)(C) serves as the most significant mechanism for implementing NEPA. These regulations set forth the requirements for EPA to carry out its obligations under NEPA.

Proposed regulations were published in the **Federal Register** on June 18, 1979 (44 FR 35158).

In view of the President's directive to CEQ to establish a single set of regulations for government-wide NEPA implementation, CEQ has directed the Federal agencies to avoid restating or paraphrasing the CEQ Regulations, even though agencies may quote or cross-reference the CEQ regulations in their implementing procedures. Therefore, it must be made clear the following regulations shall be read with the understanding that the reader has available the policy statements and definitions contained in the CEQ Regulations. In this respect, it is noted that previous nomenclature used by EPA has been adjusted to conform with the CEQ regulations. The document entitled "environmental assessment" as previously used by EPA is now referred to as an "environmental information document"; the document entitled "negative declaration" is now a "finding of no significant impact"; and the document entitled "environmental impact appraisal" is now "environmental assessment."

These regulations amend EPA regulations under 40 CFR Part 6 previously promulgated in Subparts A-H on April 14, 1975 (see 40 FR 16823) and Subpart I on January 11, 1977 (see 42 FR 2450).

Exemptions

Over the past several years there has been much controversy surrounding this Agency's preparation of EISs. Considering the nature of EPA's activities is generally concerning actions protective of the environment, the Congress and the Courts have seen fit to exempt numerous EPA activities from EIS applicability.

The Congress has provided major exemptions under the Clean Water Act and the Clean Air Act. Specifically, under section 511(c)(1) of the Clean Water Act (CWA) (PL 92-500), EPA is exempt from preparing EISs under the CWA except for the issuance of new source National Pollutant Discharge Elimination System (NPDES) permits as authorized under section 402 and the provision of Federal financial assistance for the purpose of assisting the construction of publicly owned treatment works under section 201. Under Section 7(c)(1) of the Energy Supply and Environmental Coordination Act of 1974 (PL 93-319), all activities under the Clean Air Act are exempt from the EIS requirements of NEPA. Further, the courts have found EPA to be exempt from the EIS requirements for regulatory actions under the Clean Air Act, the Federal Insecticide, Fungicide, and Rodenticide Act, and the Marine Protection, Research, and Sanctuaries Act, because major EPA actions under these statutes are undertaken with sufficient safeguards to ensure performance of a functionally equivalent analysis of NEPA's EIS requirements. See *EDF* v. *EPA*, 489 F2d 1247 (D.C. Circuit, 1973); *Wyoming* v. *Hathaway*, 525 F2d 66 (Tenth Circuit, 1975); *Maryland* v. *Train*, 415 F. Supp. 116 (District Court, Maryland, 1976). In addition, the Agency has determined that EPA regulatory activities under the Resource Conservation and Recovery Act of 1976, the Toxic Substances Control Act of 1976, the Safe Drinking Water Act, and the Noise Control Act are exempt from the EIS requirements of NEPA. Nevertheless, on May 7, 1974, Administrator Russell Train decided that the Agency would voluntarily prepare EISs on certain regulatory activities in spite of the statutory and court exemptions that existed at that time. This revised regulation does not affect those voluntary EIS procedures.

Summary of Regulation

The regulations set forth below are intended to meet the requirements for Federal agency procedures under § 1507.3 of the CEQ Regulations.

Subparts A through I describe procedures for preparing required EISs. Subparts E through I establish those classes of action and create specific criteria for preparing environmental assessments and EISs pursuant to § 1507.3(b)(2) of the CEQ Regulations. Also, numerous categorical exemptions have been created over and above those exemptions referred to above. Specifically, Subpart E relates to environmental review procedures for the Wastewater Treatment Construction Grants Program of the CWA; Subpart F relates to environmental review procedures for the new source NPDES permit program; Subpart G relates to environmental review procedures for research and development programs; Subpart H relates to environmental review procedures for solid waste demonstration projects; and Subpart I relates to environmental review procedures for EPA undertakings for construction of special purpose facilities or facility renovations. Subpart C relates to integrating the requirements of other environmental laws with the environmental review procedures set forth in subparts E through I. It is emphasized that subpart C simply provides a recitation of these other environmental laws. Detailed procedures for compliance with these laws are set forth in regulations promulgated by the responsible agency, i.e., Advisory Council on Historic Preservation regulations dated January 30, 1979, or EPA procedures implementing these laws, i.e., EPA's Statement of Procedures on Floodplain Management and Wetlands Protection dated January 5, 1979.

Regarding EPA's Statement of Procedures on Floodplain Management and Wetlands Protection, there has been some confusion as to the effective date of this interim Statement of Procedures. As described under section 7.a, further EPA program amendments implementing the Statement of Procedures were required by July 5, 1979. This was intended to be the effective date of this Statement of Procedures. In order to emphasize the importance of this Statement of Procedures, it has been made a part of this Regulation as Appendix A.

Application to Ongoing Construction Grant Activities

Applicants for wastewater treatment construction grants currently face a tremendous number of policies and regulations with which they must comply. Each time new requirements are imposed, grantees are faced with a situation that can require them to duplicate work that already has been completed. This duplication causes undue delay in the grants process and causes both delay and increased costs in water-pollution control efforts. Although these regulations impose very few new, substantive requirements, lack of phasing could cause administrative delays. To avoid this problem and to alleviate any possible burden on grantees and their consultants, § 6.102(c)(2) provides for phasing these regulations in the construction grants program.

This regulation replaces EPA's previous NEPA regulation dated April, 1975. The regulation, however, contains primarily procedural changes and few substantive changes; the regulation consolidates those policies and procedures that have been imposed since EPA's previous NEPA regulations were promulgated, as well as implementing the CEQ Regulations which were effective on July 30, 1979.

Under these circumstances, this regulation is intended to require compliance with new administrative procedures effective December 15, 1979. These procedures are those that are primarily the responsibility of EPA and include procedures such as review time for findings of no significant impact, environmental assessment and EIS format changes, etc.

No new substantive requirements will be imposed on grantees that are well along in the planning process. These grantees will be allowed to complete their environmental information documents under policies and regulations already applicable to their projects. If a case should arise in which EPA determines that environmental documents submitted by a grantee fail to comply with any new substantive requirement under this regulation, it will be the responsibility of EPA to supplement this information. It should be emphasized, however, that any lack of environmental documentation required under previously applicable procedures will still be the responsibility of the grantee.

All facility plans submitted after September 30, 1980 must comply with this regulation. This allows a nine month phase in period which permits grantees progressing in a timely manner to complete environmental review procedures under existing procedures.

Response to Comments on Proposed Regulations

EPA received 39 comment letters on the proposed regulations. As a result of external and internal input, numerous technical changes were made to improve the regulation. The following is the response to the substantive comments made on the regulation:

1. *Request for comment-period extension and public hearing on the proposed regulation.*

Several commenters requested an extension of the comment period beyond the July 18, 1979 (or 30-day) limit. Upon individual request, EPA considered comments submitted as late as August 2, 1979. However, because of the time constraints imposed by the CEQ Regulations' effective date of July 30, 1979, the comment period could not be extended any further. More importantly, we believe that the issues presented by these proponents for a time extension and a public hearing were given a thorough airing during the development of the CEQ Regulations which evolved with substantial government, industry, and citizen interaction between June, 1977 and November, 1978. Three specific issues that were raised include:

a. Limiting construction activities on a project until the environmental review process is completed; see item 5 below for further discussion.

b. Conditioning or denying new source NPDES permits based on other than water quality considerations; see item 5 below for further discussion.

c. Assuring there is no conflict of interest or financial stake in the outcome of the project by the contractor preparing the EIS under EPA's, "third party" EIS preparation method; see § 1506.5(c) of the CEQ Regulations (also, see item 6 below).

Considering the CEQ Regulations are already in effect, we believe than an appropriate comment period was created for this regulation.

2. *Establishing the need for mandatory public hearings when an EIS is prepared.* In the proposed regulation, we solicited comments on conducting mandatory public hearings attendant to a draft EIS. The response was about even. The CEQ Regulations under § 1506.6 do not require mandatory public hearings. However, EPA's public participation regulations pertaining to "Grants for Construction of Treatment Works" under 40 CFR Part 35, Subpart E, require grant applicants to undertake a Full-scale Public Participation Program whenever an EIS is required. This includes the conduct of two public meetings and one mandatory hearing during the developing of the facilities plan; the hearing is encouraged to be held in conjunction with the public hearing on the draft EIS (see 40 CFR § 35.917-5(c)(3)(viii)). EPA prepares most of its EISs on construction grant projects. Additionally, there is the view that because an EIS is generated in most cases on controversial projects, members of the local community should have the opportunity to hear testimony on the necessity of the project and personally to make comments. Therefore, we have concluded that mandatory hearings shall be conducted within 45 days after the issuance of a draft EIS; see § 6.400(c).

3. *Providing substantial exemptions for EPA programs for the application of NEPA requirements.* The "Exemptions" provision set forth above spells out the numerous statutory, court-ordered and EPA-interpreted exemptions from NEPA. Several commenters suggested that the scope of EPA's NEPA exemptions is overly broad. While taking philosophical exception to EPA's statutory and judicial exemptions, the commenters suggest that there continues to remain a duty for EPA to prepare some form of an "environmental assessment" for related actions; further, it was suggested that EISs be prepared for actions which are not environmentally protective regulations, particularly where EPA's action is potentially environmentally detrimental or highly controversial. We continue to maintain that we are exempt from the requirements of NEPA except as set forth under subparts A through I of this regulation. This view is supported by a legal opinion dated March 22, 1979 on the application of the EIS requirements to the regulatory actions taken under the Resource Conservation and Recovery Act of 1976. It is further supported by a legal opinion dated May 7, 1979 on the application of section 102(2)(E) of NEPA to NPDES permitting activities for either new dischargers or existing sources.

4. *Assuring inclusion, monitoring, and enforcement of mitigation measures in grants.* In the proposed regulation, we solicited comments on the question of whether third parties (e.g., citizens) should be able to file lawsuits to enforce compliance with grant conditions. Generally, opposition has been expressed to this concept. However, these regulations under § 6.509 entitled "Identification of mitigation measures" have been bolstered to recognize the affirmative duty of the responsible official to assure that effective mitigation measures identified in a finding of no significant impact (FNSI) or EIS are implemented by the grantee. Further, explicit monitoring and enforcement measures are set forth under § 6.510. However, the absence of an explicit third-party enforcement role in these regulations does not limit citizen enforcement authority under NEPA and the CEQ Regulations.

5. *Imposing substantive conditions on new source NPDES permitting.* Several commenters expressed concern about EPA's position that NEPA requires us to impose substantive conditions on the new source NPDES permitting process. As mentioned in item 1 above, the two major issues raised by commenters relate to limiting construction activities on a project until the environmental process is completed and conditioning or denying NPDES permits based on factors identified during the NEPA process. These issues were recently considered in the June 7, 1979 promulgation of the revised NPDES Regulations (see 44 FR 32854). That Regulation is preceded by a discussion explaining EPA's basis for setting the requirements at issue (see 44 FR 32871-32872). With regard to the preconstruction activities issue, today's rulemaking cross-references the NPDES Regulations under 40 CFR 122.47(c) (see § 6.603). However, the mitigation and monitoring requirements contained in the proposed NEPA regulations are retained in today's promulgation to clarify how the results of the NEPA process are to be made effective (see § § 6.606(b) and 6.607).

6. *Using third party EIS preparation method for new source NPDES permitting.* Several commenters raised several issues pertaining to the third party method for preparing new source EISs. This is the method whereby the applicant, aware in the early planning stages of his project of the need for an EIS, contracts directly with a consulting firm to prepare the EIS. This forecloses the need for other preliminary environmental documents (i.e., an environmental information document or environmental assessment); however, EPA must select the consultant, ensure the consultant has no conflict of interest, and oversee the preparation of the EIS (see § 6.604(g)(3)).

This method of EIS preparation created such interest on the part of some commenters that they requested a public hearing just on this issue (see comment 1 above). We determined that there was not sufficient basis for a hearing for the following reasons: Several commenters who raised this issue expressed concern over an alleged mandatory use of this method for new source EIS preparation; however, it is clear, as was set forth in the proposed regulation and now in § 6.603(g) of this regulation, that the third party method is discretionary, and is one of three methods which may be used by EPA to prepare new source EISs. A major concern raised by commenters was over the conflict of interest limitations imposed on the third party contractor. This is a primary concern to engineering consulting firm parent companies involved in performing engineering work for the applicant who desire to see their subsidiary, an environmental consulting firm, to be used as the third party contractor. This practice is generally prohibited under the EPA rules (see § 6.604(g)(3)(ii)). This standard concerning limitations on conflict of interest, as well as full disclosure pertaining to the contractor's stake in the outcome of the project, has been imposed under § 1506.5(c) of the CEQ Regulations; EPA fully supports this provision in principle. We believe that the objectivity standard that EPA has always imposed in selecting a third party contractor supports the primary NEPA concepts of full and complete disclosure in an objective way and assurance of the integrity of the environmental review process. Several commenters also noted the total exclusion of the applicant in EPA's selecting the third party contractor. Although EPA must select the contractor, a provision has been added under § 6.604(g)(3)(i) giving the applicant a consultation role in the choosing of the contractor.

7. *Deleting express reference to the adoption method for preparing construction grant EISs.* In the proposed regulation, EPA included a provision for preparing wastewater treatment construction grant EISs by a method whereby the grantee's facilities plan and NEPA-related information, to the extent they adequately addressed relevant environmental issues, and after independent EPA evaluation, could be adopted to satisfy the requirement for an EIS. Several commenters suggested that this method of EIS preparation contravened the intent of Subpart E for the identification of the need for preparation of EISs early in the facilities planning process. For instance, under the joint EIS process method (see § 6.507(h)(3)), the EIS is prepared in parallel with the facilities plan. Therefore, this adoption provision has been deleted. Nevertheless, it should be made clear that although we are discouraging use of the adoption approach, we recognize that situations may arise which require reliance on this method, which would be permissible under Subpart E.

8. *Conformity to state implementation plans.* In the proposed regulation we proposed a course of action to ensure the conformity of EPA actions to the provisions of each state air quality implementation plan (SIP) pursuant to the requirements of section 176(c) of the Clean Air Act, as amended in 1977. Subpart C of the final regulation sets forth the requirements, revised in response to public comments, for EPA to carry out its obligations under section 176(c).

The Agency intends that the responsible EPA official will consult with State and local agencies during the preparation of the environmental assessment to obtain a recommendation as to the conformity of the proposed EPA action to the SIP. Section 6.303 of the regulation has been specifically revised to require the responsible EPA official to provide a written assurance in the FNSI or draft EIS that the proposed EPA action conforms with the SIP. The final regulation now provides that the opportunity for State concurrence or nonconcurrence with EPA's conformity determination will occur during the FNSI or draft EIS review time periods.

Some comments on the proposed § 6.303(c) requested clarification of the extent to which the Agency will delay a proposed action when it has determined that the action will not be in conformity with the SIP. Accordingly, § 6.303 has been revised to provide for an explicit Agency response to any notification of State nonconcurrence with the EPA conformity determination. If EPA finds that the State nonconcurrence is unjustified, then an explanation of this finding will be included in the final FNSI or final EIS.

However, if EPA finds that the State nonconcurrence is warranted, then the Agency intends that the proposed action will not receive final approval until it has been brought into conformity with the SIP. Achieving conformity may necessitate modifications to proposed actions implemented by EPA or actions implemented by others, but subject to EPA approval. In some instances, the State may wish to revise its SIP to account for the proposed action. When the Agency has been notified of and agrees with the State nonconcurrence, the final FNSI or final EIS will detail the measures that will need to be taken, prior to final EPA approval, to assure conformity with the SIP.

Dated: October 29, 1979.

Note.—EPA has determined that because this document does not constitute a significant regulation within the meaning of Executive Order 12044, preparation of a regulatory analysis is not required.

Douglas M. Costle,
Administrator.

Accordingly, 40 CFR Part 6 is revised in its entirety to read as follows:

PART 6—IMPLEMENTATION OF PROCEDURES ON THE NATIONAL ENVIRONMENTAL POLICY ACT

Authority: Secs. 101, 102, and 103 of the National Environmental Policy Act of 1969 (42 U.S.C. 4321 *et seq.*); also, the Council on Environmental Quality Regulations dated November 29, 1978 (40 CFR Part 1500).

SUBPART A—GENERAL

§ *6.100 Purpose and policy.*

(a) The National Environmental Policy Act of 1969 (NEPA), 42 U.S.C. 4321 *et seq.*, as implemented by Executive Orders 11514 and 11991 and the Council on Environmental Quality (CEQ) Regulations of November 29, 1978 (43 FR 55978) requires that Federal agencies include in their decision-making processes appropriate and careful consideration of all environmental effects of proposed actions, analyze potential environmental effects of proposed actions and their alternatives for public understanding and scrutiny, avoid or minimize adverse effects of proposed actions, and restore and enhance environmental quality as much as possible. The Environmental Protection Agency (EPA) shall integrate these NEPA factors as early in the Agency planning processes as possible. The environmental review process shall be the focal point to assure NEPA considerations are taken into account. To the extent applicable, EPA shall prepare environmental impact statements (EISs) on those major actions determined to have significant impact on the quality of the human environment. This part takes into account the EIS exemptions set forth under section 511(c)(1) of the Clean Water Act (Pub. L. 92-500) and section 7(c)(1) of the Energy Supply and Environmental Coordination Act of 1974 (Pub. L. 93-319).

(b) This part establishes EPA policy and procedures for the identification and analysis of the environmental impacts of EPA-related activities and the preparation and processing of EISs.

§ ***6.101 Definitions.***

(a) *Terminology.* All terminology used in this part will be consistent with the terms as defined in 40 CFR Part 1508 (the CEQ Regulations). Any qualifications will be provided in the definitions set forth in each subpart of this regulation.

(b) The term "CEQ Regulations" means the regulations issued by the Council on Environmental Quality on November 29, 1978 (see 43 FR 55978), which implement Executive Order 11991. The CEQ Regulations will often be referred to throughout this regulation by reference to 40 CFR Part 1500 *et al.*

(c) The term "environmental review" means the process whereby an evaluation is undertaken by EPA to determine whether a proposed Agency action may have a significant impact on the environment and therefore require the preparation of the EIS.

(d) The term "environmental information document" means any written analysis prepared by an applicant, grantee or contractor describing the environmental impacts of a proposed action. This document will be of sufficient scope to enable the responsible official to prepare an environmental assessment as described in the remaining subparts of this regulation.

(e) The term "grant" as used in this part means an award of funds or other assistance by a written grant agreement or cooperative agreement under 40 CFR Chapter I, Subpart B.

§ ***6.102 Applicability.***

(a) *Administrative actions covered.* This part applies to the activities of EPA in accordance with the outline of the subparts set forth below. Each subpart describes the detailed environmental review procedures required for each action.

(1) Subpart A sets forth an overview of the regulation. Section 6.102(b) describes the requirements for EPA legislative proposals.

(2) Subpart B describes the requirements for the content of an EIS prepared pursuant to subparts E, F, G, H, and I.

(3) Subpart C describes the requirements for coordination of all environmental laws during the environmental review undertaken pursuant to Subparts E, F, G, H, and I.

(4) Subpart D describes the public information requirements which must be undertaken in conjunction with the environmental review requirements under Subparts E, F, G, H, and I.

(5) Subpart E describes the environmental review requirements for the wastewater treatment construction grants program under Title II of the Clean Water Act.

(6) Subpart F describes the environmental review requirements for new source

National Pollutant Discharge Elimination System (NPDES) permits under section 402 of the Clean Water Act.

(7) Subpart G describes the environmental review requirements for research and development programs undertaken by the Agency.

(8) Subpart H describes the environmental review requirements for solid waste demonstration projects undertaken by the Agency.

(9) Subpart I describes the environmental review requirements for construction of special purpose facilities and facility renovations by the Agency.

(b) *Legislative proposals.* As required by the CEQ Regulations, legislative EISs are required for any legislative proposal developed by EPA which significantly affects the quality of the human environment. A preliminary draft EIS shall be prepared by the responsible EPA office concurrently with the development of the legislative proposal and contain information required under subpart B. The EIS shall be processed in accordance with the requirements set forth under 40 CFR 1506.8.

(c) *Application to ongoing activities*—(1) *General.* The effective date for these regulations is December 5, 1979. These regulations do not apply to an EIS or supplement to that EIS if the draft EIS was filed with the Office of Environmental Review (OER) before July 30, 1979. No completed environmental documents need be redone by reason of these regulations.

(2) With regard to activities under Subpart E, these regulations shall apply to all EPA environmental review procedures effective December 15, 1979. However, for facility plans begun before December 15, 1979, the responsible official shall impose no new requirements on the grantee. Such grantees shall comply with requirements applicable before the effective date of this regulation. Notwithstanding the above, this regulation shall apply to any facility plan submitted to EPA after September 30, 1980.

§ *6.103* ***Responsibilities.***

(a) *General responsibilities.* (1) The responsible official's duties include:

(i) Requiring applicants, contractors, and grantees to submit environmental information documents and related documents and assuring that environmental reviews are conducted on proposed EPA projects at the earliest possible point in EPA's decision-making process. In this regard, the responsible official shall assure the early involvement and availability of information for private applicants and other non-Federal entities requiring EPA approvals.

(ii) When required, assuring that adequate draft EISs are prepared and distributed at the earliest possible point in EPA's decision-making process, their internal and external review is coordinated, and final EISs are prepared and distributed.

(iii) When an EIS is not prepared, assuring that findings of no significant impact (FNSIs) and environmental assessments are prepared and distributed for those actions requiring them.

(1) *The responsible official's duties include:*

(i) Requiring applicants, contractors, and grantees to submit environmental information documents and related documents and assuring that environmental reviews are conducted on proposed EPA projects at the earliest possible point in EPA's decision-making process. In this regard, the responsible official shall assure the early involvement and availability of information for private applicants and other non-Federal entities requiring EPA approvals.

(ii) When required, assuring that adequate draft EISs are prepared and distributed at the earliest possible point in EPA's decision-making process, their internal and external review is coordinated, and final EISs are prepared and distributed.

(iii) When an EIS is not prepared, assuring that findings of no significant impact (FNSIs) and environmental assessments are prepared and distributed for those actions requiring them.

(iv) Consulting with appropriate officials responsible for other environmental laws set forth in Subpart C.

(v) Consulting with the Office of Environmental Review (OER) on actions involving unresolved conflicts concerning this part or other Federal agencies.

(vi) When required, assuring that public participation requirements are met.

(2) *Office of Environmental Review duties include:*

(i) Supporting the Administrator in providing EPA policy guidance and assuring that EPA offices establish and maintain adequate administrative procedures to comply with this part.

(ii) Monitoring the overall timeliness and quality of the EPA effort to comply with this part.

(iii) Providing assistance to responsible officials as required, i.e., preparing guidelines describing the scope of environmental information required by private applicants relating to their proposed actions.

(iv) Coordinating the training of personnel involved in the review and preparation of EISs and other associated documents.

(v) Acting as EPA liaison with the Council on Environmental Quality and other Federal and State entities on matters of EPA policy and administrative mechanisms to facilitate external review of EISs, to determine lead agency and to improve the uniformity of the NEPA procedures of Federal agencies.

(vi) Advising the Administrator and Deputy Administrator on projects which involve more than one EPA office, are highly controversial, are nationally significant, or "pioneer" EPA policy, when these projects have had or should have an EIS prepared on them.

(vii) Carrying out administrative duties relating to maintaining status of EISs within EPA, i.e., publication of notices of intent in the **Federal Register** and making available to the public status reports on EISs and other elements of the environmental review process.

(3) *Office of an Assistant Administrator duties include:*

(i) Providing specific policy guidance to their respective offices and assuring that those offices establish and maintain adequate administrative procedures to comply with this part.

(ii) Monitoring the overall timeliness and quality of their respective office's efforts to comply with this part.

(iii) Acting as liaison between their offices and the OER and between their offices and other Assistant Administrators or Regional Administrators on matters of agencywide policy and procedures.

(iv) Advising the Administrator and Deputy Administrator through the OER on projects or activities within their respective areas of responsibilities which involve more than one EPA office, are highly controversial, are nationally significant, or "pioneer" EPA policy, when these projects will have or should have an EIS prepared on them.

(v) Pursuant to § 6.102(b) of this subpart, preparing legislative EISs as appropriate on EPA legislative initiatives.

(4) The Office of Planning and Evaluation shall be responsible for coordinating the preparation of EISs required on EPA legislative proposals (see § 6.102(b)).

(b) *Responsibilities for Subpart E.*

(1) *Responsible official.* The responsible official for EPA actions covered by this subpart is the Regional Administrator.

(2) *Assistant Administrator.* The responsibilities of the Office of the Assistant Administrator, as described in § 6.103(a)(3) shall be assumed by the Assistant Administrator for Water and Waste Management for EPA actions covered by this subpart.

(c) *Responsibilities for Subpart F.*

(1) *Responsible official.* The responsible official for activities covered by this subpart is the Regional Administrator.

(2) *Assistant Administrator.* The responsibilities of the Assistant Administrator, as described in section 6.103(a)(3) shall be assumed by the Assistant Administrator for Enforcement for EPA actions covered by this subpart.

(d) *Responsibilities for Subpart G.*

The Assistant Administrator for Research and Development will be the responsible official for activities covered by this subpart.

(e) *Responsibilities for Subpart H.*

The Deputy Assistant Administrator for Solid Waste will be the responsible official for activities covered by this subpart.

(f) *Responsibilities for Subpart I.*

(1) *Responsible official.* The responsible official for new construction and modification of special purpose facilities is as follows:

(i) The Chief, Facilities Management Branch, Facilities and Support Services Division, Office of Management and Agency Services, shall be the responsible official on all new construction of special purpose facilities and on all improvement and modification projects for which the Facilities Management branch has received a funding allowance.

(ii) The Regional Administrator shall be the responsible official on all improvement and modification projects for which the regional office has received the funding allowance.

(iii) The Center Directors shall be the responsible officials on all improvement and modification projects for which the National Environmental Research Centers have received the funding allowance.

§ *6.104 Early involvement of private parties.*

As required by 40 CFR 1501.2(d) and § 6.103(a)(3)(v) of this regulation, responsible officials must ensure early involvement of private applicants or other non-Federal entities in the environmental review process related to EPA grant and permit actions set forth under Subparts E, F, G, and H. The responsible official in conjunction with OER shall:

(a) Prepare, where practicable, generic guidelines describing the scope and level of environmental information required from applicants as a basis for evaluating their proposed actions, and make these guidelines available upon request.

(b) Provide such guidance on a project-by-project basis to any applicant seeking assistance.

(c) Upon receipt of an application for agency approval, or notification that an application will be filed, consult as required with other appropriate parties to initiate and coordinate the necessary environmental analyses.

§ *6.105 Synopsis of EIS procedures.*

(a) *Responsible official.* The responsible official shall utilize a systematic, interdisciplinary approach to integrate natural and social sciences as well as environmental design arts in planning programs and making decisions which are subject to environmental review. The respective staffs may be supplemented by professionals from other agencies (see 40 CFR § 1501.6) or consultants, whenever in-house capabilities are insufficiently interdisciplinary.

(b) *Environmental information documents.* Environmental information documents must be prepared by applicants, grantees, or permittees and submitted to EPA as required in Subparts E, F, G, H, and I. The environmental information document will be of sufficient scope to enable the responsible official to prepare an environmental assessment as described under § 6.105(d) below and Subparts E through I.

(c) *Environmental reviews.* Environmental reviews shall be conducted on the EPA activities outlined in § 6.102 above and set forth under Subparts E, F, G, H, and I. This process shall consist of a study of the action to identify and evaluate the related environmental impacts. The process shall include a review of any related environmental information document to determine whether any significant impacts are anticipated and whether any changes can be made in the proposed action to eliminate significant adverse impacts; when an EIS is required, EPA has overall responsibility for this review, although grantees, applicants, permittees, or contractors will contribute to the review through submission of environmental information documents.

(d) *Environmental assessments.* Environmental assessments (i.e., concise public documents for which EPA is responsible) are prepared to provide sufficient data and analysis to determine whether an EIS or finding of no significant impact is required. Where EPA determines that an EIS will be prepared, there is no need to prepare a formal environmental assessment.

(e) *Notice of intent and EISs.* When the environmental review indicates that a significant environmental impact may occur and significant adverse impacts cannot be eliminated by making changes in the project, a notice of intent to prepare an EIS shall be published in the **Federal Register,** scoping shall be undertaken in accordance with 40 CFR § 1501.7, and a draft EIS shall be prepared and distributed. After external coordination and evaluation of the comments received, a final EIS shall be prepared and disseminated. The final EIS shall list any mitigation measures necessary to make the recommended alternative environmentally acceptable.

(f) *Finding of no significant impact* (FNSI). When the environmental review indicates no significant impacts are anticipated or when the project is altered to eliminate any significant adverse impacts, a FNSI shall be issued and made available to the public. The environmental assessment shall be included as a part of the FNSI. The FNSI shall list any mitigation measures necessary to make the recommended alternative environmentally acceptable.

(g) *Record of decision.* At the time of its decision on any action for which a final EIS has been prepared, the responsible official shall prepare a concise public record of the decision. The record of decision shall describe those mitigation measures to be undertaken which will make the selected alternative environmentally acceptable. Where the final EIS recommends the alternative which is ultimately chosen by the responsible official, the record of decision may be extracted from the executive summary to the final EIS.

(h) *Monitoring.* The responsible official shall provide for monitoring to assure that decisions on any action where a final EIS has been prepared are properly implemented. Appropriate mitigation measures shall be included in actions undertaken by EPA.

§ ***6.106 Deviations.***

(a) *General.* The Director, OER, is authorized to approve deviations from these regulations. Deviation approvals shall be made in writing by the Director, OER.

(b) *Requirements.* (1) Where emergency circumstances make it necessary to take an action with significant environmental impact without observing the substantive provisions of these regulations or the CEQ Regulations, the responsible official shall notify the Director, OER, before taking such action. The responsible official shall consider to the extent possible alternative arrangements; such arrangements will be limited to actions necessary to control the immediate impacts of the emergency; other actions remain subject to the environmental review process. The Director, OER, after consulting CEQ, will inform the responsible official, as expeditiously as possible, of the disposition of his request.

(2) Where circumstances make it necessary to take action without observing procedural provisions of these regulations, the responsible official shall notify the Director, OER, before taking such action. If the Director, OER, determines such a deviation would be in the best interest of the Government, he shall inform the responsible official, as soon as possible, of his approval.

(3) The Director, OER, shall coordinate his action on a deviation under § 6.106(b)(1) or (2) above with the Director, Grants Administration Division, Office of Planning and Management, for any required grant-related deviation under 40 CFR 30.1000, as well as the appropriate Assistant Administrator.

SUBPART B—CONTENT OF EISs

§ ***6.200 The environmental impact statement.***

Preparers of EISs must conform with the requirements of 40 CFR Part 1502 in writing EISs.

§ ***6.201 Format.***

The format used for EISs shall encourage good analysis and clear presentation of alternatives, including the proposed action, and their environmental, economic, and social impacts. The following standard format for EISs should be used unless the responsible official determines that there is a compelling reason to do otherwise:

(a) Cover sheet;

(b) Executive summary;

(c) Table of contents;

(d) Purpose of and need for action;

(e) Alternatives including proposed action;

(f) Affected environment;

(g) Environmental consequences of the alternatives;

(h) Coordination (includes a list of agencies, organizations, and persons to whom copies of the EIS are sent);

(i) List of preparers;

(j) Index (commensurate with complexity of EIS);

(k) Appendices.

§ ***6.202 Executive summary.***

The executive summary shall describe in sufficient detail (10-15 pages) the critical facets of the EIS so that the reader can become familiar with the proposed project or action and its net effects. The executive summary shall focus on:

(a) The existing problem;

(b) A brief description of each alternative evaluated (including the preferred and no action alternatives) along with a listing of the environmental impacts, possible mitigation measures relating to each alternative, and any areas of controversy (including issues raised by governmental agencies and the public); and

(c) Any major conclusions.

A comprehensive summary may be prepared in instances where the EIS is unusually long in nature. In accordance with 40 CFR 1502.19, the comprehensive summary may be circulated in lieu of the EIS; however, both documents shall be distributed to any Federal, State, and local agencies who have EIS review responsibilities and also shall be made available to other interested parties upon request.

§ *6.203* ***Body of EIS.***

(a) *Purpose and need.* The EIS shall clearly specify the underlying purpose and need to which EPA is responding. If the action is a request for a permit or a grant, the EIS shall clearly specify the goals and objectives of the applicant.

(b) *Alternatives including the proposed action.* In addition to 40 CFR 1502.14, the EIS shall discuss:

(1) *Alternatives considered by the Applicant.* This section shall include a *balanced* description of each alternative considered by the applicant. These discussions shall include size and location of facilities, land requirements, operation and maintenance requirements, auxiliary structures such as pipelines or transmission lines, and construction schedules. The alternative of no action shall be discussed and the applicant's preferred alternative(s) shall be identified. For alternatives which were eliminated from detailed study, a brief discussion of the reasons for their having been eliminated shall be included.

(2) *Alternatives available to EPA.* EPA alternatives to be discussed shall include: (i) Taking an action; or (ii) taking an action on a modified or alternative project, including an action not considered by the applicant; and (iii) denying the action.

(3) *Alternatives available to other permitting agencies.* When preparing a joint EIS, and if applicable, the alternatives available to other Federal and/or State agencies shall be discussed.

(4) *Identifying preferred alternative.* In the final EIS, the responsible official shall signify the preferred alternative.

(c) *Affected environment and environmental consequences of the alternatives.* The affected environment on which the evaluation of each alternative shall be based includes, for example, hydrology, geology, air quality, noise, biology, socioeconomics, energy, land use, and archaeology/history. These subject matters shall be adapted to analyze each alternative within a project area. The discussion shall be structured so as to present the impacts of each alternative under each subject heading for easy comparison by the reader. The "no action" alternative should be described first so that the reader may relate the other alternatives to beneficial and adverse impacts related to the applicant doing nothing. Description of environmental setting for the purpose of necessary background shall be included in this discussion of the impacts of the "no action" alternative. The amount of detail in describing the affected environment shall be commensurate with the complexity of the situation and the importance of the anticipated impacts.

(d) *Coordination.* The EIS shall include: (1) The objections and suggestions made by local, State, and Federal agencies before and during the EIS review process must be given full consideration, along with the issues of public concern expressed by individual citizens and interested environmental groups. The EIS must include discussions of any such comments concerning our actions, and the author of each comment should be identified. If a comment has resulted in a change in the project or the EIS, the impact statement should explain the reason.

(2) Public participation through public hearings or scoping meetings shall also be included. If a public hearing has been held prior to the publication of the EIS, a summary of the transcript should be included in this section. For the public hearing which shall be held after the publication of the draft EIS, the date, time, place, and purpose shall be included here.

(3) In the final EIS, a summary of the coordination process and EPA responses to comments on the draft EIS shall be included.

§ *6.204 Incorporation by reference.*

In addition to 40 CFR 1502.21, material incorporated into an EIS by reference shall be organized to the extent possible into a Supplemental Information Document and be made available for review upon request. No material may be incorporated by reference unless it is reasonably available for inspection by potentially interested persons within the period allowed for comment.

§ *6.205 List of preparers.*

When the EIS is prepared by contract, either under direct contract to EPA or through an applicant's or grantee's contractor, the responsible official must independently evaluate the EIS prior to its approval and take responsibility for its sçope and contents. The EPA officials who undertake this evaluation shall also be described under the list of preparers.

SUBPART C—COORDINATION WITH OTHER ENVIRONMENTAL REVIEW AND CONSULTATION REQUIREMENTS

§ *6.300 General.*

Various Federal laws and executive orders address specific environmental concerns. The responsible official shall integrate to the greatest practicable extent the applicable procedures in this subpart during the implementation of the environmental review process under subparts E through I. This subpart presents the central requirements of these laws and executive orders. It refers to the pertinent authority and regulations or guidance that contain the procedures. These laws and executive orders establish review procedures independent of NEPA requirements. The responsible official shall be familiar with any other EPA or appropriate agency procedures implementing these laws and executive orders.

§ *6.301 Historical and archaeological sites.*

EPA is subject to the requirements of the National Historic Preservation Act of 1966, as amended, 16 U.S.C. 47 *et seq.*, the Archaeological and Historic Preservation Act of 1974, 16 U.S.C. 469 *et seq.*, and Executive Order 11593, entitled "Protection and Enhancement of the Cultural Environment." These provisions and regulations establish review procedures independent of NEPA requirements.

(a) Under section 106 of the National Historic Preservation Act and Executive Order 11593, if an EPA undertaking affects any property with historic, architectural, archaeological or cultural value that is listed on or eligible for listing on the National Register of Historic Places, the responsible official shall comply with the procedures for consultation and comment promulgated by the Advisory Council on Historic Preservation in 36 CFR Part 800. The responsible official must identify properties affected by the undertaking that are potentially eligible for listing on the National Register and shall request a determination of eligibility from the Keeper of the National Register, Department of the Interior, under the procedures in 36 CFR Part 63.

(b) Under the Archaeological and Historic Preservation Act, if an EPA activity may cause irreparable loss or destruction of significant scientific, prehistoric, historic, or archaeological data, the responsible official or the Secretary of the Interior is authorized to undertake data recovery and preservation activities. Applicable procedures are found in 36 CFR Parts 64 and 66.

§ ***6.302 Wetlands, floodplains, agricultural lands, coastal zones, wild and scenic rivers, fish and wildlife, and endangered species.***

The following procedures shall apply to EPA administrative actions in programs to which the pertinent statute or executive order applies.

(a) *Wetlands protection.* Executive Order 11990, Protection of Wetlands, requires Federal agencies conducting certain activities to avoid, to the extent possible, the adverse impacts associated with the destruction or loss of wetlands and to avoid support of new construction in wetlands if a practicable alternative exists. EPA's Statement of Procedures on Floodplain Management and Wetlands Protection (dated January 5, 1979, incorporated as Appendix A hereto) requires EPA programs to determine if proposed actions will be in or will affect wetlands. If so, the responsible official shall prepare a floodplains/wetlands assessment, which will be part of the environmental assessment or environmental impact statement. The responsible official shall either avoid adverse impacts or minimize them if no practicable alternative to the action exists.

(b) *Floodplain management.* Executive Order 11988, Floodplain Management, requires Federal agencies to evaluate the potential effects of actions they may take in a floodplain to avoid, to the extent possible, adverse effects associated with direct and indirect development of a floodplain. EPA's Statement of Procedures on Floodplain Management and Wetlands Protection (dated January 5, 1979, incorporated as Appendix A hereto), requires EPA programs to determine whether an action will be located in or will affect a floodplain. If so, the responsible official shall prepare a floodplain/wetlands assessment. The assessment will become part of the environmental assessment or environmental impact statement. The responsible official shall either avoid adverse impacts or minimize them if no practicable alternative exists.

(c) *Agricultural lands.* It is EPA's policy to consider the protection of the Nation's environmentally significant agricultural land from irreversible conversion to uses which result in its loss as an environmental or essential food production resource. Before undertaking an action, the responsible official shall determine whether there are significant agricultural lands in the planning area. If significant agricultural lands are identified, direct and indirect effects of the undertaking on the land shall be evaluated and adverse effects avoided or mitigated, to the extent possible, in accordance with EPA's Policy to Protect Environmentally Significant Agricultural Lands (September 8, 1978).

(d) *Coastal zone management.* The Coastal Zone Management Act, 16 U.S.C. 1451 *et. seq.*, requires that all Federal activities in coastal areas be consistent with approved State Coastal Zone Management Programs, to the maximum extent possible. If an EPA action may affect a coastal zone area, the responsible official shall assess the impact of the action on the coastal zone. If the action significantly affects the coastal zone area and the State has an approved coastal zone management program, a consistency determination shall be sought in accordance with procedures promulgated by the Office of Coastal Zone Management in 15 CFR 930.

(e) *Wild and scenic rivers.* Under the Wild and Scenic Rivers Act, 16 U.S.C. 1274 *et seq.*, a Federal agency may not assist, through grant, loan, license, or otherwise, the construction of a water resources project that would have a direct and adverse effect on the values for which such river was established, as determined by the Secretary charged with its administration. Nothing contained in the foregoing sentence, however, shall preclude licensing of, or assistance to, developments below or above a wild, scenic, or recreational river area or on any stream tributary thereto which will not invade the area or unreasonably diminish the scenic, recreational, and fish and wildlife values present in the area on the date of approval of the Wild and Scenic Rivers Act. The responsible official shall determine whether there are any designated rivers in the planning area. The responsible official shall not recommend authorization of any water resources project that would have a direct and adverse effect on the values for which such river was established, as determined by the Secretary charged with its administration, in request of appropriations to begin construction of any such project, whether heretofore or hereafter authorized, without advising the Secretary of Interior or the Secretary of Agriculture, as the case may be, in writing of his intention at least sixty days in advance, and without specifically reporting to the Congress in writing at the time he makes his recommendation or request in what respect construction of such project would be in conflict with the purposes of the Wild and Scenic Rivers Act and would affect the component and the values to be protected by him under the Act. Applicable consultation procedures are found in section 7 of the Act.

(f) *Fish and wildlife protection.* The Fish and Wildlife Coordination Act, 16 U.S.C. 661 *et seq.*, requires Federal agencies involved in actions that will result in the control or structural modification of any natural stream or body of water for any purpose, to take action to protect the fish and wildlife resources which may be affected by the action. The responsible official shall consult with the Fish and Wildlife Service and the appropriate State agency to ascertain the means and measures necessary to mitigate, prevent, and compensate for project-related losses of wildlife resources and to enhance the resources. Reports and recommendations of wildlife agencies should be incorporated into the environmental assessment or environmental impact statement. Consultation procedures are detailed in 16 U.S.C. 662.

(g) *Endangered species protection.* Under the Endangered Species Act, 16 U.S.C. 1531 *et seq.*, Federal agencies are prohibited from jeopardizing threatened or endangered species or adversely modifying habitats essential to their survival. The responsible official shall identify all designated endangered or threatened species or their habitats that may be affected by an EPA action. If listed species or their habitats may be affected, formal consultation must be undertaken with the Fish and Wildlife Service or the National Marine Fisheries Service, as appropriate. If the consultation reveals that the EPA activity may jeopardize a listed species or habitat, mitigation measures should be considered. Applicable consultation procedures are found in 50 CFR 402.

§ ***6.303 Air quality.***

(a) The Clean Air Act, as amended in 1977, 42 U.S.C. 7476(c), requires all Federal projects, licenses, permits, plans, and financial assistance activities to conform to any State Air Quality Implementation Plan (SIP) approved or promulgated under section 110 of the Act. For proposed EPA actions that may significantly affect air quality, the responsible official shall assess the extent of the direct or indirect increases in emissions and the resultant change in air quality.

(b) If the proposed action may have a significant direct or indirect adverse effect on air quality, the responsible official shall consult with the appropriate State and local agencies as to the conformity of the proposed action with the SIP. Such agencies shall include the State agency with primary responsibility for the SIP, the agency designated under section 174 of the Clean Air Act and, where appropriate, the metropolitan planning organization (MPO). This consultation should include a request for a recommendation as to the conformity of the proposed action with the SIP.

(c) The responsible official shall provide an assurance in the FNSI or the draft EIS that the proposed action conforms with the SIP.

(d) The assurance of conformity shall be based on a determination of the following:

(1) The proposed action will be in compliance with all applicable Federal and State air pollution emission limitations and standards;

(2) The direct and indirect air pollution emissions resulting from the proposed action have been expressly quantified in the emissions growth allowance of the SIP; or if a case-by-case offset approach is included in the SIP, that offsets have been obtained for the proposed action's air quality impacts;

(3) The proposed action conforms to the SIP's provisions for demonstrating reasonable further progress toward attainment of the national ambient air quality standards by the required date;

(4) The proposed action complies with all other provisions and requirements of the SIP.

(e) During the 30-day FNSI and 45-day draft EIS review time periods, EPA shall provide an opportunity for the State agency with primary responsibility for the SIP to concur or nonconcur with the determination of conformity. All State notifications of concurrence or nonconcurrence with the EPA conformity determination shall include a record of consultation with the appropriate section 174 agency and, where different, the MPO. There shall be a presumption of State concurrence if no objection is received by EPA during the review time period.

(f) The responsible official shall provide in the FNSI or the final EIS a response to a notification of state nonconcurrence with the EPA conformity determination. This response shall include the basis on which the conformity of the proposed action to the SIP will be assured. If the responsible official finds that the State nonconcurrence with the EPA conformity determination is unjustified, then an explanation of this finding shall be included in the FNSI or the final EIS.

SUBPART D—PUBLIC AND OTHER FEDERAL AGENCY INVOLVEMENT

§ *6.400* ***Public involvement.***

(a) *General.* EPA shall make diligent efforts to involve the public in the environmental review process consistent with program regulations and EPA policies on public participation. The responsible official shall ensure that public notice is provided for in accordance with 40 CFR § 1506.6(b) and shall ensure that public involvement is carried out in accordance with EPA Public Participation Regulations, 40 CFR Part 25, and other applicable EPA public participation procedures.

(b) *Publication of notices of intent.* As soon as practicable after his decision to prepare an EIS and before the scoping process, the responsible official shall send the notice of intent to interested and affected members of the public and shall request the OER to publish the notice of intent in the **Federal Register.** The responsible official shall send to OER the signed original notice of intent for **Federal Register** publication purposes. The scoping process should be initiated as soon as practicable in accordance with the requirements of 40 CFR § 1501.7. Participants in the scoping process shall be kept informed of substantial changes which evolve during the EIS drafting process.

(c) *Public meetings or hearings.* Public meetings or hearings shall be conducted consistent with Agency program requirements. There shall be a presumption that a scoping meeting will be conducted whenever a notice of intent has been published. The responsible official shall conduct a public hearing on a draft EIS. The responsible official shall ensure that the draft EIS is made available to the public at least 30 days in advance of the hearing.

(d) *Findings of no significant impact.* The responsible official shall allow for sufficient public review of a FNSI before it becomes final. The FNSI and attendant publication must state that interested persons disagreeing with the decision may submit comments to EPA. The responsible official shall not take administrative action on the project for at least thirty (30) calendar days after release of the FNSI and may allow more time for response. The responsible official shall consider fully comments submitted before taking administrative action. The FNSI shall be made available to the public in accordance with the requirements of 40 CFR § 1506.6. One copy shall be submitted to OER.

(e) *Record of decision.* The responsible official shall disseminate the record of decision to those parties which commented on the draft or final EIS. One copy shall be submitted to OER.

§ ***6.401 Official filing requirements.***

(a) *General.* OER is responsible for the conduct of the official filing system for EISs. This system was established as a central repository for all EISs which serves not only as means of advising the public of the availability of each EIS but provides a uniform method for the computation of minimum time periods for the review of EISs. OER publishes a weekly notice in the **Federal Register** listing all EISs received during a given week. The 45-day and 30-day review periods for draft and final EISs, respectively, are computed from the Friday following a given reporting week. Pursuant to 40 CFR § 1506.9, responsible officials shall comply with the guidelines established by OER on the conduct of the filing system.

(b) *Minimum time periods.* No decision on EPA actions shall be made until the later of the following dates: (1) Ninety (90) days after the date established in § 6.401(a) above from which the draft EIS review time period is computed.

(2) Thirty (30) days after the date established in § 6.401(a) above from which the final EIS review time period is computed.

(c) *Filing of EISs.* All EISs, including supplements, must be officially filed with OER. Responsible officials shall transmit each EIS in five (5) copies to the Director, Office of Environmental Review, EIS Filing Section (A-104). OER will provide CEQ with one copy of each EIS filed. No EIS will be officially filed by OER unless the EIS has been made available to the public. OER will not accept unbound copies of EISs for filing.

(d) *Extensions or waivers.* The responsible official may independently extend review periods. In such cases, the responsible official shall notify OER as soon as possible so that adequate notice may be published in the weekly **Federal Register** report. OER upon a showing of compelling reasons of national policy may reduce the prescribed review periods. Also, OER upon a showing by any other Federal agency of compelling reasons of national policy may extend prescribed review periods, but only after consultation with the responsible official. If the responsible official does not concur with the extension of time, OER may not extend a prescribed review period more than 30 days beyond the minimum prescribed review period.

(e) *Rescission of filed EISs.* The responsible official shall file EISs with OER at the same time they are transmitted to commenting agencies and made available to the public. The responsible official is required to reproduce an adequate supply of EISs to satisfy these distribution requirements prior to filing an EIS. If the EIS is not made available, OER will consider retraction of the EIS or revision of the prescribed review periods based on the circumstances.

§ *6.402* ***Availability of documents.***

(a) *General.* The responsible official will ensure sufficient copies of the EIS are distributed to interested and affected members of the public and are made available for further public distribution. EISs, comments received, and any underlying documents should be available to the public pursuant to the provisions of the Freedom of Information Act (5 U.S.C. section 552(b)), without regard to the exclusion for interagency memoranda where such memoranda transmit comments of Federal agencies on the environmental impact of the proposed actions. To the extent practicable, materials made available to the public shall be provided without charge; otherwise, a fee may be imposed which is not more than the actual cost of reproducing copies required to be sent to another Federal agency.

(b) *Public information.* Lists of all notices of intent, EISs, FNSIs, and records of decision prepared by EPA shall be maintained by OER for the public. Each responsible official will maintain a similar monthly status report for all environmental documents prepared. In addition, OER will make available for public inspection copies of EPA EISs; the responsible official shall do the same for any prepared EIS.

§ *6.403* ***The commenting process.***

(a) *Inviting comments.* After preparing a draft EIS and before preparing a final EIS, the responsible official shall obtain the comments of Federal agencies, other governmental entities and the public in accordance with 40 CFR § 1503.1.

(b) *Response to comments.* The responsible official shall respond to comments in the final EIS in accordance with 40 CFR § 1503.4.

§ *6.404* ***Supplements.***

(a) *General.* The responsible official shall consider preparing supplements to draft and final EISs in accordance with 40 CFR § 1502.9(c). A supplement shall be prepared, circulated and filed in the same fashion (exclusive of scoping) as draft and final EISs.

(b) *Alternative procedures.* In the case where the responsible official wants to deviate from existing procedures, OER shall be consulted. OER shall consult with CEQ on any alternative arrangements.

SUBPART E—ENVIRONMENTAL REVIEW PROCEDURES FOR WASTEWATER TREATMENT CONSTRUCTION GRANTS PROGRAM

§ *6.500* ***Purpose.***

This subpart amplifies the procedures described in Subparts A through D with detailed environmental review procedures for the wastewater treatment works construction grants program under Title II of the Clean Water Act.

§ ***6.501 Definitions.***

(a) "Step 1 grant" means grant assistance for a project for preparation of a facilities plan as described in 40 CFR 35.930-1(a)(1).

(b) "Step 2 grant" means grant assistance for a project for preparation of construction drawings and specifications as described in 40 CFR 35.930-1(a)(2).

(c) "Step 3 grant" means grant assistance for a project for erection and building of a publicly owned treatment works as described in 40 CFR 35.930-1(a)(3).

(d) "Step 2 plus step 3 grant" means grant assistance for a project which combines the grants set forth in section 6.501 (b) and (c) above as described in 40 CFR 35.930-1(a)(4).

(e) "Applicant" means any individual, agency, or entity which has filed an application for grant assistance under 40 CFR Part 35.

(f) "Grantee" means any individual, agency, or entity which has been awarded assistance under 40 CFR 35.930-1.

§ ***6.502 Applicability.***

(a) *Administrative actions covered.* This subpart applies to the administrative actions listed below (except as provided in § 6.502(c) below):

(1) Approval of a facilities plan; and

(2) Award of grant assistance for a project involving step 2 or step 3 when the responsible official determines that a significant change has occurred in the project or its impact from that described in the facilities plan.

(b) *Administrative actions excluded.* The actions listed below are not subject to the requirements of this subpart:

(1) Approval of State priority lists;

(2) Award of a step 1 grant;

(3) Approval or award of a section 208 planning grant;

(4) Award of grant assistance for a step 2 or step 3 project unless the responsible official determines that a significant change has occurred in the project or its impact from that described in the facilities plan (see § 6.502(a)(2) above);

(5) Approval of issuing an invitation for bids or awarding a construction contract;

(6) Actual physical commencement of building construction;

(7) Award of a section 206 grant for reimbursement;

(8) Award of a grant increase unless the responsible official determines that a significant change has occurred in a project or its impact as described in the approved facilities plan; and

(9) Awards of training assistance under section 109(b) of the Clean Water Act;

(10) Approval of user charge system or industrial cost recovery system.

§ ***6.503 Consultation during the environmental review process.***

When there are overriding considerations of cost or impaired program effectiveness, the responsible official may award a step 2 or step 3 grant for a discrete segment of the project plans or construction before the environmental review is completed. The project segment must be noncontroversial, necessary to correct water quality or other immediate environmental problems and cannot, by its completion, foreclose any reasonable options being considered in the environmental review. The remaining portion of the project shall be evaluated to determine if an EIS is required. In applying the criteria for this determination, the entire project shall be considered, including those parts permitted to proceed. In no case may these types of grant assistance for step 2 or step 3 projects be awarded unless both the OER and CEQ have been consulted, a FNSI has been issued on the segments permitted to proceed at least 30 days prior to grant award, and the grant award contains a specific agreement prohibiting action on the segment of planning or construction for which the environmental review is not complete. The Director, OER, is responsible for consulting with CEQ and the Assistant Administrator for Water and Waste Management.

§ ***6.504 Public participation.***

(a) *General.* It is EPA policy that optimum public participation be achieved during the environmental review process as deemed appropriate by the responsible official under 40 CFR Part 25 and implementing provisions of Part 35, Subpart E of this Chapter. Compliance with Part 25 and implementing provisions constitutes compliance with public participation requirements under this part.

(b) *Full-scale public participation.* In accordance with 40 CFR 35.917-5(c), the responsible official shall assure that a full-scale public participation program shall be undertaken where EPA prepares or requires the preparation of an EIS during the facility planning process. If the need for an EIS is identified late in the facility planning process, the responsible official shall determine on an individual project basis what elements are necessary to ensure full-scale public participation.

(c) Public participation activities undertaken in connection with the environmental review process should be coordinated with the facility planning public participation program wherever possible.

(d) The responsible official may institute such additional NEPA-related public participation procedures as he deems necessary during the environmental review process.

§ ***6.505 Limitations on actions during environmental review process.***

No administrative action under § 6.502(a) shall be taken until the environmental review process has been completed except as provided under § 6.502(c) above. The responsible official shall ensure compliance in accordance with 40 CFR § 1506.1 and subparts A, C, and D of this regulation, and all policies, guidance and regulations adopted to implement the requirements under 42 U.S.C. 7616 of the Clean Air Act.

§ ***6.506 Criteria for preparing EISs.***

(a) The responsible official shall assure that an EIS will be issued when he determines that any of the following conditions exists:

(1) The treatment works will induce significant changes (either absolute changes or increases in the rate of change) in industrial, commercial, agricultural, or residential land use concentrations or distributions. Factors that should be considered in determining if these changes are significant include but are not limited to: (i) The vacant land subject to increased development pressure as a result of the treatment works; (ii) the increases in population which may be induced; (iii) the faster rate of change of population; changes in population density; (iv) the potential for overloading sewage treatment works; (v) the extent to which landowners may benefit from the areas subject to increased development; (vi) the nature of land use regulations in the affected area and their potential effects on development; and (vii) deleterious changes in the availability or demand for energy.

(2) The treatment works or collector system will have significant adverse effects on wetlands, including indirect effects, or any major part of the treatment works will be located on wetlands.

(3) The treatment works or collector system will significantly affect a habitat identified on the Department of the Interior's or a State's threatened and endangered species lists, or the treatment works will be located on the habitat.

(4) Implementation of the treatment works or plan may directly cause or induce changes that significantly:

(i) Displace population;

(ii) Alter the character of an existing residential area;

(iii) Adversely affect a floodplain; or

(iv) Adversely affect significant amounts of prime or unique agricultural land, or agricultural operations on this land as defined in EPA's Policy to Protect Environmentally Significant Agricultural Land.

(5) The treatment works will have significant adverse direct or indirect effects on parklands, other public lands, or areas of recognized scenic, recreational, archaeological, or historic value.

(6) The treatment works may directly or through induced development have a significant adverse effect upon local ambient air quality, local ambient noise levels, surface or groundwater quality or quantity, fish, wildlife, and their natural habitats.

(7) The treated effluent is being discharged into a body of water where the present classification is too lenient or is being challenged as too low to protect present or recent uses, and the effluent will not be of sufficient quality or quantity to meet the requirements of these uses.

(b) When the treatment works shall threaten a violation of Federal, State, or local law or requirements imposed for the protection of the environment, the responsible official shall consider preparing an EIS.

(c) When full-scale public participation is required under 40 CFR 35.917-5(c), the responsible official shall consider preparing an EIS.

§ ***6.507 Environmental review process.***

Consistent with 40 CFR 1501.2, EPA shall integrate the environmental review process throughout the construction grants program facilities planning process (Step 1). Critical decision-making points and the scope of review recommended include:

(a) *Award of a facilities planning grant (Step 1).* Prior to award of Step 1 assistance, or within no more than 30 days thereafter, EPA may review, or request that the State review, if the facilities plan review is delegated under section 205(g) of the Clean Water Act, the existence of environmentally sensitive areas in the facilities planning area. This review is intended to be brief and concise, drawing on existing information and knowledge of EPA, State agencies, regional planning agencies, areawide water quality management agencies, and grantees. This review may be used to determine the scope of the environmental information document prepared by the grantee. It may also be used to make an early determination of the need for an EIS. Whenever possible, this initial review should be discussed at the first conference held with the potential grantee.

(b) *Mid-course reviews.* A review of environmental information developed by the grantee should be conducted to the extent practicable whenever meetings are held to assess the progress of facilities plan development. These meetings should be held after completion and submission to EPA and the State of the majority of the environmental information document and before a preferred alternative is selected. When the program is delegated, the state shall forward to EPA the required preliminary environmental assessment to enable EPA to make decisions with respect to the need for an EIS. Although the decision whether to prepare an EIS must be made before a facilities plan can be approved, a decision to prepare an EIS is encouraged earlier during the facilities planning process. Following any mid-course review meeting, EPA should inform interested parties as to the following:

(1) The preliminary nature of the agency's position on preparing an EIS;

(2) The relationship between the facilities planning and environmental review processes;

(3) The desirability of further public input; and

(4) A contact person for further information.

(c) *Review of completed facilities plan.* EPA, or the State when the program is delegated, shall review any completed facilities plan with particular attention to the environmental information document and its utilization in the development of alternatives and the selection of a preferred alternative. An adequate environmental information document should be an integral part of any facilities plan submitted to EPA or to a State. The environmental information document shall be of sufficient scope to enable the responsible official to prepare an environmental assessment. For those States where the review of facilities plans has been delegated, State personnel will be required to prepare a preliminary environmental assessment which serves as an adequate basis for EPA's decision to issue an FNSI or an EIS. The environmental assessment shall cover all potentially significant environmental impacts and related factors. Each of the following subjects shall be critically reviewed to identify potentially significant environmental concerns and shall be addressed in the environmental assessment.

(1) *Description of the existing environment.* For the delineated facilities planning area, the existing environmental conditions relevant to the analysis of alternatives or determinations of the environmental impacts (especially indirect) of the proposed action shall be considered. The description may include those environmental factors potentially affected by the alternatives under consideration, such as: surface and groundwater quality; water supply and use; general hydrology; air quality; noise levels, energy production and consumption; land use trends, including probable development of regional shopping centers; population projections; wetlands, floodplains, coastal zones, prime agricultural lands, and other environmentally sensitive areas; historic and archaeological sites; other related Federal or State projects in the area; plant and animal communities which may be affected, especially those containing threatened or endangered species.

(2) *Description of the future environment without the project.* The relevant future environmental conditions shall be described. The no action alternative must be adequately evaluated.

(3) *Purpose and need.* This should include a summary discussion and demonstration of the need for wastewater treatment in the facilities planning area, with particular emphasis on existing public health or water quality problems and their severity and extent.

(4) *Documentation.* Sources of information used to describe the existing environment and to assess future environmental impacts should be clearly referenced. These sources should include regional, State, and Federal agencies with responsibility or interest in the types of impacts listed in § 6.506(a)(1) above and in Subpart C.

(5) *Evaluation of Alternatives.* This discussion shall include a comparative analysis of feasible alternatives (including the no action alternative) throughout the study area. The alternatives shall be screened with respect to capital and operating costs; significant direct and indirect environmental effects; physical, legal, or institutional constraints; and compliance with regulatory requirements. Special attention should be given to long term impacts, irreversible impacts, and induced impacts such as development. The reasons for rejecting any alternatives shall be presented in addition to any significant environmental benefits precluded by rejection of an alternative. The analysis should consider, when relevant to the project:

(i) Flow and waste reduction measures, including infiltration/inflow reduction;

(ii) Appropriate water conservation measures;

(iii) Alternative locations, capacities, and construction phasing of facilities;

(iv) Alternative waste management techniques, including treatment and discharge, wastewater reuse, land application, and individual systems;

(v) Alternative methods for management of sludge, other residual materials, including utilization options such as land application, composting, and conversion of sludge for marketing as a soil conditioner or fertilizer.

(vi) Improving effluent quality through more efficient operation and maintenance;

(vii) Appropriate energy reduction measures;

(viii) Multiple use, including recreation and education.

(6) *Environmental consequences.* Relevant direct and indirect impacts of the proposed action shall be considered, giving special attention to unavoidable impacts, steps to mitigate adverse impacts, any irreversible or irretrievable commitments of resources to the project and the relationship between local short term uses of the environment and the maintenance and enhancement of long term productivity. The significance of land use impacts shall be considered, based on the analysis required under Appendix A to 40 CFR Part 35, Subpart E. Any specific land use controls (including grant conditions and areawide waste treatment management plan requirements) should be identified and referenced. In addition to these items, the responsible official may require that other analyses and data, which are needed to satisfy environmental review requirements, be included with the facilities plan. Such requirements should be discussed during initial conferences with potential grantees or mid-course review meetings. The responsible official also may require submission of supplementary information either before or after award of grant assistance for a step 2 project or before a step 3 project if needed for compliance with environmental review requirements. Requests for supplementary information shall be made in writing.

(7) *Steps to minimize adverse effects.* (i) This section shall describe structural and nonstructural measures, if any, in the facilities plan, or additional measures identified during the review, to mitigate or eliminate significant adverse effects on the human and natural environments. Structural provisions include changes in facility design, size, and location; nonstructural provisions include staging facilities as well as developing and enforcing land use regulations and environmental protection regulations.

(ii) The Responsible official shall not award step 2 or step 3 grant assistance if the grantee has not made, or agreed to make, pertinent changes in the project, in accordance with determinations made in a FNSI or EIS. He shall condition a grant to ensure that the grantee will comply, or seek to obtain compliance, with such environmental review determinations.

(d) *Environmental review.* The environmental review shall apply the criteria under § 6.506 above. This review shall be conducted by the responsible official and based on any of the following:

(1) A complete facilities plan and the environmental information document, whenever review of facilities plan has not been delegated;

(2) A complete facilities plan, the applicant's environmental information document, and the preliminary environmental assessment prepared by the State, for a State which has been delegated authority for facilities plan review; or

(3) Other documentation, deemed necessary by the responsible official or submitted by a State with delegated review authority, adequate to make an EIS determination by EPA. Where EPA determines that an EIS is to be prepared, there is no need to prepare a formal environmental assessment.

If deficiencies exist in the environmental information document, preliminary environmental assessment, or other supporting documentation, they may be identified by EPA, and necessary corrections shall be made before the facilities plan is approved.

(e) *Finding of No Significant Impact.* If, after completion of the environmental review, a determination is made that an EIS will not be required, the responsible official shall prepare and distribute a FNSI in accordance with § 6.104 and subpart D of this Chapter. The FNSI will be based on EPA's independent review and environmental assessment finalized by EPA, which will either be incorporated into or attached to the FNSI. The FNSI shall list any mitigation measures necessary to eliminate significant adverse environmental effects and make the recommended alternative environmentally acceptable. Once a FNSI and environmental assessment have been prepared for the facilities plan for a certain area, grant awards may proceed without preparation of additional FNSIs, unless the responsible official has determined that the project has changed significantly from that described in the facilities plan.

(f) *Notice of intent.* If, after completion of the environmental review, or subsequent to any of the steps described in § 6.507 (a), (b), or (c) above, a determination is made that an EIS will be required, the responsible official shall prepare and distribute a notice of intent in accordance with § 6.104 and subpart D.

(g) *Scoping.* As soon as possible, after the publication of the notice of intent, the responsible official will convene a meeting of affected Federal, State, and local agencies, the grantee, and other interested parties (e.g. Advisory Group members under 40 CFR 25.7) to determine the scope of the EIS. A notice of this scoping meeting will meet the requirements of subpart D. As part of the scoping meeting EPA will as a minimum:

(1) Determine the scope and the significant issues to be analyzed in depth in the EIS;

(2) Identify those issues which are not significant;

(3) Determine what information is needed from cooperating agencies or other parties;

(4) Discuss the method for EIS preparation and the public participation strategy;

(5) Identify consultation requirements of other environmental laws, in accordance with subpart C; and

(6) Determine the relationship between the EIS and the completion of the facilities plan and any necessary coordination arrangements between the preparers of both documents.

(h) *EIS method.* EPA shall prepare the EIS by any one of the following means:

(1) Directly by its own staff;

(2) By contracting directly with a qualified consulting firm; or

(3) By utilizing a joint EIS process, whereby the grantee contracts directly with a qualified consulting firm. In this case the draft EIS serves the purpose of and satisfies the requirement for an environmental information document. In this instance, the following selection requirements shall be fulfilled:

(i) A Memorandum of Understanding shall be developed between EPA, the grantee, and where possible, the State, outlining the responsibilities of each party and their relationship to the EIS consultant.

(ii) EPA shall approve evaluation criteria to be used in the consultant selection process.

(iii) EPA shall review and approve the selection process.

(iv) EPA shall approve the consultant selected for EIS preparation.

(v) The detailed Scope of Work prepared by the EIS consultant must be approved by EPA.

(vi) The EIS consultant shall execute a disclosure statement prepared by EPA indicating that the consultant has no financial or other interest in the outcome of the project.

§ 6.508 Limits on delegation to States.

(a) *General.* In cases where the authority for facilities plan review has been delegated to the State under section 205(g) of the Clean Water Act, EPA shall, as a minimum, retain the following responsibilities:

(1) The determination of whether or not to prepare an EIS shall be solely that of EPA. EPA may consider a State's recommendation, but the ultimate decision under NEPA shall not be delegated.

(2) Findings of no significant impact and the environmental assessment shall be approved, finalized, and issued by EPA.

(3) Notices of intent shall be prepared and issued by EPA.

(b) *Elimination of duplication.* The responsible official shall assure that maximum efforts are undertaken to minimize duplication within the limits described under §§ 6.506 and 6.507(a) above. In carrying out requirements under this subpart, maximum consideration should be given to eliminating duplication in accordance with 40 CFR § 1506.2, where there are State or local procedures comparable to NEPA, and entering into Memoranda of Understanding with a State concerning workload distribution and responsibilities for implementing the facilities planning process.

§ 6.509 Identification of mitigation measures.

(a) *Record of decision.* When a final EIS has been issued, the responsible official shall prepare a record of decision in accordance with 40 CFR 1505.2 prior to the approval of the facilities plan. The record of decision shall include identification of mitigation measures derived from the EIS process which are necessary to make the recommended alternative environmentally acceptable.

(b) *Specific mitigation measures.* Prior to the award of step 2 or step 3 grant assistance, the responsible official must ensure that effective mitigation measures identified in the FNSI, final EIS, or record of decision are implemented by the grantee. This should be done by revising the facilities plan, initiating other steps to mitigate adverse effects, or agreeing to conditions in grants requiring actions to minimize effects. Care should be exercised if a condition is to be imposed in a grant document to assure that the applicant possesses the authority to fulfill the conditions.

§ 6.510 Monitoring.

(a) *General.* The responsible official shall ensure there is adequate monitoring of mitigation measures and other grant conditions which are identified in the FNSI, final EIS, and record of decision.

(b) *Enforcement.* The responsible official may consider taking the following actions consistent with 40 CFR 35.965 if the grantee fails to comply with grant conditions:

(1) Terminating or annulling the grant;

(2) Disallowing project costs related to noncompliance;

(3) Withholding project payments;

(4) Suspending work;

(5) Finding the grantee to be nonresponsible or ineligible for future Federal assistance or for approval for future contract awards under EPA grants;

(6) Seeking an injunction against the grantee; or

(7) Instituting such other administrative or judicial action as may be legally available and appropriate.

SUBPART F—ENVIRONMENTAL REVIEW PROCEDURES FOR THE NEW SOURCE NPDES PROGRAM

§ *6.600* ***Purpose.***

(a) *General.* This subpart provides procedures for carrying out the environmental review process for the issuance of new source National Pollutant Discharge Elimination System (NPDES) discharge permits authorized under section 306, section 402, and section 511(c)(1) of the Clean Water Act.

(b) *Permit regulations.* All references in this subpart to the "permit regulations" shall mean Parts 122, 124, and 125 of Title 40 of the Code of Federal Regulations relating to the NPDES program.

§ *6.601* ***Definitions.***

(a) The term "administrative action" for the sake of this subpart means the issuance by EPA of an NPDES permit to discharge as a new source, pursuant to 40 CFR 124.61.

(b) The term "applicant" for the sake of this subpart means any person who applies to EPA for the issuance of an NPDES permit to discharge as a new source.

§ *6.602* ***Applicability.***

(a) *General.* The procedures set forth under subparts A, B, C, and D and this subpart shall apply to the issuance of new source NPDES permits, except for the issuance of a new source NPDES permit from any State which has an approved NPDES program in accordance with section 402(b) of the Clean Water Act.

(b) *New source determination.* An NPDES permittee must be determined a "new source" before these procedures apply. New source determinations will be undertaken pursuant to the provisions of the permit regulations under 40 CFR 122.47 (a) and (b) and 124.12.

§ *6.603* ***Limitations on actions during environmental review process.***

The processing and review of an applicant's NPDES permit application shall proceed concurrently with the procedures within this subpart. Actions undertaken by the applicant or EPA shall be performed consistent with the requirements of 40 CFR 122.47(c).

§ *6.604* ***Environmental review process.***

(a) *New source.* If EPA's initial determination under § 6.602(b) is that the facility is a new source, the responsible official shall evaluate any environmental information to determine if any significant impacts are anticipated and an EIS is necessary. If the permit applicant requests, the responsible official shall establish time limits for the completion of the environmental review process consistent with 40 CFR 1501.8.

(b) *Information needs.* Information necessary for a proper environmental review shall be provided by the permit applicant in an environmental information document. The responsible official shall consult with the applicant to determine the scope of an environmental information document. In doing this the responsible official shall consider the size of the new source and the extent to which the applicant is capable of providing the required information. The responsible official shall not require the applicant to gather data or perform analyses which unnecessarily duplicate either existing data or the results of existing analyses available to EPA. The responsible official shall keep requests for data to the minimum consistent with his responsibilities under NEPA.

(c) *Environmental assessment.* The responsible official shall prepare a written environmental assessment based on an environmental review of either the environmental information document and/or any other available environmental information.

(d) *EIS determination.* (1) When the environmental review indicates that a significant environmental impact may occur and that the significant adverse impacts cannot be eliminated by making changes in the proposed new source project, a notice of intent shall be issued, and a draft EIS prepared and distributed. When the environmental review indicates no significant impacts are anticipated or when the proposed project is changed to eliminate the significant adverse impacts, a FNSI shall be issued which lists any mitigation measures necessary to make the recommended alternative environmentally acceptable.

(2) The FNSI together with the environmental assessment that supports the finding shall be distributed in accordance with section 6.400(d) of this regulation.

(e) *Lead agency.* (1) If the environmental review reveals that the preparation of an EIS is required, the responsible official shall determine if other Federal agencies are involved with the project. The responsible official shall contact all other involved agencies and together the agencies shall decide the lead agency based on the criteria set forth in 40 CFR 1501.5.

(2) If, after the meeting of involved agencies, EPA has been determined to be the lead agency, the responsible official may request that other involved agencies be cooperating agencies. Cooperating agencies shall be chosen and shall be involved in the EIS preparation process in the manner prescribed in the 40 CFR 1501.6(a). If EPA has been determined to be a cooperating agency, the responsible official shall be involved in assisting in the preparation of the EIS in the manner prescribed in 40 CFR 1501.6(b).

(f) *Notice of intent.* (1) If EPA is the lead agency for the preparation of an EIS, the responsible official shall arrange through OER for the publication of the notice of intent in the **Federal Register,** distribute the notice of intent, and arrange and conduct a scoping meeting as outlined in 40 CFR 1501.7.

(2) If the responsible official and the permit applicant agree to a third party method of EIS preparation, pursuant to § 6.604(g)(3) below, the responsible official shall insure that a notice of intent is published and that a scoping meeting is held before the third party contractor begins work which may influence the scope of the EIS.

(g) *EIS method.* EPA shall prepare EISs by one of the following means:

(1) Directly by its own staff;

(2) By contracting directly with a qualified consulting firm; or

(3) By utilizing a third party method, whereby the responsible official enters into a "third party agreement" for the applicant to engage and pay for the services of a third party contractor to prepare the EIS. Such an agreement shall not be initiated unless both the applicant and the responsible official agree to its creation. A third party agreement will be established prior to the applicant's environmental information document and eliminate the need for that document. In proceeding under the third party agreement, the responsible official shall carry out the following practices:

(i) In consultation with the applicant, choose the third party contractor and manage that contract.

(ii) Select the consultant based on his ability and an absence of conflict of interest. Third party contractors will be required to execute a disclosure statement prepared by the responsible official signifying they have no financial or other conflicting interest in the outcome of the project.

(iii) Specify the information to be developed and supervise the gathering, analysis, and presentation of the information. The responsible official shall have sole authority for approval and modification of the statements, analyses, and conclusions included in the third party EIS.

(h) *Documents for the administrative record.* Pursuant to 40 CFR § § 124.35(a)(7) and 124.122, any environmental assessment, FNSI, EIS, or supplement to an EIS shall be made a part of the administrative record related to permit issuance.

§ *6.605 Criteria for preparing EISs.*

(a) *General guidelines.* (1) When determining the significance of a proposed new source's impact, the responsible official shall consider both its short term and long term effects as well as its direct and indirect effects and beneficial and adverse environmental impacts as defined in 40 CFR 1508.8.

(2) If EPA is proposing to issue a number of new source NPDES permits during a limited time span and in the same general geographic area, the responsible official shall examine the possibility of tiering EISs. If the permits are minor and environmentally insignificant when considered separately, the responsible official may determine that the cumulative impact of the issuance of all these permits may have a significant environmental effect and require an EIS for the area. Each separate decision to issue an NPDES permit shall then be based on the information in this areawide EIS. Site specific EISs may be required in certain circumstances in addition to the areawide EIS.

(b) *Specific criteria.* An EIS will be prepared when:

(1) The new source will induce or accelerate significant changes in industrial, commercial, agricultural, or residential land use concentrations or distributions which have the potential for significant environmental effects. Factors that should be considered in determining if these changes are environmentally significant include but are not limited to: the nature and extent of the vacant land subject to increased development pressure as a result of the new source; the increases in population or population density which may be induced and the ramifications of such changes; the nature of land use regulations in the affected area and their potential effects on development and the environment; and the changes in the availability or demand for energy and the resulting environmental consequences.

(2) The new source will directly, or through induced development, have significant adverse effect upon local ambient air quality, local ambient noise levels, floodplains, surface or groundwater quality or quantity, fish, wildlife, and their natural habitats.

(3) Any major part of the new source will have significant adverse effect on the habitat of threatened or endangered species on the Department of the Interior's or a State's lists of threatened and endangered species.

(4) The environmental impact of the issuance of a new source NPDES permit will have significant direct and adverse effect on a property listed in or eligible for listing in the National Register of Historic Places.

(5) Any major part of the source will have significant adverse effects on parklands, wetlands, wild and scenic rivers, reservoirs or other important bodies of water, navigation projects, or agricultural lands.

§ *6.606 Record of decision.*

(a) *General.* At the time of permit award, the responsible official shall prepare a record of decision in those cases where a final EIS was issued in accordance with 40 CFR 1505.2 and pursuant to the provisions of the permit regulations under 40 CFR 124.61 and 124.122. The record of decision shall list any mitigation measures necessary to make the recommended alternative environmentally acceptable.

(b) *Mitigation measures.* The mitigation measures derived from the EIS process shall be incorporated as conditions of the permit; ancillary agreements shall not be used to require mitigation.

§ ***6.607 Monitoring.***

In accordance with 40 CFR 1505.3 and pursuant to 40 CFR 122.47(c) and 122.17, the responsible official shall ensure that there is adequate monitoring of compliance with all NEPA related requirements contained in the permit.

SUBPART G—ENVIRONMENTAL REVIEW PROCEDURES FOR RESEARCH AND DEVELOPMENT PROGRAMS

§ ***6.700 Purpose.***

This subpart amplifies the requirements described in subparts A through D by providing more specific environmental review procedures on research and development programs undertaken by the Office of Research and Development (ORD).

§ ***6.701 Definition.***

The term "appropriate program official" means the official at each decision level within ORD to whom the Assistant Administrator has delegated responsibility for carrying out the environmental review process.

§ ***6.702 Applicability.***

The requirements of this subpart apply to administrative actions undertaken to approve intramural and extramural programs under the purview of ORD.

§ ***6.703 Criteria for preparing EISs.***

(a) The responsible official shall assure that an EIS will be prepared when it is determined that any of the conditions under § 6.506(a) (1) through (6) and (8) exist and when:

(1) The project consists of field tests involving the introduction of significant quantities of toxic or polluting agricultural chemicals, animal wastes, pesticides, radioactive materials, or other hazardous substances into the environment by ORD, its grantees, or its contractors;

(2) The action may involve the introduction of species or subspecies not indigenous to an area;

(3) There is a high probability of an action ultimately being implemented on a large scale, and this implementation may result in significant environmental impacts; or

(4) The project involves commitment to a new technology which is significant and may restrict future viable alternatives;

(b) An EIS will not usually be needed when:

(1) The project is conducted completely within any laboratory or other facility, and external environmental effects have been eliminated by methods for disposal of laboratory wastes and safeguards to prevent hazardous materials entering the environment accidentally; or

(2) The project is a relatively small experiment or investigation that is part of a non-Federally funded activity of the private sector, and it makes no significant new or additional contribution to existing pollution.

§ ***6.704 Environmental review process.***

Environmental review activities will be integrated into the decision levels of ORD's research planning system to assure managerial control.

(a) *Environmental information.* (1) Environmental information documents shall be submitted with all grant applications and all unsolicited contract proposals. The documents shall contain the same information required for EISs under Subpart B. Guidance on environmental information documents shall be included in all grant application kits and attached to instructions for the submission of unsolicited proposals.

(2) In the case of competitive contracts, environmental information documents need not be submitted by potential contractors since the environmental review procedures must be completed before a request for proposal (RFP) is issued. If there is a question concerning the need for an environmental information document, the potential contractor should contact the official responsible for the contract.

(b) *Environmental review.* (1) At the start of the planning year, an environmental review will be performed for each program plan with its supporting substructures (work plans and projects) before incorporating them into the ORD program planning system, unless they are excluded from review by existing legislation. This review is an evaluation of the potentially adverse environmental effects of the efforts required by the program plan. The criteria in § 6.703 above shall be used in conducting this review. Each program plan with its supporting substructures which does not have significant adverse impacts may be dismissed from further current year environmental considerations with a single FNSI. Any supporting substructures of a program plan which cannot be dismissed with the parent plan shall be reviewed at the appropriate subordinate levels of the planning system.

(i) All continuing program plans and supporting substructures, including those previously dismissed from consideration, will be reevaluated annually. An environmental review will coincide with the annual planning cycle and whenever a major redirection of a parent plan is undertaken. All environmental documents will be updated as appropriate.

(ii) Later plans and/or projects, added to fulfill the mission objectives but not identified at the time program plans were approved, will be subjected to the same environmental review.

(2) The responsible official shall assure completion of the EPA Form 5300-23 for each extramural project subject to an environmental review. If the project consists of literature studies, computer studies, or studies in which essentially all work is performed within the confines of the laboratory, the Form 5300-23 may be issued as a finding of no significant impact.

(c) *Notice of intent and EIS.* (1) If the reviews conducted according to § 6.704(b) above reveal a potential significant adverse effect on the environment and the adverse impact cannot be eliminated by replanning, the appropriate program official shall issue a notice of intent and through proper organizational channels shall request the Regional Administrator to assist him in the preparation and distribution of the EIS.

(2) As soon as possible after release of the notice of intent, the appropriate program official shall prepare a draft EIS in accordance with subpart B and distribute the draft EIS in accordance with subpart D.

(3) All draft and final EISs shall be sent through the proper organizational channels to the Assistant Administrator for ORD for approval.

(d) *Finding of no significant impact.* If an environmental review conducted according to § 6.704(b) above reveals that proposed actions will not have significant adverse environmental impacts, the appropriate program official shall prepare a FNSI which lists any mitigation measures necessary to make the recommended alternative environmentally acceptable.

(e) *Timing of action.* Pursuant to § 6.401(b), in no case shall a contract or grant be awarded or intramural activity undertaken until the prescribed 30-day review period for a final EIS has elapsed. Similarly, no action shall be taken until the 30-day comment period for a FNSI is completed.

§ *6.705* ***Record of decision.***

The responsible official shall prepare a record of decision in any case where final EIS has been issued in accordance with 40 CFR 1505.2. It shall be prepared at the time of contract or grant award or before the undertaking of the intramural activity. The record of decision shall list any mitigation measures necessary to make the recommended alternative environmentally acceptable.

SUBPART H—ENVIRONMENTAL REVIEW PROCEDURES FOR SOLID WASTE DEMONSTRATION PROJECTS

§ *6.800* ***Purpose.***

This subpart amplifies the procedures described in subparts A through D by providing more specific environmental review procedures for demonstration projects undertaken by the Office of Solid Waste (OSW).

§ *6.801* ***Applicability.***

The requirements of this subpart apply to solid waste demonstration projects for resource recovery systems and improved solid waste disposal facilities undertaken pursuant to § 8006 of the Resource Conservation and Recovery Act of 1976.

§ *6.802* ***Criteria for preparing EISs.***

The responsible official shall assure that an EIS will be prepared when it is determined that any of the conditions in § 6.506(a) (1) through (6) and (8) exist.

§ *6.803* ***Environmental review process.***

(a) *Environmental information.* (1) Environmental information documents shall be submitted to EPA by grant applicants or contractors. If there is a question concerning the need for a document, the potential contractor or grantee should consult with the appropriate project officer for the grant or contract.

(2) The environmental information document shall contain the same sections specified for EISs in Subpart B. Guidance alerting potential grantees and contractors of the environmental information documents shall be included in all grant application kits, attached to letters concerning the submission of unsolicited proposals, and included with all requests for proposal.

(b) *Environmental review.* An environmental review will be conducted before a grant or contract award is made. This review will include the preparation of an environmental assessment by the responsible official; the appropriate Regional Administrator's input will include his recommendations on the need for an EIS.

(c) *Notice of intent and EIS.* Based on the environmental review if the criteria in § 6.802 above apply, the responsible official will assure that a notice of intent and a draft EIS are prepared. The responsible official may request the appropriate Regional Administrator to assist him in the preparation and distribution of the environmental documents.

(d) *Finding of no significant impact.* If the environmental review indicated no significant environmental impacts, the responsible official will assure that a FNSI is prepared which lists any mitigation measures necessary to make the recommended alternative environmentally acceptable.

(e) *Timing of action.* Pursuant to § 6.401(b), in no case shall a contract or grant be awarded until the prescribed 30-day review period for a final EIS has elapsed. Similarly, no action shall be taken until the 30-day comment period for a FNSI is completed.

§ 6.804 Record of decision.

The responsible official shall prepare a record of decision in any case where final EIS has been issued in accordance with 40 CFR 1505.2. It shall be prepared at the time of contract or grant award. The record of decision shall list any mitigation measures necessary to make the recommended alternative environmentally acceptable.

SUBPART I—ENVIRONMENTAL REVIEW PROCEDURES FOR EPA FACILITY SUPPORT ACTIVITIES

§ 6.900 Purpose.

This subpart amplifies the general requirements described in subparts A through D by providing environmental procedures for the preparation of EISs on construction and renovation of special purpose facilities.

§ 6.901 Definitions.

(a) The term "special purpose facility" means a building or space, including land incidental to its use, which is wholly or predominantly utilized for the special purpose of an agency and not generally suitable for other uses, as determined by the General Services Administration.

(b) The term "program of requirements" means a comprehensive document (booklet) describing program activities to be accomplished in the new special purpose facility or improvement. It includes architectural, mechanical, structural, and space requirements.

(c) The term "scope of work" means a document similar in content to the program of requirements but substantially abbreviated. It is usually prepared for small-scale projects.

§ 6.902 Applicability.

(a) *Actions covered.* These procedures apply to all new special-purpose facility construction, activities related to this construction (e.g., site acquisition and clearing), and any improvements or modifications to facilities having potential environmental effects external to the facility, including new construction and improvements undertaken and funded by the Facilities Management Branch, Facilities and Support Services Division, Office of Management and Agency Services; by a regional office; or by a National Environmental Research Center.

(b) *Actions excluded.* This subpart does not apply to those activities of the Facilities Management Branch, Facilities and Support Services Division, for which the branch does not have full fiscal responsibility for the entire project. This includes pilot plant construction, land acquisition, site clearing and access road construction where the Facilities Management Branch's activity is only supporting a project financed by a program office. Responsibility for considering the environmental impacts of such projects rests with the office managing and funding the entire project. Other subparts of this regulation apply, depending on the nature of the project.

§ 6.903 Criteria for preparing EISs.

(a) *Preliminary information.* The responsible official shall request an environmental information document from a construction contractor or consulting architect/engineer employed by EPA if he is involved in the planning, construction, or modification of special purpose facilities when his activities have potential environmental effects external to the facility. Such modifications include but are not limited to facility additions, changes in central heating systems or wastewater treatment systems, and land clearing for access roads and parking lots.

(b) *EIS preparation criteria.* The responsible official shall conduct an environmental review of all actions involving construction of special purpose facilities. The responsible official shall assure that an EIS will be prepared when it is determined that any of the conditions in § 6.506(a) (1) through (6) and (8) above exist.

§ *6.904 Environmental review process.*

(a) *Environmental review.* (1) An environmental review shall be conducted when the program of requirements or scope of work has been completed for the construction, improvements, or modification of special purpose facilities. For special purpose facility construction, the Chief, Facilities Management Branch, shall request the assistance of the appropriate program office and Regional Administrator in the review. For modifications and improvement, the appropriate responsible official shall request assistance in making the review from other cognizant EPA offices.

(2) Any environmental information documents requested shall contain the same sections listed for EISs in subpart B. Contractors and consultants shall be notified in contractual documents when an environmental information document must be prepared.

(b) *Notice of intent, EIS, and FNSI.* The responsible official shall decide at the completion of the Environmental review whether there may be any significant environmental impacts. If there could be significant environmental impacts, a notice of intent and an EIS shall be prepared according to the procedures under subparts A, B, C, and D. If there are not any significant environmental impacts, a FNSI shall be prepared according to the procedures in subparts A and D. The FNSI shall list any mitigation measures necessary to make the recommended alternative environmentally acceptable.

(c) *Timing of action.* Pursuant to § 6.401(b), in no case shall a contract be awarded or construction activities begun until the prescribed 30-day wait period for a final EIS has elapsed. Similarly, under § 6.400(d), no action shall be taken until the 30-day comment period for FNSIs is completed.

§ *6.905 Record of decision.*

At the time of contract award, the responsible official shall prepare a record of decision in those cases where a final EIS has been issued in accordance with 40 CFR 1505.2. The record of decision shall list any mitigation measures necessary to make the recommended alternative environmentally acceptable.

APPENDIX A—STATEMENT OF PROCEDURES ON FLOODPLAIN MANAGEMENT AND WETLANDS PROTECTION

Contents:

Section 1 General

a. Executive Order 11988 entitled "Floodplain Management" dated May 24, 1977, requires Federal agencies to evaluate the potential effects of actions it may take in a floodplain to avoid adversely impacting floodplains wherever possible, to ensure that its planning programs and budget requests reflect consideration of flood hazards and floodplain management, including the restoration and preservation of such land areas as natural undeveloped floodplains, and to prescribe procedures to implement the policies and procedures of this Executive Order. Guidance for implementation of the Executive Order has been provided by the U.S. Water Resources Council in its Floodplain Management Guidelines dated February 10, 1978 (see 40 FR 6030).

b. Executive Order 11990 entitled "Protection of Wetlands," dated May 24, 1977, requires Federal agencies to take action to avoid adversely impacting wetlands wherever possible, to minimize wetlands destruction and to preserve the values of wetlands, and to prescribe procedures to implement the policies and procedures of this Executive Order.

c. It is the intent of these Executive Orders that, wherever possible, Federal agencies implement the floodplains/wetlands requirements through existing procedures, such as those internal procedures established to implement the National Environmental Policy Act (NEPA) and OMB A-95 review procedures. In those instances where the environmental impacts of a proposed action are not significant enough to require an environmental impact statement (EIS) pursuant to section 102(2)(C) of NEPA, or where programs are not subject to the requirements of NEPA, alternative but equivalent floodplain/wetlands evaluation and notice procedures must be established.

Section 2 Purpose

a. The purpose of this Statement of Procedures is to set forth Agency policy and guidance for carrying out the provisions of Executive Orders 11988 and 11990.

b. EPA program offices shall amend existing regulations and procedures to incorporate the policies and procedures set forth in this Statement of Procedures.

c. To the extent possible, EPA shall accommodate the requirements of Executive Orders 11988 and 11990 through the Agency NEPA procedures contained in 40 CFR Part 6.

Section 3 Policy

a. The Agency shall avoid wherever possible the long and short term impacts associated with the destruction of wetlands and the occupancy and modification of floodplains and wetlands, and avoid direct and indirect support of floodplain and wetlands development wherever there is a practicable alternative.

b. The Agency shall incorporate floodplain management goals and wetlands protection considerations into its planning, regulatory, and decision-making processes. It shall also promote the preservation and restoration of floodplains so that their natural and beneficial values can be realized. To the extent possible EPA shall:

(1) Reduce the hazard and risk of flood loss and wherever it is possible, avoid direct or indirect adverse impact on floodplains;

(2) Where there is no practical alternative to locating in a floodplain, minimize the impact of floods on human safety, health, and welfare, as well as the natural environment;

(3) Restore and preserve natural and beneficial values served by floodplains;

(4) Require the construction of EPA structures and facilities to be in accordance with the standards and criteria of the regulations promulgated pursuant to the National Flood Insurance Program;

(5) Identify floodplains which require restoration and preservation and recommend management programs necessary to protect these floodplains and to include such considerations as part of ongoing planning programs; and

(6) Provide the public with early and continuing information concerning floodplain management and with opportunities for participating in decision-making, including the (evaluation of) tradeoffs among competing alternatives.

c. The Agency shall incorporate wetlands protection considerations into its planning, regulatory, and decision-making processes. It shall minimize the destruction, loss, or degradation of wetlands and preserve and enhance the natural and beneficial values of wetlands. Agency activities shall continue to be carried out consistent with the Administrator's Decision Statement No. 4 dated February 21, 1973 entitled "EPA Policy to Protect the Nation's Wetlands."

Section 4 Definitions

a. "Base Flood" means that flood which has a one percent chance of occurrence in any given year (also known as a 100-year flood). This term is used in the National Flood Insurance Program (NFIP) to indicate the minimum level of flooding to be used by a community in its floodplain management regulations.

b. "Base Floodplain" means the 100-year floodplain (one percent chance floodplain). Also see definition of floodplain.

c. "Flood or Flooding" means a general and temporary condition of partial or complete inundation of normally dry land areas from the overflow of inland and/or tidal waters, and/or the unusual and rapid accumulation or runoff of surface waters from any source, or flooding from any other source.

d. "Floodplain" means the lowland and relatively flat areas adjoining inland and coastal waters and other floodprone areas such as offshore islands, including at a minimum, that area subject to a one percent or greater chance of flooding in any given year. The base floodplain shall be used to designate the 100-year floodplain (one percent chance floodplain). The critical action floodplain is defined as the 500-year floodplain (0.2 percent chance floodplain).

e. "Floodproofing" means modification of individual structures and facilities, their sites, and their contents to protect against structural failure, to keep water out, or to reduce effects of water entry.

f. "Minimize" means to reduce to the smallest possible amount or degree.

g. "Practicable" means capable of being done within existing constraints. The test of what is practicable depends upon the situation and includes consideration of the pertinent factors such as environment, community welfare, cost, or technology.

h. "Preserve" means to prevent modification to the natural floodplain environment or to maintain it as closely as possible to its natural state.

i. "Restore" means to re-establish a setting or environment in which the natural functions of the floodplain can again operate.

j. "Wetlands" means those areas that are inundated by surface or ground water with a frequency sufficient to support and under normal circumstances does or would support a prevalence of vegetative or aquatic life that requires saturated or seasonally saturated soil conditions for growth and reproduction. Wetlands generally include swamps, marshes, bogs, and similar areas such as sloughs, potholes, wet meadows, river overflows, mud flats, and natural ponds.

Section 5 Applicability

a. The Executive Orders apply to activities of Federal agencies pertaining to (1) acquiring, managing, and disposing of Federal lands and facilities, (2) providing Federally undertaken, financed, or assisted construction and improvements, and (3) conducting Federal activities and programs affecting land use, including but not limited to water and related land resources planning, regulating, and licensing activities.

b. These procedures shall apply to EPA's programs as follows: (1) All Agency actions involving construction of facilities or management of lands or property. This will require amendment of the EPA Facilities Management Manual (October 1973 and revisions thereafter).

(2) All Agency actions where the NEPA process applies. This would include the programs under section 306/402 of the Clean Water Act pertaining to new source permitting and section 201 of the Clean Water Act pertaining to wastewater treatment construction grants.

(3) All agency actions where there is sufficient independent statutory authority to carry out the floodplain/wetlands procedures.

(4) In program areas where there is no EIS requirement nor clear statutory authority for EPA to require procedural implementation, EPA shall continue to provide leadership and offer guidance so that the value of floodplain management and wetlands protection can be understood and carried out to the maximum extent practicable in these programs.

c. These procedures shall not apply to any permitting or source review programs of EPA once such authority has been transferred or delegated to a State. However, EPA shall, to the extent possible, require States to provide equivalent effort to assure support for the objectives of these procedures as part of the state assumption process.

Section 6 Requirements

a. Floodplain/Wetlands review of proposed Agency actions.

(1) *Floodplain/Wetlands Determination*—Before undertaking an Agency action, each program office must determine whether or not the action will be located in or affect a floodplain or wetlands. The Agency shall utilize maps prepared by the Federal Insurance Administration (Flood Insurance Rate Maps or Flood Hazard Boundary Maps), Fish and Wildlife Service (National Wetlands Inventory Maps), and other appropriate agencies to determine whether a proposed action is located in or will likely affect a floodplain or wetlands. If there is no adverse floodplain/wetlands impact identified, the action may proceed without further consideration of the remaining procedures set forth below.

(2) *Early Public Notice*—When it is apparent that a proposed or potential agency action is likely to impact a floodplain or wetlands, the public should be informed through appropriate public notice procedures.

(3) *Floodplain/Wetlands Assessment*—If the Agency determines a proposed action is located in or affects a floodplain or wetlands, a floodplain/wetlands assessment shall be undertaken. For those actions where an environmental assessment (EA) or environmental impact statement (EIS) is prepared pursuant to 40 CFR Part 6, the floodplain/wetlands assessment shall be prepared concurrently with these analyses and shall be included in the EA or EIS. In all other cases, a "floodplain/wetlands assessment" shall be prepared. Assessments shall consist of a description of the proposed action, a discussion of its effect on the floodplain/wetlands, and shall also describe the alternatives considered.

(4) *Public Review of Assessments*—For proposed actions impacting floodplain/wetlands where an EA or EIS is prepared, the opportunity for public review will be provided through the EIS provisions contained in 40 CFR Parts 6, 25, or 35, where appropriate. In other cases, an equivalent public notice of the floodplain/wetlands assessment shall be made consistent with the public involvement requirements of the applicable program.

(5) *Minimize, Restore or Preserve*—If there is no practicable alternative to locating in or affecting the floodplain or wetlands, the Agency shall act to minimize potential harm to the floodplain or wetlands. The Agency shall also act to restore and preserve the natural and beneficial values of floodplains and wetlands as part of the analysis of all alternatives under consideration.

(6) *Agency Decision*—After consideration of alternative actions, as they have been

modified in the preceding analysis, the Agency shall select the desired alternative. For all Agency actions proposed to be in or affecting a floodplain/wetlands, the Agency shall provide further public notice announcing this decision. This decision shall be accompanied by a Statement of Findings, not to exceed three pages. This Statement shall include: (i) The reasons why the proposed action must be located in or affect the floodplain or wetlands; (ii) a description of significant facts considered in making the decision to locate in or affect the floodplain or wetlands including alternative sites and actions; (iii) a statement indicating whether the proposed action conforms to applicable State or local floodplain protection standards; (iv) a description of the steps taken to design or modify the proposed action to minimize potential harm to or within the floodplain or wetlands; and (v) a statement indicating how the proposed action affects the natural or beneficial values of the floodplain or wetlands. If the provisions of 40 CFR Part 6 apply, the Statement of Findings may be incorporated in the final EIS or in the environmental assessment. In other cases, notice should be placed in the **Federal Register** or other local medium and copies sent to Federal, State, and local agencies and other entities which submitted comments or are otherwise concerned with the floodplain/wetlands assessment. For floodplain actions subject to Office of Management and Budget (OMB) Circular A-95, the Agency shall send the Statement of Findings to State and areawide A-95 clearinghouse in the geographic area affected. At least 15 working days shall be allowed for public and interagency review of the Statement of Findings.

(7) *Authorizations/Appropriations*—Any requests for new authorizations or appropriations transmitted to OMB shall include, a floodplain/wetlands assessment and, for floodplain impacting actions, a Statement of Findings, if a proposed action will be located in a floodplain or wetlands.

b. *Lead agency concept.* To the maximum extent possible, the Agency shall rely on the lead agency concept to carry out the provisions set forth in section 6.a. above. Therefore, when EPA and another Federal agency have related actions, EPA shall work with the other agency to identify which agency shall take the lead in satisfying these procedural requirements and thereby avoid duplication of efforts.

c. *Additional floodplain management provisions relating to Federal property and facilities.*

(1) *Construction Activities*—EPA controlled structures and facilities must be constructed in accordance with existing criteria and standards set forth under the NFIP and must include mitigation of adverse impacts wherever feasible. Deviation from these requirements may occur only to the extent NFIP standards are demonstrated as inappropriate for a given structure or facility.

(2) *Flood Protection Measures*—If newly constructed structures or facilities are to be located in a floodplain, accepted floodproofing and other flood protection measures shall be undertaken. To achieve flood protection, EPA shall, wherever practicable, elevate structures above the base flood level rather than filling land.

(3) *Restoration and Preservation*—As part of any EPA plan or action, the potential for restoring and preserving floodplains and wetlands so that their natural and beneficial values can be realized must be considered and incorporated into the plan or action wherever feasible.

(4) *Property Used by Public*—If property used by the public has suffered damage or is located in an identified flood hazard area, EPA shall provide on structures, and other places where appropriate, conspicuous indicators of past and probable flood height to enhance public knowledge of flood hazards.

(5) *Transfer of EPA Property*—When property in flood plains is proposed for lease, easement, right-of-way, or disposal to non-Federal public or private parties, EPA shall reference in the conveyance those uses that are restricted under Federal, State, and local floodplain regulations and attach other restrictions to uses of the property as may be deemed appropriate. Notwithstanding, EPA shall consider withholding such properties from conveyance.

Section 7 Implementation.

a. Pursuant to section 2, the EPA program offices shall amend existing regulations, procedures, and guidance, as appropriate, to incorporate the policies and procedures set forth in this Statement of Procedures. Such amendments shall be made within six months of the date of these Procedures.

b. The Office of Federal Activities (OFA) is responsible for the oversight of the implementation of this Statement of Procedures and shall be given advanced opportunity to review amendments to regulations, procedures, and guidance. OFA shall coordinate efforts with the program offices to develop necessary manuals and more specialized supplementary guidance to carry out this Statement of Procedures.

[FR Doc. 79-34157 Filed 11-5-79; 8:45 am]

APPENDIX H
EPA–EIS Preparation Regulations

EPA Preparation of Environmental Impact Statements Final Regulations

TITLE 40—PROTECTION OF ENVIRONMENT

CHAPTER I—ENVIRONMENTAL PROTECTION AGENCY

[FRL 327-5]

PART 6—PREPARATION OF ENVIRONMENTAL IMPACT STATEMENTS

Final Regulation

The National Environmental Policy Act of 1969 (NEPA), implemented by Executive Order 11514 of March 5, 1970, and the Council on Environmental Quality's (CEQ's) Guidelines of August 1, 1973, requires that all agencies of the Federal Government prepare detailed environmental impact statements on proposals for legislation and other major Federal actions significantly affecting the quality of the human environment. NEPA requires that agencies include in their decision-making process an appropriate and careful consideration of all environmental aspects of proposed actions, an explanation of potential environmental effects of proposed actions and their alternatives for public understanding, a discussion of ways to avoid or minimize adverse effects of proposed actions and a discussion of how to restore or enhance environmental quality as much as possible.

On January 17, 1973, the Environmental Protection Agency (EPA) published a new Part 6 in interim form in the **Federal Register** (33 FR 1696), establishing EPA policy and procedures for the identification and analysis of environmental impacts and the preparation of environmental impact statements (EIS's) when significant impacts on the environment are anticipated.

On July 17, 1974, EPA published a notice of proposed rulemaking the **Federal Register** (39 FR 26254). The rulemaking provided detailed procedures for applying NEPA to EPA's nonregulatory programs only. A separate notice of administrative procedure published in the October 21, 1974, **Federal Register** (39 FR 37419) gave EPA's procedures for voluntarily preparing EIS's on certain regulatory activities. EIS procedures for another regulatory activity, issuing National Pollutant Discharge Elimination System (NPDES) discharge permits to new sources, will appear in 40 CFR 6. Associated amendments to the NPDES operating regulations, covering permits to new sources, will appear in 40 CFR 125.

The proposed regulation on the preparation of EIS's for nonregulatory programs was published for public review and comment. EPA received comments on this proposed regulation from environmental groups; Federal, State and local governmental agencies: industry; and private individuals. As a result of the comments received, the following changes have been made:

(1) Coastal zones, wild and scenic rivers, prime agricultural land and wildlife habitat were included in the criteria to be considered during the environmental review.

The Coastal Zone Management Act and the Wild and Scenic Rivers Act are intended to protect these environmentally sensitive areas; therefore, EPA should consider the effects of its projects on these areas. Protection of prime agricultural lands and wildlife habitat has become an important concern as a result of the need to further increase food production from domestic sources as well as commercial harvesting of fish and other wildlife resources and from the continuing need to preserve the diversity of natural resources for future generations.

(2) Consideration of the use of floodplains as required by Executive Order 11296 was added to the environmental review process.

Executive Order 11296 requires agencies to consider project alternatives which will preclude the uneconomic, hazardous or unnecessary use of floodplains to minimize the exposure of facilities to potential flood damage, lessen the need for future Federal expenditures for flood protection and flood disaster relief and preserve the unique and significant public value of the floodplain as an environmental resource.

(3) Statutory definitions of coastal zones and wild and scenic rivers were added to § 6.214(b).

These statutes define sensitive areas and require states to designate areas which must be protected.

(4) The review and comment period for negative declarations was extended from 15 days to 15 working days.

Requests for negative declarations and comments on negative declarations are not acted on during weekends and on holidays. In addition, mail requests often take two or three days to reach the appropriate office and several more days for action and delivery of response. Therefore, the new time frame for review and response to a negative declaration is more realistic without adding too much delay to a project.

(5) Requirements for more data in the negative declaration to clarify the proposed action were added in § 6.212(b).

Require a summary of the impacts of a project and other data to support the negative declaration in this document improves its usefulness as a tool to review the decision not to prepare a full EIS on a project.

(6) The definitions of primary and secondary impacts in § 6.304 were clarified.

The definitions were made more specific, especially in the issue areas of induced growth and growth rates, to reduce subjectivity in deciding whether an impact is primary or secondary.

(7).Procedures for EPA public hearings in Subpart D were clarified.

Language was added to this subpart to distinguish EPA public hearings from applicant hearings required by statute or regulation, such as the facilities plan hearings.

(8) The discussion of retroactive application (§ 6.504) was clarified and abbreviated.

The new language retains flexibility in decision-making for the Regional Administrator while eliminating the ambiguity of the language in the interim regulation.

(9) The criteria for writing an EIS if wetlands may be affected were modified in § 6.510(b).

The new language still requires an EIS on a project which will be located on wetlands but limits the requirements for an EIS on secondary wetland effects to those which are significant and adverse.

(10) A more detailed explanation of the data required in environmental assessments (§ 6.512) was added.

Requiring more specific data in several areas, including energy production and consumption as well as land use trends and population projections, from the applicant will provide a more complete data base for the environmental review. Documentation of the applicant's data will allow EPA to evaluate the validity of this data.

(11) Subpart F, Guidelines for Compliance with NEPA in Research and Development Programs and Activities, was revised.

ORD simplified this subpart by removing the internal procedures and assignments of responsibility for circulation in internal memoranda. Only the general application of this regulation to ORD programs was retained.

(12) The discussions of responsibilities and document distribution procedures were moved to appendices attached to the regulations.

These sections were removed from the regulatory language to improve the readability of the regulation and because these discussions are more explanatory and do not need to have the legal force of regulatory language.

(13) Consideration of the Endangered Species Act of 1973 was incorporated into the regulation.

EPA recognizes its responsibility to assist with implementing legislation which will help preserve or improve our natural resources.

The major issues raised on this regulation were on new and proposed criteria for determining when to prepare an EIS and the retroactive application of the criteria to projects started before July 1, 1975. In addition to the new criteria which were added, CEQ requested the addition of several quantitative criteria for which parameters have not been set. These new criteria are being discussed with CEQ and may be added to the regulation at a future date. Changes in the discussion of retroactive application of the criteria are described in item 8 above.

EPA believes that Agency compliance with the regulations of Part 6 will enhance the present quality of human life without endangering the quality of the natural environment for future generations.

*

Effective date: This regulation will become effective April 14, 1975.

Dated: April 3, 1975.

RUSSELL E. TRAIN,
Administrator.

Subpart A—General

Sec.
6.100 Purpose and policy.
6.102 Definitions.
6.104 Summary of procedures for implementing NEPA.
6.106 Applicability.
6.108 Completion of NEPA procedures before start of administrative action.
6.110 Responsibilities.

Subpart B—Procedures

6.200 Criteria for determining when to prepare an environmental impact statement.
6.202 Environmental assessment.
6.204 Environmental review.
6.206 Notice of intent.
6.208 Draft environmental impact statements.
6.210 Final environmental impact statements.
6.212 Negative declarations and environmental impact appraisals.
6.214 Additional procedures.
6.216 Availability of documents.

Subpart C—Content of Environmental Impact Statements

6.300 Cover sheet.
6.302 Summary sheet.

Sec.
6.304 Body of statement.
6.306 Documentation.

Subpart D—EPA Public Hearings on Impact Statements

6.400 General.
6.402 Public hearing process.

Subpart E—Guidelines for Compliance with NEPA in the Title II Wastewater Treatment Works Construction Grants Program and the Areawide Waste Treatment Management Planning Program

6.500 Purpose.
6.502 Definitions.
6.504 Applicability.
6.506 Completion of NEPA procedures before start of administrative actions.
6.510 Criteria for preparation of environmental impact statements.
6.512 Procedures for implementing NEPA.
6.514 Content of environmental impact statements.

Subpart F—Guidelines for Compliance with NEPA in Research and Development Programs and Activities

6.600 Purpose.
6.602 Definitions.
6.604 Applicability.
6.608 Criteria for determining when to prepare environmental impact statements.
6.610 Procedures for compliance with NEPA.

Subpart G—Guidelines for Compliance with NEPA in Solid Waste Management Activities

6.700 Purpose.
6.702 Criteria for the preparation of environmental assessments and EIS's.
6.704 Procedures for compliance with NEPA.

Subpart H—Guidelines for Compliance with NEPA in Construction of Special Purpose Facilities and Facility Renovations

6.800 Purpose.
6.802 Definitions.
6.804 Applicability.
6.808 Criteria for the preparation of environmental assessments and EIS's.
6.810 Procedures for compliance with NEPA.

EXHIBITS

1. (Page 1.) Notice of Intent Transmittal Memorandum Suggested Format.
   (Page 2.) Notice of Intent Suggested Format.
2. Public Notice and News Release Suggested Format.
3. Negative Declaration Suggested Format.
4. Environmental Impact Appraisal Suggested Format.
5. Cover Sheet Format for Environmental Impact Statements.
6. Summary Sheet Format for Environmental Impact Statements.
7. Flowchart for Solid Waste Management Program Operations.

Appendix A—Checklist for Environmental Reviews.
Appendix B—Responsibilities.
Appendix C—Availability and Distribution of Documents.

Authority: Secs. 102, 103 of 83 Stat. 854 (42 U.S.C. 4321 et seq.)

SUBPART A—GENERAL

§ *6.100* *Purpose and policy.*

(a) The National Environmental Policy Act (NEPA) of 1969, implemented by Executive Order 11514 and the Council on Environmental Quality's (CEQ's) Guidelines of August 1, 1973 (38 FR 20550), requires that all agencies of the Federal Government prepare detailed environmental impact statements on proposals for legislation and other major Federal actions significantly affecting the quality of the human environment. NEPA requires that agencies include in the decision-making process appropriate and careful consideration of all environmental effects of proposed actions, explain potential environmental effects of proposed actions and their alternatives for public understanding, avoid or minimize adverse effects of proprosed actions and restore or enhance environmental quality as much as possible.

(b) This part establishes Environmental Protection Agency (EPA) policy and procedures for the identification and analysis of the environmental impacts of EPA nonregulatory actions and the preparation and processing of environmental impact statements (EIS's) when significant impacts on the environment are anticipated.

§ *6.102* *Definitions.*

(a) "Environmental assessment" is a written analysis submitted to EPA by its grantees or contractors describing the environmental impacts of proposed actions undertaken with the financial support of EPA. For facilities or section 203 plans as defined in § 6.102 (j) and (k), the assessment must be an integral, though indentifiable, part of the plan submitted to EPA for review.

(b) "Environmental review" is a formal evaluation undertaken by EPA to determine whether a proposed EPA action may have a significant impact on the environment. The environmental assessment is one of the major sources of information used in this review.

(c) "Notice of intent" is a memorandum, prepared after the environmental review, announcing to Federal, regional, State, and local agencies, and to interested persons, that a draft EIS will be prepared.

(d) "Environmental impact statement" is a report, prepared by EPA, which identifies and analyzes in detail the environmental impacts of a proposed EPA action and feasible alternatives.

(e) "Negative declaration" is a written announcement, prepared after the environmental review, which states that EPA has decided not to prepare an EIS and summarizes the environmental impact appraisal.

(f) "Environmental impact appraisal" is based on an environmental review and supports a negative declaration. It describes a proposed EPA action, its expected environmental impact, and the basis for the conclusion that no significant impact is anticipated.

(g) "NEPA-associated documents" are any one or combination of: notices of intent, negative declarations, exemption certifications, environmental impact appraisals, news releases, EIS's, and environmental assessments.

(h) "Responsible official" is an Assistant Administrator, Deputy Assistant Administrator, Regional Administrator or their designee.

(i) "Interested persons" are individuals, citizen groups, conservation organizations, corporations, or other nongovernmental units, including applicants for EPA contracts or grants, who may be interested in, affected by, or technically competent to comment on the environmental impacts of the proposed EPA action.

(j) "Section 208 plan" is an areawide waste treatment management plan prepared under section 208 of the Federal Water Pollution Control Act (FWPCA), as amended, under 40 CFR Part 126 and 40 CFR Part 35, Subpart F.

(k) "Facilities plan" is a preliminary plan prepared as the basis for construction of publicly owned waste treatment works under Title II of FWPCA, as amended, under 40 CFR 35.917.

(l) "Intramural project" is an in-house project undertaken by EPA personnel.

(m) "Extramural project" is a project undertaken by grant or contract.

§ ***6.104 Summary of procedures for implementing NEPA.***

(a) *Responsible official.* The responsible official shall utilize a systematic, interdisciplinary approach to integrate natural and social sciences as well as environmental design arts in planning programs and making decisions which are subject to NEPA review. His staff may be supplemented by professionals from other agencies, universities or consultants whenever in-house capabilities are insufficiently interdisciplinary.

(b) *Environmental assessment.* Environmental assessments must be submitted to EPA by its grantees and contractors, as required in Subparts E, F, G, and H of this part. The assessment is used by EPA to decide if an EIS is required and to prepare one if necessary.

(c) *Environmental review.* Environmental reviews shall be made of proposed and certain ongoing EPA actions as required in § 6.106(c). This process shall consist of a study of the action to identify and evaluate the environmental impacts of the action. Types of grants, contracts and other actions requiring study are listed in the subparts following Subpart D. The process shall include a review of any environmental assessment received to determine whether any significant impacts are anticipated, whether any changes can be made in the proposed action to eliminate significant adverse impacts, and whether an EIS is required. EPA has overall responsibility for this review, although its grantees and contractors will contribute to the review through their environmental assessments.

(d) *Notice of intent and EIS's.* When an environmental review indicates that a significant environmental impact may occur and the significant adverse impacts cannot be eliminated by making changes in the project, a notice of intent shall be published, and a draft EIS shall be prepared and distributed. After external coordination and evaluation of the comments received, a final EIS shall be prepared and distributed. EIS's should be prepared first on those proposed actions with the most adverse effects which are scheduled for earliest implementation and on other proposed actions according to priorities assigned by the responsible official.

(e) *Negative declaration and environmental impact appraisal.* When the environmental review indicates no significant impacts are anticipated or when the project is changed to eliminate the significant adverse impacts, a negative declaration shall be issued. For the cases in Subparts E, F, G, and H of this part, an environmental impact appraisal shall be prepared which summarizes the impacts, alternatives and reasons an EIS was not prepared. It shall remain on file and be available for public inspection.

§ ***6.106 Applicability.***

(a) *Administrative actions covered.* This part applies to the administrative actions listed below. The subpart referenced with each action lists the detailed NEPA procedures associated with the action. Administrative actions are:

(1) Development of EPA legislative proposals;

(2) Development of favorable reports on legislation initiated elsewhere and not

accompanied by an EIS, when they relate to or affect matters within EPA's primary areas of responsibility;

(3) For the programs under Title II of FWPCA, as amended, those administrative actions in § 6.504;

(4) For the Office of Research and Development, those administrative actions in § 6.604;

(5) For the Office of Solid Waste Management Programs, those administrative actions in § 6.702;

(6) For construction of special purpose facilities and facility renovations, those administrative actions in § 6.804; and

(7) Development of an EPA project in conjunction with or located near a project or complex of projects started by one or more Federal agencies when the cumulative effects of all the projects will be major allocations of resources or foreclosures of future land-use options.

(b) *Administrative actions excluded.* The requirements of this part do not apply to environmentally protective regulatory activities undertaken by EPA, nor to projects exempted in § 6.504, § 6.604, and § 6.702.

(c) *Application to ongoing actions.* This regulation shall apply to uncompleted and continuing EPA actions initiated before the promulgation of these procedures when modifications of or alternatives to the EPA action are still available, except for the Title II construction grants program. Specific application for the construction grants program is in § 6.504(c). An EIS shall be prepared for each project found to have significant environmental effects as described in § 6.200.

(d) *Application to legislative proposals.* (1) As noted in paragraphs (a) (1) and (2) of this section, EIS's of negative declarations shall be prepared for legislative proposals or favorable reports relating to legislation which may significantly affect the environment. Because of the nature of the legislative process, EIS's for legislation must be prepared and reviewed according to the procedures followed in the development and review of the legislative matter. These procedures are described in Office of Management and Budget (OMB) Circular No. A-19.

(2) A working draft EIS shall be prepared by the EPA office responsible for preparing the legislative proposal or report on legislation. It shall be prepared concurrently with the development of the legislative proposal or report and shall contain the information required in § 6.304. The EIS shall be circulated for internal EPA review with the legislative proposal or report and other supporting documentation. The working draft EIS shall be modified to correspond with changes made in the proposal or report during the internal review. All major alternatives developed during the formulation and review of the proposal or report should be retained in the working draft EIS.

(i) The working draft EIS shall accompany the legislative proposal or report to OMB. EPA shall revise the working draft EIS to respond to comments from OMB and other Federal agencies.

(ii) Upon transmittal of the legislative proposal or report to Congress, the working draft EIS will be forwarded to CEQ and the Congress as a formal legislative EIS. Copies will be distributed according to procedures described in Appendix C.

(iii) Comments received by EPA on the legislative EIS shall be forwarded to the appropriate Congressional Committees. EPA also may respond to specific comments and forward its responses with the comments. Because legislation undergoes continuous changes in Congress beyond the control of EPA, no final EIS need be prepared by EPA.

§ *6.108 Completion of NEPA procedures before starting administrative action.*

(a) No administrative action shall be taken until the environmental review process, resulting in an EIS or a negative declaration with environmental appraisal, has been completed.

(b) *When an EIS will be prepared.* Except when requested by the responsible official in writing and approved by CEQ, no administrative action shall be taken sooner than ninety (90) calendar days after a draft EIS has been distributed or sooner than thirty (30) calendar days after the final EIS has been made public. If the final text of an EIS is filed within ninety (90) days after a draft EIS has been circulated for comment, furnished to CEQ, and made public, the minimum thirty (30) day period and the ninety (90) day period may run concurrently if they overlap. The minimum periods for review and advance availability of EIS's shall begin on the date CEQ publishes the notice of receipt of the EIS in the **Federal Register**. In addition, the proposed action should be modified to conform with any changes EPA considers necessary before the final EIS is published.

(c) *When an EIS will not be prepared.* If EPA decides not to prepare an EIS on any action listed in this part for which a negative declaration with environmental appraisal has been prepared, no administrative action shall be taken for at least fifteen (15) working days after the negative declaration is issued, to allow public review of the decision. If significant environmental issues are raised during the review period, the decision may be changed and a new environmental appraisal or an EIS may be prepared.

§ *6.110 Responsibilities.*

See Appendix B for responsibilities of this part.

SUBPART B—PROCEDURES

§ *6.200 Criteria for determining when to prepare an EIS.*

The following general criteria shall be used when reviewing a proposed EPA action to determine if it will have a significant impact on the environment and therefore require an EIS:

(a) *Significant environmental effects.* (1) An action with both beneficial and detrimental effects should be classified as having significant effects on the environment, even if EPA believes that the net effect will be beneficial. However, preference should be given to preparing EIS's on proposed actions which, on balance, have adverse effects.

(2) When determining the significance of a proposed action's impacts, the responsible official shall consider both its short term and long term effects as well as its primary and secondary effects as defined in § 6.304(c). Particular attention should be given to changes in land use patterns; changes in energy supply and demand; increased development in floodplains; significant changes in ambient air and water quality or noise levels; potential violations of air quality, water quality, and noise level standards; significant changes in surface of groundwater quality or quantity; and encroachments on wetlands, coastal zones, or fish and wildlife habitat, especially when threatened or endangered species may be affected.

(3) Minor actions which may set a precedent for future major actions with significant adverse impacts or a number of actions with individually insignificant but cumulatively significant adverse impacts shall be classified as having significant environmental impacts. If EPA is taking a number of minor, environmentally insignificant actions that are similar in execution and purpose, during a limited time span and in the same general geographic area, the cumulative environmental impact of all of these actions shall be evaluated.

(4) In determining the significance of a proposed action's impact, the unique characteristics of the project area should be carefully considered. For example, proximity to historic sites, parklands, or wild and scenic rivers may make the impact significant. A project discharging into a drinking water aquifer may make the impact significant.

(5) A proposed EPA action which will have direct and significant adverse effects on a property listed in or eligible for listing in the National Register of Historic Places or will cause irreparable loss or destruction of significant scientific, prehistoric, historic, or archaeological data shall be classified as having significant environmental impacts.

(b) *Controversial actions.* An EIS shall be prepared when the environmental impact of a proposed EPA action is likely to be highly controversial.

(c) *Additional criteria for specific programs.* Additional criteria for various EPA programs are in Subpart E (Title II Wastewater Treatment Works Construction Grants Program), Subpart F (Research and Development Programs), Subpart G (Solid Waste Management Programs), and Subpart H (Construction of Special Facilities and Facility Renovations).

§ *6.202 Environmental assessment.*

Environmental assessments must be submitted to EPA by its grantees and contractors as required in Subparts E, F, G, and H of this part. The assessment is to ensure that the applicant considers the environmental impacts of the proposed action at the earliest possible point in his planning process. The assessment and other relevant information are used by EPA to decide if an EIS is required. While EPA is responsible for ensuring that EIS's are factual and comprehensive, it expects assessments and other data submitted by grantees and contractors to be accurate and complete. The responsible official may request additional data and analyses from grantees or other sources any time he determines they are needed to comply adequately with NEPA.

§ *6.204 Environmental review.*

Proposed EPA actions, as well as ongoing EPA actions listed in § 6.106(c), shall be subjected to an environmental review. This review shall be a continuing one, starting at the earliest possible point in the development of the project. It shall consist of a study of the proposed action, including a review of any environmental assessments received, to identify and evaluate the environmental impacts of the proposed action and feasible alternatives. The review will determine whether significant impacts are anticipated from the proposed action, whether any feasible alternatives can be adopted or changes can be made in project design to eliminate significant adverse impacts, and whether an EIS or a negative declaration is required. The responsible official shall determine the proper scope of the environmental review. The responsible official may delay approval of related projects until the proposals can be reviewed together to allow EPA to properly evaluate their cumulative impacts.

§ *6.206 Notice of intent.*

(a) *General.* (1) When an environmental review indicates a significant impact may occur and significant adverse impacts cannot be eliminated by making changes in the project, a notice of intent, announcing the preparation of a draft EIS, shall be issued by the responsible official. The notice shall briefly describe the EPA action, its location, and the issues involved (Exhibit 1).

(2) The purpose of a notice of intent is to involve other government agencies and interested persons as early as possible in the planning and evaluation of EPA actions which may have significant environmental impacts. This notice should encourage

agency and public input to a draft EIS and assure that environmental values will be identified and weighed from the outset rather than accommodated by adjustments at the end of the decision-making process.

(b) *Specific actions.* The specific actions to be taken by the responsible official on notices of intent are:

(1) When the review process indicates a significant impact may occur and significant adverse impacts cannot be eliminated by making changes in the project, prepare a notice of intent immediately after the environmental review.

(2) Distribute copies of the notice of intent as required in Appendix C.

(3) Publish in a local newspaper, with adequate circulation to cover the area affected by the project, a brief public notice stating that an EIS will be prepared on a particular project, and the public may participate in preparing the EIS (Exhibit 2). News releases also may be submitted to other media.

(c) *Regional office assistance to program offices.* Regional offices will provide assistance to program offices in taking these specific actions when the EIS originates in a program office.

§ ***6.208 Draft EIS's.***

(a) *General.* (1) The responsible official shall assure that a draft EIS is prepared as soon as possible after the release of the notice of intent. Before releasing the draft EIS to CEQ, a preliminary version may be circulated for review to other offices within EPA with interest in or technical expertise related to the action. Then the draft EIS shall be sent to CEQ and circulated to Federal, State, regional, and local agencies with special expertise or jurisdiction by law, and to interested persons. If the responsible official determines that a public hearing on the proposed action is warranted, the hearing will be held after the draft EIS is prepared, according to the requirements of § 6.402.

(2) Draft EIS's should be prepared at the earliest possible point in the project development. If the project involves a grant applicant or potential contractor, he must submit any data EPA requests for preparing the EIS. Where a plan or program has been developed by EPA or submitted to EPA for approval, the relationship between the plan and the later projects encompassed by it shall be evaluated to determine the best time to prepare an EIS. Whenever possible, an EIS will be drafted for the total program at the initial planning stage. Then later component projects included in the plan will not require individual EIS's unless they differ substantially from the plan, or unless the overall plan did not provide enough detail to fully assess significant impacts of individual projects. Plans shall be reevaluated by the responsible official to monitor the cumulative impact of the component projects and to preclude the plan's obsolescence.

(b) *Specific actions.* The specific actions to be taken by the responsible official on draft EIS's are:

(1) Distribute the draft EIS according to the procedures in Appendix C.

(2) Inform the agencies to reply directly to the originating EPA office. Commenting agencies shall have at least forty-five (45) calendar days to reply, starting from the date of publication in the **Federal Register** of lists of statements received by CEQ. If no comments are received during the reply period and no time extension has been requested, it shall be presumed that the agency has no comment to make. EPA may grant extensions of fifteen (15) or more calendar days. The time limits for review and extensions for State and local agencies; State, regional, and metropolitan clearinghouses; and interested persons shall be the same as those available to Federal agencies.

(3) Publish a notice in local newspapers stating that the draft EIS is available for

comment and listing where copies may be obtained (Exhibit 2), and submit news releases to other media.

(4) Include in the draft EIS a notice stating that only those Federal, State, regional, and local agencies and interested persons who make substantive comments on the draft EIS or request a copy of the final EIS will be sent a copy.

(c) *Regional office assistance to program office.* If requested, regional offices will provide assistance to program offices in taking these specific actions when the EIS originates in a program office.

§ ***6.210 Final EIS's.***

(a) Final EIS's shall respond to all substantive comments raised through the review of the draft EIS. Special care should be taken to respond fully to comments disagreeing with EPA's position. (See also § 6.304(g).)

(b) Distribution and other specific actions are described in Appendix C. If there is an applicant, he shall be sent a copy. When the number of comments on the draft EIS is so large that distribution of the final EIS to all commenting entities appears impractical, the program or regional office preparing the EIS shall consult with OFA, which will consult with CEQ about alternative arrangements for distribution of the EIS.

§ ***6.212 Negative declaration and environmental impact appraisals.***

(a) *General.* When an environmental review indicates there will be no significant impact or significant adverse impacts have been eliminated by making changes in the project, the responsible official shall prepare a negative declaration to allow public review of his decision before it becomes final. The negative declaration and news release must state that interested persons disagreeing with the decision may submit comments for consideration by EPA. EPA shall not take administrative action on the project for at least fifteen (15) working days after release of the negative declaration and may allow more time for response. The responsible official shall have an environmental impact appraisal supporting the negative declaration available for public review when the negative declaration is released for those cases given in Subparts E, F, G, and H.

(b) *Specific actions.* The responsible official shall take the following specific actions on those projects for which both a negative declaration and an impact appraisal will be prepared.

(1) *Negative declaration.* (i) Prepare a negative declaration immediately after the environmental review. This document shall briefly summarize the purpose of the project, its location, the nature and extent of the land use changes related to the project, and the major primary and secondary impacts of the project. It shall describe how the more detailed environmental impact appraisal may be obtained at cost. (See Exhibit 3.)

(ii) Distribute the negative declaration according to procedures in Appendix C. In addition, submit to local newspapers and other appropriate media a brief news release with a negative declaration attached, informing the public that a decision not to prepare an EIS has been made and a negative declaration and environmental impact appraisal are available for public review and comment (Exhibit 2).

(2) *Environmental impact appraisal.* (i) Prepare an environmental impact appraisal concurrently with the negative declaration. This document shall briefly describe the proposed action and feasible alternatives, environmental impacts of the proposed action, unavoidable adverse impacts of the proposed action, the relationship between short term uses of man's environment and the maintenance and enhancement of long term productivity, steps to minimize harm to the environment, irreversible and irretrievable commitments of resources to implement the action, comments and consultations

on the project, and reasons for concluding there will be no significant impacts. (See Exhibit 4.)

(ii) Distribute the environmental impact appraisal according to procedures in Appendix C.

§ ***6.214 Additional procedures.***

(a) *Historical and archaeological sites.* EPA is subject to the requirements of section 106 of the National Historic Preservation Act of 1966, 16 U.S.C. 470 *et seq.*, Executive Order 11593, the Archaeological and Historic Preservation Act of 1974, 16 U.S.C. 469 *et seq.*, and the regulations promulgated under this legislation. These statutes and regulations establish environmental review procedures which are independent of the NEPA requirements.

(1) If an EPA action may affect properties with historic, architectural, archaeological, or cultural value which are listed in the National Register of Historic Places (published in the **Federal Register** each February with supplements on the first Tuesday of each month), the responsible official shall comply with the procedures of the Advisory Council on Historic Preservation (36 CFR 800), including determining the need for a Memorandum of Agreement among EPA, the State Historic Preservation Officer, and the Advisory Council. If a Memordandum of Agreement is executed, it shall be included in an EIS whenever one is prepared on a proposed action. See § 6.512(c) of this part for additional procedures for the construction grants program under Title II of the FWPCA, as amended.

(2) If an EPA action may cause irreparable loss or destruction of significant scientific, prehistoric, historic, or archaeological data, the responsible official shall consult with the State Historic Preservation Officer in compliance with the Archaeological and Historic Preservation Act (P.L. 93-291).

(b) *Wetlands, floodplains, coastal zones, wild and scenic rivers, fish and wildlife.* The following procedures shall be applied to all EPA administrative actions covered by this part that may affect these environmentally sensitive resources.

(1) If an EPA action may affect wetlands, the responsible official shall consult with the appropriate offices of the Department of the Interior, Department of Commerce, and the U.S. Army Corps of Engineers during the environmental review to determine the probable impact of the action on the pertinent fish and wildlife resources and land use of these areas.

(2) If an EPA action may directly cause or induce the construction of buildings or other facilities in a floodplain, the responsible official shall evaluate flood hazards in connection with these facilities as required by Executive Order 11296 and shall, as far as practicable, consider alternatives to preclude the uneconomic, hazardous or unnecessary use of floodplains to minimize the exposure of facilities to potential flood damage, lessen the need for future Federal expenditures for flood protection and flood disaster relief and preserve the unique and significant public value of the floodplain as an environmental resource.

(3) If an EPA action may affect coastal zones or coastal waters as defined in Title III of the Coastal Zone Management Act of 1972 (Pub. L. 92-583), the responsible official shall consult with the appropriate State offices and with the appropriate office of the Department of Commerce during the environmental review to determine the probable impact of the action on coastal zone or coastal water resources.

(4) If an EPA action may affect portions of rivers designated wild and scenic or being considered for this designation under the Wild and Scenic Rivers Act (Pub. L. 90-542), the responsible official shall consult with appropriate State offices and with the Secretary of the Interior or, where national forest lands are involved, with the Secretary of

Agriculture during the environmental review to determine the status of an affected river and the probable impact of the action on eligible rivers.

(5) If an EPA action will result in the control or structural modification of any stream or other body of water for any purpose, including navigation and drainage, the responsible official shall consult with the United States Fish and Wildlife Service (Department of the Interior), the National Marine Fisheries Service of the National Oceanic and Atmospheric Administration (Department of Commerce), the U.S. Army Corps of Engineers and the head of the agency administering the wildlife resources of the particular State in which the action will take place with a view to the conservation of wildlife resources. This consultation shall follow the procedures in the Fish and Wildlife Coordination Act (Pub. L. 85-624) and shall occur during the environmental review of an action.

(6) If an EPA action may affect threatened or endangered species defined under section 4 of the Endangered Species Act of 1973 (Pub. L. 93-205), the responsible official shall consult with the Secretary of the Interior or the Secretary of Commerce, according to the procedures in Section 7 of that act.

(7) Requests for consultation and the results of consultation shall be documented in writing. In all cases where consultation has occurred, the agencies consulted should receive copies of either the notice of intent and EIS or the negative declaration and environmental appraisal prepared on the proposed action. If a decision has already been made to prepare an EIS on a project and wetlands, floodplains, coastal zones, wild and scenic rivers, fish or wildlife may be affected, the required consultation may be deferred until the preparation of the draft EIS.

§ ***6.216 Availability of documents.***

(a) EPA will print copies of draft and final EIS's for agency and public distribution. A nominal fee may be charged for copies requested by the public.

(b) When EPA no longer has copies of an EIS to distribute, copies shall be made available for public inspection at regional and headquarters Offices of Public Affairs. Interested persons also should be advised of the availability (at cost) of the EIS from the Environmental Law Institute, 1356 Connecticut Avenue NW., Washington, D.C. 20036.

(c) Lists of EIS's prepared or under preparation and lists of negative declarations prepared will be available at both the regional and headquarters Offices of Public Affairs.

SUBPART C—CONTENT OF ENVIRONMENTAL IMPACT STATEMENTS

§ ***6.300 Cover sheet.***

The cover sheet shall indicate the type of EIS (draft or final), the official project name and number, the responsible EPA office, the date, and the signature of the responsible official. The format is shown in Exhibit 5.

§ ***6.302 Summary Sheet.***

The summary sheet shall conform to the format in Exhibit 6, based on Appendix I of the August 1, 1973, CEQ Guidelines, or the latest revision of the CEQ Guidelines.

§ ***6.304 Body of EIS.***

The body of the EIS shall identify, develop, and analyze the pertinent issues discussed in the seven sections below; each section need not be a separate chapter. This analysis should include, but not be limited to, consideration of the impacts of the proposed project on the environmental areas listed in Appendix A which are relevant to the project. The EIS shall serve as a means for the responsible official and the public to assess the environmental impacts of a proposed EPA action, rather than as a justification for decisions already made. It shall be prepared using a systematic, interdisciplinary approach and shall incorporate all relevant analytical disciplines to provide meaningful and factual data, information, and analyses. The presentation of data should be clear and concise, yet include all facts necessary to permit independent evaluation and appraisal of the beneficial and adverse environmental effects of alternative actions. The amount of detail provided should be commensurate with the extent and expected impact of the action and the amount of information required at the particular level of decision making. To the extent possible, an EIS shall not be drafted in a style which requires extensive scientific or technical expertise to comprehend and evaluate the environmental impact of a proposed EPA action.

(a) *Background and description of the proposed action.* The EIS shall describe the recommended or proposed action, its purpose, where it is located, and its time setting. When a decision has been made not to favor an alternative until public comments on a proposed action have been received, the draft EIS may treat all feasible alternatives at similar levels of detail; the final EIS should focus on the alternative the draft EIS and public comments indicate is the best. The relationship of the proposed action to other projects and proposals directly affected by or stemming from it shall be discussed, including not only other EPA activities, but also those of other governmental and private organizations. Land use patterns and population trends in the project area and the assumptions on which they are based also shall be included. Available maps, photos, and artist's sketches should be incorporated when they help depict the environmental setting.

(b) *Alternatives to the proposed action.* The EIS shall develop, describe, and objectively weigh feasible alternatives to any proposed action, including the options of taking no action or postponing action. The analysis should be detailed enough to show EPA's comparative evaluation of the environmental impacts, commitments of resources, costs, and risks of the proposed action and each feasible alternative. For projects involving construction, alternative sites must be analyzed in enough detail for reviewers independently to judge the relative desirability of each site. For alternatives involving regionalization, the effects of varying degrees of regionalization should be addressed. If a cost-benefit analysis is prepared, it should be appended to the EIS and referenced in the body of the EIS. In addition, the reasons why the proposed action is believed by EPA to be the best course of action shall be explained.

(c) *Environmental impacts of the proposed action.* (1) The positive and negative effects of the proposed action as it affects both the national and international environment should be assessed. The attention given to different environmental factors will vary according to the nature, scale, and location of proposed actions. Primary attention should be given to those factors most evidently affected by the proposed action. The factors shall include, where appropriate, the proposed action's effect on the resource base, including land, water quality and quantity, air quality, public services and energy supply. The EIS shall describe primary and secondary environmental impacts, both beneficial and adverse, anticipated from the action. The description shall include short term and long term impacts on both the natural and human environments.

(2) Primary impacts are those that can be attributed directly to the proposed action. If the action is a field experiment, materials introduced into the environment which might damage certain plant communities or wildlife species would be a primary impact. If the action involves construction of a facility, such as a sewage treatment works, an office building or a laboratory, the primary impacts of the action would include the environmental impacts related to construction and operation of the facility and land use changes at the facility site.

(3) Secondary impacts are indirect or induced changes. If the action involves construction of a facility, the secondary impacts would include the environmental impacts related to:

(i) induced changes in the pattern of land use, population density and related effects on air and water quality or other national resources;

(ii) increased growth at a faster rate than planned for or above the total level planned by the existing community.

(4) A discussion of how socioeconomic activities and land use changes related to the proposed action conform or conflict with the goals and objectives of approved or proposed Federal, regional, State, and local land use plans, policies and controls for the project area should be included in the EIS. If a conflict appears to be unresolved in the EIS, EPA should explain why it has decided to proceed without full reconciliation.

(d) *Adverse impacts which cannot be avoided should the proposal be implemented and steps to minimize harm to the environment.* The EIS shall describe the kinds and magnitudes and adverse impacts which cannot be reduced in severity or which can be reduced to an acceptable level but not eliminated. These may include water or air pollution, undesirable land use patterns, damage to fish and wildlife habitats, urban congestion, threats to human health, or other consequences adverse to the environmental goals in section 101(b) of NEPA. Protective and mitigative measures to be taken as part of the proposed action shall be identified. These measures to reduce or compensate for any environmentally detrimental aspect of the proposed action may include those of EPA, its contractors and grantees and others involved in the action.

(e) *Relationship between local short term uses of man's environment and the maintenance and enhancement of long term productivity.* The EIS shall describe the extent to which the proposed action involves tradeoffs between short term environmental gains at the expense of long term gains or vice-versa and the extent to which the proposed action forecloses future options. Special attention shall be given to effects which narrow the range of future uses of land and water resources or pose long term risks to health or safety. Consideration should be given to windfall gains or significant decreases in current property values from implementing the proposed action. In addition, the reasons the proposed action is believed by EPA to be justified now, rather than reserving a long term option for other alternatives, including no action, shall be explained.

(f) *Irreversible and irretrievable commitments of resources to the proposed action should it be implemented.* The EIS shall describe the extent to which the proposed action requires commitment of construction materials, person-hours, and funds to design and implement the project, as well as curtails the range of future uses of land and water resources. For example, induced growth in undeveloped areas may curtail alternative uses of that land. Also, irreversible environmental damage can result from equipment malfunctions or industrial accidents at the project site. Therefore, the need for any irretrievable and significant commitments of resources shall be explained fully.

(g) *Problems and objections raised by other Federal, State, and local agencies and by interested persons in the review process.* Final EIS's (and draft EIS's if appropriate) shall summarize the comments and suggestions made by reviewing organizations and shall describe the disposition of issues raised, e.g., revisions to the proposed action to mitigate anticipated impacts or objections. In particular, the EIS shall address the major issues raised when the EPA position differs from most recommendations and explain the factors of overriding importance overruling the adoption of suggestions. Reviewer's statements should be set forth in a "comment" and discussed in a "response." In addition, the source of all comments should be clearly identified, and copies of the comments should be attached to the final EIS. Summaries of comments should be attached when a response has been exceptionally long or the same comments were received from many reviewers.

§ ***6.306 Documentation.***

All books, research reports, field study reports, correspondence, and other documents which provided the data base for evaluating the proposed action and alternatives discussed in the EIS shall be used as references in the body of the EIS and shall be included in a bibliography attached to the EIS.

SUBPART D—EPA PUBLIC HEARINGS ON EIS's

§ ***6.400 General.***

While EPA is not required by statute to hold public hearings on EIS's, the responsible official should hold a public hearing on a draft EIS whenever a hearing may facilitate the resolution of conflicts or significant public controversy. This hearing may be in addition to public hearings held on facilities plans or section 209 plans. The responsible official may take special measures to involve interested persons through personal contact.

§ ***6.402 Public hearing process.***

(a) When public hearings are to be held, EPA shall inform the public of the hearing, for example, with a notice in the draft EIS. The notice should follow the summary sheet at the beginning of the EIS. The draft EIS shall be available for public review at least thirty (30) days before the public hearing. Public notice shall be given at least fifteen (15) working days before the public hearing and shall include:

(1) Publication of a public notice in a newspaper which covers the project area, identifying the project, announcing the date, time, and place of the hearing, and announcing the availability of detailed information on the proposed action for public inspection at one or more locations in the area in which the project will be located. "Detailed information" shall include a copy of the project application and the draft EIS.

(2) Notification of appropriate State and local agencies and appropriate State, regional, and metropolitan clearinghouses.

(3) Notification of interested persons.

(b) A written record of the hearing shall be made. A stenographer may be used to record the hearing. As a miminum, the record shall contain a list of witnesses with the text of each presentation. A summary of the record, including the issues raised, conflicts resolved and unresolved, and any other significant portions of the record, shall be appended to the final EIS.

(c) When a public hearing has been held by another Federal, State, or local agency on an EPA action, additional hearings are not necessary. The responsible official shall decide if additional hearings are needed.

(d) When a program office is the originating office, the appropriate regional office will provide assistance to the originating office in holding any public hearing if assistance is requested.

SUBPART E—GUIDELINES FOR COMPLIANCE WITH NEPA IN THE TITLE II WASTEWATER TREATMENT WORKS CONSTRUCTION GRANTS PROGRAM AND THE AREAWIDE WASTE TREATMENT MANAGEMENT PLANNING PROGRAM

§ ***6.500 Purpose.***

This subpart amplifies the general EPA policies and procedures described in Subparts A through D with detailed procedures for compliance with NEPA in the wastewater treatment works construction grants program and the areawide waste treatment management planning program.

§ *6.502 Definitions.*

(a) "Step 1 grant." A grant for preparation of a facilities plan as described in 40 CFR 35. 930-1.

(b) "Step 2 grant." A grant for preparation of construction drawings and specifications as described in 40 CFR 35.930-1.

(c) "Step 3 grant." A grant for fabrication and building of a publicly owned treatment works as described in 40 CFR 35.930-1.

§ *6.504 Applicability.*

(a) *Administrative actions covered.* This subpart applies to the administrative actions listed below:

(1) Approval of all section 208 plans according to procedures in 40 CFR 35.1067-2;

(2) Approval of facilities plans except those listed in paragraph (a) (5) of this section;

(3) Award of step 2 and step 3 grants, if an approved facilities plan was not required;

(4) Award of a step 2 or step 3 grant when either the project or its impact has changed significantly from that described in the approved facilities plan, except when the situation in paragraph (a) (5) of this section exists;

(5) Consultation during the NEPA review process. When there are overriding considerations of cost or impaired program effectiveness, the Regional Administrator may award a step 2 or a step 3 grant for a discrete segment of the project plans or construction before the NEPA review is completed if this project segment is noncontroversial. The remaining portion of the project shall be evaluated to determine if an EIS is required. In applying the criteria for this determination, the entire project shall be considered, including those parts permitted to proceed. In no case may these types of step 2 or step 3 grants be awarded unless both the Office of Federal Activities and CEQ have been consulted, a negative declaration has been issued on the segments permitted to proceed, and the grant award contains a specific agreement prohibiting action on the segment of planning or construction for which the NEPA review is not complete. Examples of consultation during the NEPA review process are: award of a step 2 grant for preparation of plans and specifications for a large treatment plant, when the only unresolved NEPA issue is where to locate the sludge disposal site; or award of a step 3 grant for site clearance for a large treatment plant, when the unresolved NEPA issue is whether sludge from the plant should be incinerated at the site or disposed of elsewhere by other means.

(b) *Administrative actions excluded.* The actions listed below are not subject to the requirements of this part:

(1) Approval of State priority lists;

(2) Award of a step 1 grant;

(3) Award of a section 208 planning grant;

(4) Award of a step 2 or step 3 grant when no significant changes in the facilities plan have occurred;

(5) Approval of issuing an invitation for bid or awarding a construction contract;

(6) Actual physical commencement of building or fabrication;

(7) Award of a section 206 grant for reimbursement;

(8) Award of grant increases whenever § 6.504(a)(4) does not apply;

(9) Awards of training assistance under FWPCA, as amended, section 109(b).

(c) *Retroactive application.* The new criteria in § 6.510 of this subpart do not apply to step 2 or step 3 grants awarded before July 1, 1975. However, the Regional Administrator may apply the new criteria of this subpart when he considers it appropriate. Any negative declarations issued before the effective date of this regulation shall remain in effect.

§ *6.506* *Completion of NEPA procedures before start of administrative actions.*
See § 6.108 and § 6.504.

§ *6.510* *Criteria for preparation of environmental impact statements.*
In addition to considering the criteria in § 6.200, the Regional Administrator shall assure that an EIS will be prepared on a treatment works facilities plan, 208 plan, or other appropriate water quality management plan when:

(a) The treatment works or plan will induce significant changes (either absolute changes or increases in the rate of change) in industrial, commercial, agricultural, or residential land use concentrations or distributions. Factors that should be considered in determining if these changes are significant include but are not limited to: the vacant land subject to increased development pressure as a result of the treatment works; the increases in population which may be induced; the faster rate of change of population; changes in population density; the potential for overloading sewage treatment works; the extent to which landowners may benefit from the areas subject to increased development; the nature of land use regulations in the affected area and their potential effects on development; and deleterious changes in the availability or demand for energy.

(b) Any major part of the treatment works will be located on productive wetlands or will have significant adverse effects on wetlands, including secondary effects.

(c) Any major part of the treatment works will be located on or significantly affect the habitat of wildlife on the Department of Interior's threatened and endangered species lists.

(d) Implementation of the treatment works or plan may directly cause or induce changes that significantly;

(1) Displace population;

(2) Deface an existing residential area; or

(3) Adversely affect significant amounts of prime agricultural land or agricultural operations on this land.

(e) The treatment works or plan will have significant adverse effects on parklands, other public lands or areas of recognized scenic, recreational, archaeological, or historic value.

(f) The works or plan may directly or through induced development have a significant adverse effect upon local ambient air quality, local ambient noise levels, surface or groundwater quantity or quality, fish, wildlife, and their natural habitats.

(g) The treated effluent is being discharged into a body of water where the present classification is too lenient or is being challenged as too low to protect present or recent uses, and the effluent will not be of sufficient quality to meet the requirements of these uses.

§ *6.512* *Procedures for implementing NEPA.*
(a) *Environmental assessment.* An adequate environmental assessment must be an integral, though identifiable, part of any facilities of section 208 plan submitted to EPA. (See § 6.202 for a general description.) The information in the facilities plan, particularly the environmental assessment, will provide the substance of an EIS and shall be submitted by the applicant. The analyses that constitute an adequate environmental assessment shall include:

(1) *Description of the existing environment without the project.* This shall include for the delineated planning area a description of the present environmental conditions

relevant to the analysis of alternatives or determinations of the environmental impacts of the proposed action. The description shall include, but not be limited to, discussions of whichever areas are applicable to a particular study: surface and groundwater quality; water supply and use; general hydrology; air quality; noise levels, energy production and consumption; land use trends; population projections, wetlands, floodplains, coastal zones and other environmentally sensitive areas; historic and archaeological sites; other related Federal or State projects in the area; and plant and animal communities which may be affected, especially those containing threatened or endangered species.

(2) *Description of the future environment without the project.* The future environmental conditions with the no project alternative shall be forecast, covering the same areas listed in § 6.512 (a) (1).

(3) *Documentation.* Sources of information used to describe the existing environment and to assess future environmental impacts should be documented. These sources should include regional, State and Federal agencies with responsibility or interest in the types of impacts listed in § 6.512 (a) (1). In particular, the following agencies should be consulted:

(i) Local and regional land use planning agencies for assessments of land use trends and population projections, especially those affecting size, timing, and location of facilities, and planning activities funded under section 701 of the Housing and Community Development Act of 1974 (Pub. L. 93-383);

(ii) The HUD Regional Office if a project involves a flood risk area identified under the Flood Disaster Protection Act of 1973 (Pub. L. 93-234);

(iii) The State coastal zone management agency, if a coastal zone is affected;

(iv) The Secretary of the Interior or Secretary of Agriculture, if a wild and scenic river is affected;

(v) The Secretary of the Interior or Secretary of Commerce, if a threatened or endangered species is affected;

(vi) The Fish and Wildlife Service (Department of Interior), the Department of Commerce, and the U.S. Army Corps of Engineers, if a wetland is affected.

(4) *Evaluation of alternatives.* This discussion shall include a comparative analysis of feasible options and a systematic development of wastewater treatment alternatives. The alternatives shall be screened with respect to capital and operating costs; significant primary and secondary environmental effects; physical, legal or institutional constraints; and whether or not they meet regulatory requirements. Special attention should be given to long term impacts, irreversible impacts and induced impacts such as development. The reasons for rejecting any alternatives shall be presented in addition to any significant environmental benefits precluded by rejection of an alternative. The analysis should consider, when relevant to the project:

(i) Flow and waste reduction measures, including infiltration/inflow reduction;

(ii) Alternative locations, capacities, and construction phasing of facilities;

(iii) Alternative waste management techniques, including treatment and discharge, wastewater reuse, and land application;

(iv) Alternative methods for disposal of sludge and other residual waste, including process options and final disposal options;

(v) Improving effluent quality through more efficient operation and maintenance;

(vi) For assessments associated with section 208 plans, the analysis of options shall include in addition;

(A) Land use and other regulatory controls, fiscal controls, non-point source controls, and institutional arrangements; and

(B) Land management practices.

(5) *Environmental impacts of the proposed action.* Primary and secondary impacts of the proposed action shall be described, giving special attention to unavoidable impacts, steps to mitigate adverse impacts, any irreversible or irretrievable commitments of resources to the project, and the relationship between local short term uses of the environment and the maintenance and enhancement of long term productivity. See § 6.304 (c), (d), (e), and (f) for an explanation of these terms and examples. The significance of land use impacts shall be evaluated, based on current population of the planning area; design year population for the service area; percentage of the service area currently vacant; and plans for staging facilities. Special attention should be given to induced changes in population patterns and growth, particularly if a project involves some degree of regionalization. In addition to these items, the Regional Administrator may require that other analyses and data, which he determines are needed to comply with NEPA, be included with the facilities or section 208 plan. Such requirements should be discussed during preapplication conferences. The Regional Administrator also may require submission of supplementary information either before or after a step 2 grant or before a step 3 grant award if he determines it is needed for compliance with NEPA. Requests for supplementary information shall be made in writing.

(6) *Steps to minimize adverse effects.* This section shall describe structural and nonstructural measures, if any, in the facilities plan to mitigate or eliminate significant adverse effects on the human and natural environments. Structural provisions include changes in facility design, size, and location; nonstructural provisions include staging facilities as well as developing and enforcing land use regulations and environmentally protective regulations.

(b) *Public hearing.* The applicant shall hold at least one public hearing before a facilities plan is adopted, unless waived by the Regional Administrator before completion of the facilities plan according to § 35.917-5 of the Title II construction grants regulations. Hearings should be held on section 208 plans. A copy of the environmental assessment should be available for public review before the hearing and at the hearing, since these hearings provide an opportunity to accept public input on the environmental issues associated with the facilities plan or the 203 water quality management strategy. In addition, a Regional Administrator may elect to hold an EPA hearing if environmental issues remain unresolved. EPA hearings shall be held according to procedures in § 6.402.

(c) *Environmental review.* An environmental review of a facilities plan or section 208 plan shall be conducted according to the procedures in § 6.204 and applying the criteria of § 6.510. If deficiencies exist in the environmental assessment, they shall be identified in writing by the Regional Administrator and must be corrected before the plan can be approved.

(d) *Additional procedures.* (1) Historic and archaeological sites. If a facilities or section 208 plan may affect properties with historic, architectural, archaeological, or cultural value which are listed in or eligible for listing in the National Register of Historic Places or may cause irreparable loss or destruction of significant scientific, prehistoric, historic or archaeological data, the applicant shall follow the procedures in § 6.214(a).

(2) If the facilities or section 208 plan may affect wetlands, floodplains, coastal zones, wild and scenic rivers, fish, or wildlife, the Regional Administrator shall follow the appropriate procedures described in § 6.214(b).

(e) *Notice of intent.* The notice of intent on a facilities plan or section 208 plan shall be issued according to § 6.206.

(f) *Scope of EIS.* It is the Regional Administrator's responsibility to determine the scope of the EIS. He should determine if an EIS should be prepared on a facilities plan(s) or section 208 plan and which environmental areas should be discussed in greatest detail in the EIS. Once an EIS has been prepared for the designated section 208 area, another need not be prepared unless the significant impacts of individual facilities or other plan elements were not adequately treated in the EIS. The Regional Administrator should document his decision not to prepare an EIS on individual facilities.

(g) *Negative declaration.* A negative declaration on a facilities plan or section 208 plan shall be prepared according to § 6.212. Once a negative declaration and environmental appraisal have been prepared for the facilities plan for a certain area, grant awards may proceed without preparation of additional negative declarations, unless the project has changed significantly from that described in the facilities plan.

§ *6.514 Content of environmental impact statements.*

EIS's for treatment works or plans shall be prepared according to § 6.304.

SUBPART F—GUIDELINES FOR COMPLIANCE WITH NEPA IN RESEARCH AND DEVELOPMENT PROGRAMS AND ACTIVITIES

§ *6.600 Purpose.*

This subpart amplifies the general EPA policies and procedures described in Subparts A through D by providing procedures for compliance with NEPA on actions undertaken by the Office of Research and Development (ORD).

§ *6.602 Definitions.*

(a) "Work plan." A document which defines and schedules all projects required to fulfill the objectives of the program plan.

(b) "Program plan." An overall planning document for a major research area which describes one or more research objectives, including outputs and target completion dates, as well as person-year and dollar resources.

(c) "Appropriate program official." The official at each decision level within ORD to whom the Assistant Administrator delegates responsibility for NEPA compliance.

(d) "Exemption certification." A certified statement delineating those actions specifically exempted from NEPA compliance by existing legislation.

§ *6.604 Applicability.*

The requirements of this subpart are applicable to administrative actions undertaken to approve program plans, work plans, and projects, except those plans and projects excluded by existing legislation. However, no administrative actions are excluded from the additional procedures in § 6.214 of this part concerning historic sites, wetlands, coastal zones, wild and scenic rivers, floodplains, or fish and wildlife.

§ *6.608 Criteria for determining when to prepare EIS's.*

(a) An EIS shall be prepared by ORD when any of the criteria in § 6.200 apply or when:

(1) The action will have significant adverse impacts on public parks, wetlands, floodplains, coastal zones, wildlife habitats, or areas of recognized scenic or recreational value.

(2) The action will significantly deface an existing residential area.

(3) The action may directly or through induced development have a significant adverse effect upon local ambient air quality, local ambient noise levels, surface or groundwater quality; and fish, wildlife, or their natural habitats.

(4) The treated effluent is being discharged into a body of water where the present classification is being challenged as too low to protect present or recent uses, and the effluent will not be of sufficient quality to meet the requirements of these uses.

(5) The project consists of field tests involving the introduction of significant quantities of: toxic or polluting agricultural chemicals, animal wastes, pesticides, radioactive materials, or other hazardous substances into the environment by ORD, its grantees, or its contractors.

(6) The action may involve the introduction of species or subspecies not indigenous to an area.

(7) There is a high probability of an action ultimately being implemented on a large scale, and this implementation may result in significant environmental impacts.

(8) The project involves commitment to a new technology which is significant and may restrict future viable alternatives.

(b) An EIS will not usually be needed when:

(1) The project is conducted completely within a laboratory or other facility, and external environmental effects have been minimized by methods for disposal of laboratory wastes and safeguards to prevent hazardous materials entering the environment accidentally; or

(2) The project is a relatively small experiment or investigation that is part of a non-Federally funded activity of the private sector, and it makes no signficant new or additional contribution to existing pollution.

§ ***6.610 Procedures for compliance with NEPA.***

EIS related activities for compliance with NEPA will be integrated into the decision levels of ORD's research planning system to assure managerial control. This control includes those administrative actions which do not come under the applicability of this subpart by assuring that they are made the subject of an exemption certification and filed with the Office of Public Affairs (OPA). ORD's internal procedures provide details for NEPA compliance.

(a) *Environmental assessment.* (1) Environmental assessments shall be submitted with all grant applications and all unsolicited contract proposals. The assessment shall contain the same information required for EIS's in § 6.304. Copies of § 6.304 (or more detailed guidance when available) and a notice of the requirement for assessment shall be included in all grant application kits and attached to letters concerning the submission of unsolicited proposals.

(2) In the case of competitive contracts, assessments need not be submitted by potential contractors since the NEPA procedures must be completed before a request for proposal (RFP) is issued. If there is a question concerning the need for an assessment, the potential contractor should contact the official responsible for the contract.

(b) *Environmental review.* (1) At the start of the planning year, an environmental review will be performed for each program plan with its supporting substructures (work plans and projects) before incorporating them into the ORD program planning system, unless they are excluded from review by existing legislation. This review is an evaluation of the potentially adverse environmental effects of the efforts required by the program plan. The criteria in § 6.608 shall be used in conducting this review. Each program plan with its supporting substructures which does not have significant adverse impacts may be dismissed from further current year environmental considerations with a single negative declaration. Any supporting substructures of a program plan which cannot be dismissed with the parent plan shall be reviewed at the appropriate subordinate levels of the planning system for NEPA compliance.

(i) All continuing program plans and supporting substructures, including those previously dismissed from consideration, will be reevaluated annually for NEPA compliance. An environmental review will coincide with the annual planning cycle and whenever a major redirection of a parent plan is undertaken. All NEPA-associated documents will be updated as appropriate.

(ii) All approved program plans and supporting substructures, less budgetary data, will be filed in the OPA with a notice of intent or negative declaration and environmental appraisal.

(iii) Later plans and/or projects, added to fulfill the mission objectives but not identified at the time the program plans were approved, will be subjected to the same NEPA requirements for environmental assessments and/or reviews.

(iv) Those projects subjected to environmental assessments as outlined in paragraph (a) of this section and not exempt under existing legislation also shall undergo an environmental review before work begins.

(c) *Notice of intent and EIS.*

(1) If the reviews conducted according to paragraph (b) of this section reveal a potentially significant adverse effect on the environment and the adverse impact cannot be eliminated by replanning, the appropriate program official shall, after making sure the project is to be funded, issue a notice of intent according to § 6.206, and through proper organizational channels, shall request the Regional Administrator to assist him in the preparation and distribution of the EIS.

(2) As soon as possible after release of the notice of intent, the appropriate program official shall prepare a draft EIS using the criteria in Subpart B, § 6.208 and Subpart C. Through proper organizational channels, he shall request the Regional Administrator to assist him in the preparation and distribution of the the draft EIS.

(3) The appropriate program official shall prepare final EIS's according to criteria in Subpart B, § 6.210 and Subpart C.

(4) All draft and final EIS's shall be sent through the proper organizational channels to the Assistant Administrator for ORD for approval. The approved statements then will be distributed according to the procedures in Appendix C.

(d) *Negative declaration and environmental impact appraisal.* If an environmental review conducted according to paragraph (b) of this section reveals that proposed actions will not have significant adverse environmental impacts, the appropriate program official shall prepare a negative declaration and environmental impact appraisal according to Subpart B, § 6.212. Upon assurance that the program will be funded, the appropriate program official shall distribute the negative declaration as described in § 6.212 and make copies of the negative declaration and appraisal available in the OPA.

(e) *Project start.* As required by § 6.108, a contract or grant shall not be awarded for an extramural project, nor for continuation of what was previously an intramural project, until at least fifteen (15) working days after a negative declaration has been issued or thirty (30) days after forwarding the final EIS to the Council on Environmental Quality.

SUBPART G—GUIDELINES FOR COMPLIANCE WITH NEPA IN SOLID WASTE MANAGEMENT ACTIVITIES

§ *6.700* ***Purpose.***

This subpart amplifies the general policies and procedures described in Subparts A through D by providing additional procedures for compliance with NEPA on actions undertaken by the Office of Solid Waste Management Programs (OSWMP).

§ ***6.702 Criteria for the preparation of environmental assessments of EIS's.***

(a) *Assessment preparation criteria.* An environmental assessment need not be submitted with all grant applications and contract proposals. Studies and investigations do not require assessments. The following sections describe when an assessment is or is not required for other actions:

(1) *Grants.* (i) *Demonstration projects.* Environmental assessments must be submitted with all applications for demonstration grants that will involve construction, land use (temporary or permanent), transport, sea disposal, any discharges into the air or water, or any other activity having any direct or indirect effects on the environment external to the facility in which the work will be conducted. Preapplication proposals for these grants will not require environmental assessments.

(ii) *Training.* Grant applications for training of personnel will not require assessments.

(iii) *Plans.* Grant applications for the development of comprehensive State, interstate, or local solid waste management plans will not require environmental assessments. A detailed analysis of environmental problems and effects should be part of the planning process, however.

(2) *Contracts.* (i) *Sole-source contract proposals.* Before a sole-source contract can be awarded, an environmental assessment must be submitted with a bid proposal for a contract which will involve construction, land use (temporary or permanent), sea disposal, any discharges into the air or water, or any other activity that will directly or indirectly affect the environment external to the facility in which the work will be performed.

(ii) *Competitive contract proposals.* Assessments generally will not be required on competitive contract proposals.

(b) *EIS preparation criteria.* The responsible official shall conduct an environmental review on those OSWMP projects on which an assessment is required or which may have effects on the environment external to the facility in which the work will be performed. The criteria in § 6.200 shall be utilized in determining whether an EIS need be prepared.

§ ***6.704 Procedures for compliance with NEPA.***

(a) *Environmental assessment.* (1) Environmental assessments shall be submitted to EPA according to procedures in § 6.702. If there is a question concerning the need for an assessment, the potential contractor or grantee should consult with the appropriate project officer for the grant or contract.

(2) The assessment shall contain the same sections specified for EIS's in § 6.304. Copies of § 6.304 (or more detailed guidance when available) and a notice alerting potential grantees and contractors of the assessment requirements in § 6.702 shall be included in all grant application kits, attached to letters concerning the submission of unsolicited proposals, and included with all RFP's.

(b) *Environmental review.* An environmental review will be conducted on all projects which require assessments or which will affect the environment external to the facility in which the work will be performed. This review must be conducted before a grant or contract award is made on an extramural project or before an intramural project begins. The guidelines in § 6.200 will be used to determine if the project will have any significant environmental effects. This review will include an evaluation of the assessment by both the responsible official and the appropriate Regional Administrator. The Regional Administrator's comments will include his recommendations on the need for an EIS. No detailed review or documentation is required on projects for which assessments are not required and which will not affect the environment external to a facility.

(c) *Notice of intent and EIS.* If any of the criteria in § 6.200 apply, the responsible official will assure that a notice of intent and a draft EIS are prepared. The responsible official may request the appropriate Regional Administrator to assist him in the distribution of the NEPA-associated documents. Distribution procedures are listed in Appendix C.

(d) *Negative declaration and environmental impact appraisal.* If the environmental review indicated no significant environmental impacts, the responsible official will assure that a negative declaration and environmental appraisal are prepared. These documents need not be prepared for projects not requiring an environmental review.

(e) The EIS process for the Office of Solid Waste Management Programs is shown graphically in Exhibit 7.

SUBPART H—GUIDELINES FOR COMPLIANCE WITH NEPA IN CONSTRUCTION OF SPECIAL PURPOSE FACILITIES AND FACILITY RENOVATIONS

§ *6.800* ***Purpose.***

This subpart amplifies general EPA policies and procedures described in Subparts A through D by providing detailed procedures for the preparation of EIS's on construction and renovation of special purpose facilities.

§ *6.802* ***Definitions.***

(a) "Special purpose facility." A building or space, including land incidental to its use, which is wholly or predominantly utilized for the special purpose of an agency and not generally suitable for other uses, as determined by the General Services Administration.

(b) "Program of requirements." A comprehensive document (booklet) describing program activities to be accomplished in the new special purpose facility or improvement. It includes architectural, mechanical, structural, and space requirements.

(c) "Scope of work." A document similar in content to the program of requirements but substantially abbreviated. It is usually prepared for small-scale projects.

§ *6.804* ***Applicability.***

(a) *Actions covered.* These guidelines apply to all new special purpose facility construction, activities related to this construction (e.g., site acquisition and clearing), and any improvements or modifications to facilities having potential environmental effects external to the facility, including new construction and improvements undertaken and funded by the Facilities Management Branch, Facilities and Support Services Division, Office of Administration; by a regional office; or by a National Environmental Research Center.

(b) *Actions excluded.* This subpart does not apply to those activities of the Facilities Management Branch, Facilities and Support Services Division, for which the branch does not have full fiscal responsibility for the entire project. This includes pilot plant construction, land acquisition, site clearing, and access road construction where the Facilities Management Branch's activity is only supporting a project financed by a program office. Responsibility for considering the environmental impacts of such projects rests with the office managing and funding the entire project. Other subparts of this regulation apply depending on the nature of the project.

§ *6.808* ***Criteria for the preparation of environmental assessments and EIS's.***

(a) *Assessment preparation criteria.* The responsible official shall request an environmental assessment from a construction contractor or consulting architect/engineer employed by EPA if he is involved in the planning, construction or modification of special purpose facilities when his activities have potential environmental effects external to the facility. Such modifications include but are not limited to: facility additions, changes in central heating systems or wastewater treatment systems, and land clearing for access roads and parking lots.

(b) *EIS preparation criteria.* The responsible official shall conduct an environmental review of all actions involving construction of special purpose facilities and improvements to these facilities. The guidelines in § 6.200 shall be used to determine whether an EIS shall be prepared.

§ *6.810* *Procedures for compliance with NEPA.*

(a) *Environmental review and assessment.* (1) An environmental review shall be conducted when the program of requirements or scope of work has been completed for the construction, improvement, or modification of special purpose facilities. For special purpose facility construction, the Chief, Facilities Management Branch, shall request the assistance of the appropriate program office and Regional Administrator in the review. For modifications and improvements, the appropriate responsible official shall request assistance in making the review from other cognizant EPA offices.

(2) Any assessments requested shall contain the same sections listed for EIS's in § 6.304. Contractors and consultants shall be notified in contractual documents when an assessment must be prepared.

(b) *Notice of intent, EIS, and negative declaration.* The responsible official shall decide at the completion of the environmental review whether there may be any significant environmental impacts. If there could be significant environmental impacts, a notice of intent and an EIS shall be prepared according to the procedures in § 6.206. If there may not be any significant environmental impacts, a negative declaration and environmental impact appraisal shall be prepared according to the procedures in § 6.212.

(c) *Project start.* As required by § 6.108, a contract shall not be awarded or construction-related activities begun until at least fifteen (15) working days after release of a negative declaration, or until thirty (30) days after forwarding the final EIS to the Council on Environmental Quality.

Seven exhibits follow.

EXHIBIT 1

NOTICE OF INTENT TRANSMITTAL MEMORANDUM SUGGESTED FORMAT

____
(Date)

ENVIRONMENTAL PROTECTION AGENCY

____
(Appropriate Office)

____
(Address, City, State, Zip Code)

To All Interested Government Agencies and Public Groups:

As required by guidelines for the preparation of environmental impact statements (EIS's), attached is a notice of intent to prepare an EIS for the proposed EPA action described below:

____
(Official Project Name and Number)

____
(City, State)

If your organization needs additional information or wishes to participate in the preparation of the draft EIS, please advise the (appropriate office, city, State).

Very truly yours,

____
(Appropriate EPA Official)

(List Federal, State, and local agencies to be solicited for comment.)
(List public action groups to be solicited for comment.)

NOTICE OF INTENT SUGGESTED FORMAT

NOTICE OF INTENT—ENVIRONMENTAL PROTECTION AGENCY

1. Project location:

   City____

   County______________________________________

   State_______________________________________
2. Proposed EPA action:

   ____

   ____
3. Issues involved:

   ____

   ____
4. Estimated project costs:

   Federal share (total)____________________________ $_______

   Contract $_______________ Grant $_______________ Other $_______

   Applicant share (if any):

   (Name)____________________________________ $_______

   Other (specify)_______________________________ $_______

   Total______________________________________ $_______
5. Period covered by project:

   Start date:______________________________________
   (Original date, if project covers more than one year)

   Dates of different project phases:________________________

   Approximate end date:______________________________
6. Estimated application filing date:________________________

EXHIBIT 2

PUBLIC NOTICE AND NEWS RELEASE SUGGESTED FORMAT

PUBLIC NOTICE

The Environmental Protection Agency (originating office) (will prepare, will not prepare, has prepared) a (draft, final) environmental impact statement on the following project:

(Official Project Name and Number)

(Purpose of Project)

(Project Location, City, County, State)

(Where EIS or negative declaration and environmental impact appraisal can be obtained)

This notice is to implement EPA's policy of encouraging public participation in the decision-making process on proposed EPA actions. Comments on this document may be submitted to (full address of originating office).

EXHIBIT 3

NEGATIVE DECLARATION SUGGESTED FORMAT

____
(Date)

ENVIRONMENTAL PROTECTION AGENCY

____
(Appropriate Office)

____
(Address, City, State, Zip Code)

To All Interested Government Agencies and Public Groups:

As required by guidelines for the preparation of environmental impact statements (EIS's), an environmental review has been performed on the proposed EPA action below:

____
(Official Project Name and Number)

____
(Potential Agency Financial Share)

____
(Project Location: City, County, State)

____
(Other Funds Included)

PROJECT DESCRIPTION, ORIGINATOR, AND PURPOSE

(Include a map of the project area and a brief narrative summarizing the growth the project will serve, the percent of vacant land the project will serve, major primary and secondary impacts of the project, and the purpose of the project.)

The review process did not indicate significant environmental impacts would result from the proposed action or significant adverse impacts have been eliminated by making changes in the project. Consequently, a preliminary decision not to prepare an EIS has been made.

This action is taken on the basis of a careful review of the engineering report, environmental impact assessment, and other supporting data, which are on file in the above office with the environmental impact appraisal and are available for public scrutiny upon request. Copies of the environmental impact appraisal will be sent at cost on your request.

Comments supporting or disagreeing with this decision may be submitted for consideration by EPA. After evaluating the comments received, the Agency will make a final decision; however, no administrative action will be taken on the project for at least fifteen (15) working days after release of the negative declaration.

Sincerely,

____
(Appropriate EPA Official)

EXHIBIT 4

ENVIRONMENTAL IMPACT APPRAISAL SUGGESTED FORMAT

A. Identify Project.

Name of Applicant: ______

Address: ______

Project Number: ______

B. Summarize Assessment.

1. Brief description of project: ______

2. Probable impact of the project on the environment: ______

3. Any probable adverse environmental effects which cannot be avoided: ______

4. Alternatives considered with evaluation of each: ______

5. Relationship between local short-term uses of man's environment and maintenance and enhancement of long-term productivity: ______

6. Steps to minimize harm to the environment: ______

7. Any irreversible and irretrievable commitment of resources: ______

8. Public objections to project, if any, and their resolution: ______

9. Agencies consulted about the project: ______

State representative's name: ______

Local representative's name: ______

Other: ______

C. Reasons for concluding there will be no significant impacts.

______

(Signature of appropriate official)

(Date)

EXHIBIT 5

COVER SHEET FORMAT FOR ENVIRONMENTAL IMPACT STATEMENTS

(Draft, Final)

ENVIRONMENTAL IMPACT STATEMENT

____

____

(Describe title of project plan and give identifying number)

Prepared by: ______________________________

(Responsible Agency Office)

Approved by: ______________________________

(Responsible Agency Official)

____

(Date)

EXHIBIT 6

SUMMARY SHEET FORMAT FOR ENVIRONMENTAL IMPACT STATEMENTS

(Check One)

( ) Draft
( ) Final

ENVIRONMENTAL PROTECTION AGENCY

____

(Responsible Agency Office)

1. Name of action. (Check one)
   ( ) Administrative action.
   ( ) Legislative action.
2. Brief description of action indicating what States (and counties) are particularly affected.
3. Summary of environmental impact and adverse environmental effects.
4. List alternatives considered.
5. a. (for draft statements) List all Federal, State, and local agencies and other comments which have been requested.
   b. (for final statements) List all Federal, State, and local agencies and other sources from which written comments have been received.
6. Dates draft statement and final statement made available to Council on Environmental Quality and public.

Exhibit 7

Flowchart For OSWMP

Procedures For Determining If An EIS Is Required On OSWMPT Projects

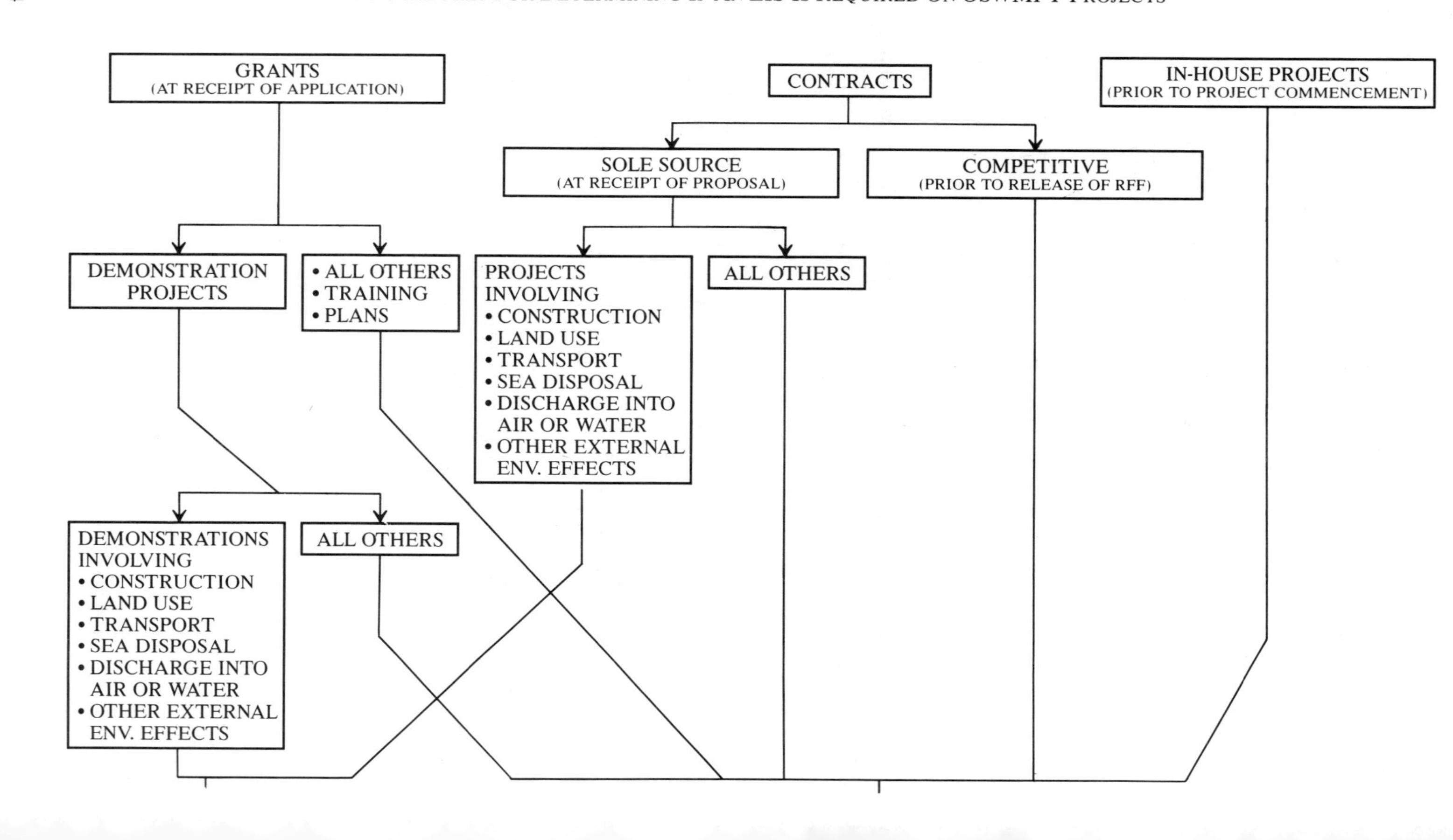

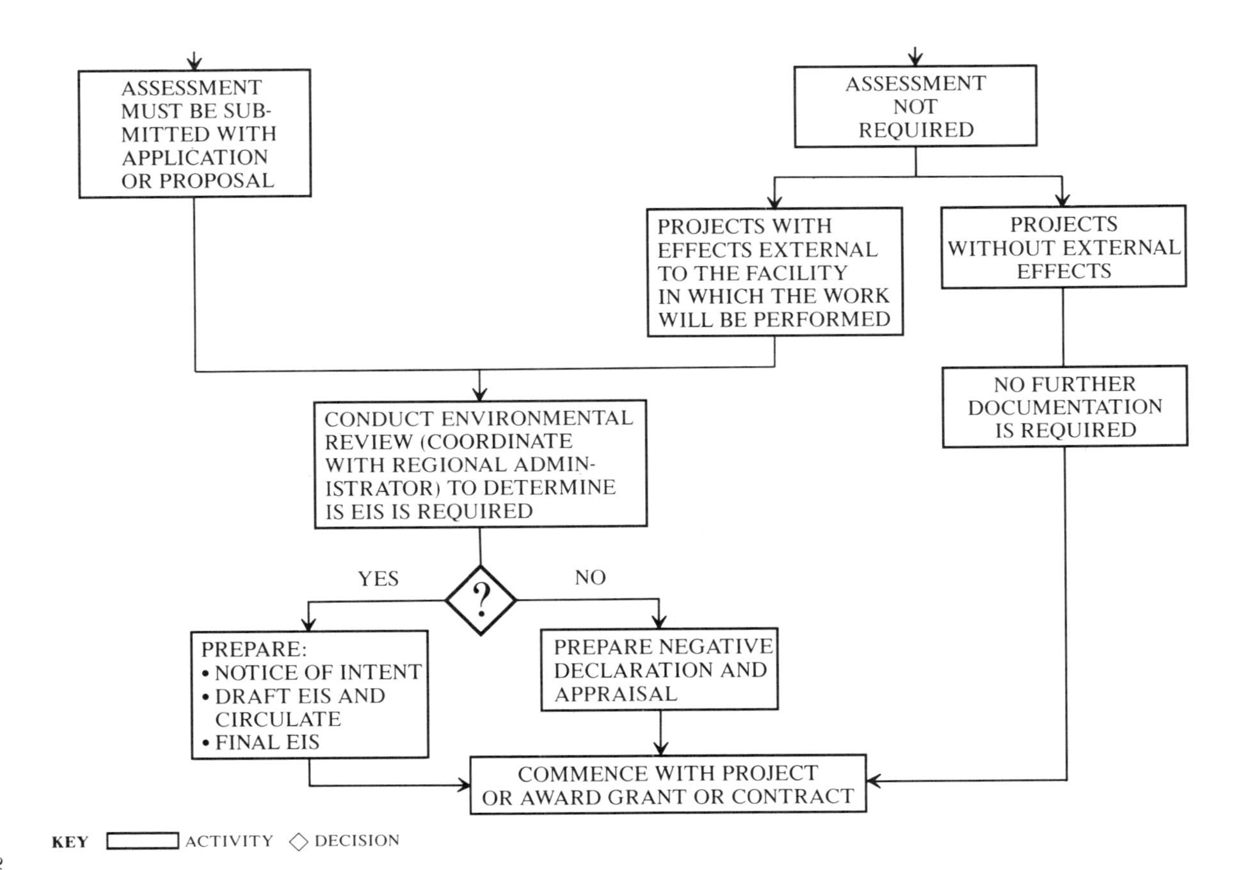
ASSESSMENT MUST BE SUBMITTED WITH APPLICATION OR PROPOSAL
ASSESSMENT NOT REQUIRED
PROJECTS WITH EFFECTS EXTERNAL TO THE FACILITY IN WHICH THE WORK WILL BE PERFORMED
PROJECTS WITHOUT EXTERNAL EFFECTS
NO FURTHER DOCUMENTATION IS REQUIRED
CONDUCT ENVIRONMENTAL REVIEW (COORDINATE WITH REGIONAL ADMINISTRATOR) TO DETERMINE IS EIS IS REQUIRED
YES
?
NO
PREPARE:
• NOTICE OF INTENT
• DRAFT EIS AND CIRCULATE
• FINAL EIS
PREPARE NEGATIVE DECLARATION AND APPRAISAL
COMMENCE WITH PROJECT OR AWARD GRANT OR CONTRACT
KEY
ACTIVITY
DECISION

APPENDIX A

CHECKLIST FOR ENVIRONMENTAL REVIEWS

Areas to be considered, when appropriate, during an environmental review include, but are not limited to, the items on this checklist, based on Appendix II of the CEQ guidelines for the preparation of environmental impact statements which appeared in the **Federal Register** August 1, 1973. The classification of items is not mandatory.

I. *Natural environment.* Consider the impacts of a proposed action on air quality, water supply and quality, soil conservation and hydrology, fish, and wildlife populations, fish and wildlife habitats, solid waste disposal, noise levels, radiation, and hazardous substances use and disposal.

II. *Land use planning and management.* Consider the impacts of a proposed action on energy supply and natural resources development; protection of environmentally critical areas, such as floodplains, wetlands, beaches and dunes, unstable soils, steep slopes and aquifer recharge areas, coastal area land use, and redevelopment and construction in built-up areas.

III. *Socioeconomic environment.* Consider the impacts of a proposed action on population density changes, congestion mitigation, neighborhood character and cohesion, low income populations, outdoor recreation, industrial/commercial/residential development and tax ratables, and historic, architectural, and archaeological preservation.

APPENDIX B

RESPONSIBILITIES

I. *General responsibilities.* (a) *Responsible official.* (1) Requires contractors and grantees to submit environmental assessments and related documents needed to comply with NEPA, and assures environmental reviews are conducted on proposed EPA projects at the earliest possible point in EPA's decision-making process.

(2) When required, assures that draft EIS's are prepared and distributed at the earliest possible point in EPA's decision-making process, their internal and external review is coordinated, and final EIS's are prepared and distributed.

(3) When an EIS is not prepared, assures that negative declarations and environmental appraisals are prepared and distributed for those actions requiring them.

(4) Consults with appropriate officials identified in § 6.214 of this part.

(5) Consults with the Office of Federal Activities on actions involving unresolved conflicts with other Federal agencies.

(b) *Office of Federal Activities.* (1) Provides EPA with policy guidance and assures that EPA offices establish and maintain adequate administrative procedures to comply with this part.

(2) Monitors the overall timeliness and quality of the EPA effort to comply with this part.

(3) Provides assistance to responsible officials as required.

(4) Coordinates the training of personnel involved in the review and preparation of EIS's and other NEPA-associated documents.

(5) Acts as EPA liaison with the Council of Environmental Quality and other Federal and State entities on matters of EPA policy and administrative mechanisms to facilitate external review of EIS's, to determine lead agency and to improve the uniformity of the NEPA procedures of Federal agencies.

(6) Advises the Administrator and Deputy Administrator on projects which involve more than one EPA office, are controversial, are nationally significant, or "pioneer" EPA policy, when these projects have had or should have an EIS prepared on them.

(c) *Office of Public Inquiries.* Assists the Office of Federal Activities and responsible officials by answering the public's queries on the EIS process and on specific EIS's and by directing requests for copies of specific documents to the appropriate regional office or program.

(d) *Office of Public Affairs.* Analyzes the present procedures for public participation, and develops and recommends to the Office of Federal Activities a program to improve those procedures and increase public participation.

(e) *Regional Office Division of Public Affairs.* (1) Assists the responsible official or his designee on matters pertaining to negative declarations, notices of intent, press releases, and other public notification procedures.

(2) Assists the responsibility official or his designee by answering the public's queries on the EIS process and on specific EIS's, and by filling requests for copies of specific documents.

(f) *Offices of the Assistant Administrators and Regional Administrators.* (1) Provides specific policy guidance to their respective offices and assures that those offices establish and maintain adequate administrative procedures to comply with this part.

(2) Monitors the overall timeliness and quality of their respective office's efforts to comply with this part.

(3) Acts as liaison between their offices and the Office of Federal Activities and between their offices and other Assistant Administrators or Regional Administrators on matters of agencywide policy and procedures.

(4) Advises the Administrator and Deputy Administrator through the Office of Federal Activities on projects or activities within their respective areas of responsibilities which involve more than one EPA office, are controversial, are nationally significant, or "pioneer" EPA policy, when these projects have had or should have an EIS prepared on them.

(g) *The Office of Legislation.* (1) Provides the necessary liaison with Congress.

(2) Coordinates the preparation of EIS's required on reports on legislation originating outside EPA. (See § 6.106(d)).

(h) *The Office of Planning and Evaluation.* Coordinates the preparation of EIS's required on EPA legislative proposals. (See § 6.106 (d)).

II. *Responsibilities for Title II Construction Grants Program (Subpart E).* (a) *Responsible official.* The responsible official for EPA actions covered by this subpart is the Regional Administrator. The responsibilities of the Regional Administrator in addition to those in Appendix B.I. are to:

(1) Assist the Office of Federal Activities in coordinating the training of personnel involved in the review and preparation of NEPA-associated documents.

(2) Require grant applicants and those who have submitted plans for approval to provide the information the regional office requires to comply with these guidelines.

(3) Consult with the Office of Federal Activities concerning works or plans which significantly affect more than one regional office, are controversial, are of national significance or "pioneer" EPA policy, when these works have had or should have had an EIS prepared on them.

(b) *Assistant Administrator.* The responsibilities of the Office of the Assistant Administrator, as described in Appendix B.I, shall be assumed by the Assistant Administrator for Water and Hazardous Materials for EPA actions covered by this subpart.

(c) Oil and Special Materials Control Division, Office of Water Program Operations, coordinates all activities and responsibilities of the Office of Water Program Operations concerned with preparation and review of EIS's. This includes providing technical assistance to the Regional Administrators on EIS's and assisting the Office of Federal

Activities in coordinating the training of personnel involved in the review and preparation of NEPA-associated documents.

(d) *Public Affairs Division, Regional Offices.* The responsibilities of the regions' Public Affairs Divisions, in addition to those in Appendix B.I, are to:

(1) Assist the Regional Administrator in the preparation and dissemination of NEPA-associated documents.

(2) Collaborate with the Headquarters Office of Public Affairs to analyze procedures in the regions for public participation and to develop and recommend to the Office of Federal Activities a program to improve those procedures.

III. *Responsibilities for Research and Development Programs (Subpart F).* The Assistant Administrator for Research and Development, in addition to those responsibilities outlined in Appendix B.I.(a), will also assume the responsibilities described in Appendix B.I(f).

IV. *Responsibilities for Solid Waste Management Programs (Subpart G).* (a) *Responsible Official.* The responsible official for EPA actions covered by this subpart is the Deputy Assistant Administrator for Solid Waste Management Programs. The responsibilities of this official, in addition to those in Appendix B.I.(a), are to:

(1) Assist the Office of Federal Activities in coordinating the training of personnel involved in the review and preparation of all NEPA-associated documents.

(2) Advise the Assistant Administrator for Air and Waste Management concerning projects which significantly affect more than one regional office, are controversial, and nationally significant, or "pioneer" EPA policy.

V. *Responsibilities for Special Purpose Facilities and Facility Renovation Programs (Subpart H).*

(a) *Responsible official.* The responsible official for new construction and modification or special purpose facilities is as follows:

(1) The Chief, Facilities Management Branch, Data and Support Systems Division, shall be the responsible official on all new construction of special purpose facilities and on all improvement and modification projects for which the Facilities Management Branch has received a funding allowance.

(2) The Regional Administrator shall be the responsible official on all improvement and modification projects for which the regional office has received the funding allowance.

(3) The Center Directors shall be the responsible officials on all improvement and modification projects for which the National Environmental Research Centers have received the funding allowance.

(b) The responsibilities of the responsible officials, in addition to those in Appendix B.I, are to:

(1) Ensure that environmental assessments are submitted when requested, that environmental reviews are conducted on all projects, and EIS's are prepared and circulated when there will be significant impacts.

(2) Assist the Office of Federal Activities in coordinating the training of personnel involved in the review and preparation of NEPA-associated documents.

APPENDIX C

DISTRIBUTION AND AVAILABILITY OF DOCUMENTS

I. *Negative Declaration.* (a) The responsible official shall distribute two copies of each negative declaration to:

(1) The appropriate Federal, State, and local agencies and to the appropriate State and areawide clearinghouses.

(2) The Office of Legislation, the Office of Public Affairs and the Office of Federal Activities.

(3) The headquarters EIS coordinator for the program office originating the document. When the originating office is a regional office and the action is related to water quality management, one copy should be forwarded to the Oil and Special Materials Control Division, Office of Water Program Operations.

(b) The responsible official shall distribute one copy of each negative declaration to:

(1) Local newspapers and other local mass media.

(2) *Interested persons on request.* If it is not practical to send copies to all interested persons, make the document available through local libraries or post offices, and notify individuals that this action has been taken.

(c) The responsible official shall have a copy of the negative declaration and any documents supporting the negative declaration available for public review at the originating office.

II. *Environmental Impact Appraisal.* (a) The responsible official shall have the environmental impact appraisal available when the negative declaration is distributed and shall forward one copy to the headquarters EIS coordinator for the program office originating the document and to any other Federal or State agency which requests a copy.

(b) The responsible official shall have a copy of the environmental impact appraisal available for public review at the originating office and shall provide copies at cost to persons who request them.

III. *Notice of Intent.* (a) The responsible official shall forward one copy of the notice of intent to:

(1) The appropriate Federal, State, and local agencies and to the appropriate State, regional, and metropolitan clearinghouses.

(2) Potentially interested persons.

(3) The Offices of Federal Activities, Public Affairs and Legislation.

(4) The headquarters Grants Administration Division, Grants Information Branch.

(5) The headquarters EIS coordinator for the program office originating the notice. When the originating office is a regional office and the action is related to water quality management, one copy should be forwarded to the Oil and Special Materials Control Division, Office of Water Program Operations.

IV. *Draft EIS's.* (a) The responsible official shall send two copies of the draft EIS to:

(1) The Office of Federal Activities.

(2) The headquarters EIS coordinator for the program office originating the document. When the originating office is a regional office and the project is related to water quality management, send two copies to the Oil and Special Materials Control Division, Office of Water Program Operations.

(b) If none of the above offices requests any changes within ten (10) working days after notification, the responsible official shall:

(1) Send five copies of the draft EIS to CEQ.

(2) Send two copies of the draft EIS to the Office of Public Affairs and to the Office of Legislation.

(3) Send two copies of the draft EIS to the appropriate offices of reviewing Federal agencies that have special expertise or jurisdiction by law with respect to any impacts involved. CEQ's guidelines (40 CRF 1500.9 and Appendices II and III) list those agencies to which draft EIS's will be sent for official review and comment.

(4) Send two copies of the draft EIS to the appropriate Federal, State, regional, and metropolitan clearinghouses.

(5) Send one copy of the draft EIS to the public libraries in the project area and interested persons. Post offices, city halls or courthouses may be used as distribution points if public library facilities are not available.

(c) The responsible official shall make a copy of the draft EIS available for public review at the originating office and at the Office of Public Affairs.

V. *Final EIS.* (a) The responsible official shall distribute the final EIS to the following offices, agencies and interested persons:

(1) Five copies to CEQ.

(2) Two copies to the Office of Public Affairs, Legislation and Federal Activities.

(3) Two copies to the headquarters EIS coordinator for the program office originating the document.

(4) One copy to Federal, State, and local agencies and interested persons who made substantive comments on the draft EIS or requested a copy of the final EIS.

(5) One copy to a grant applicant.

(b) The responsible official shall make a copy of the final EIS available for public review at the originating office and at the Office of Public Affairs.

VI. *Legislative EIS.* Copies of the legislative EIS shall be distributed by the responsible official according to the procedures in section IV(b) of this appendix. In addition, the responsible official shall send two copies of the EIS to the Office of Federal Activities and the EIS coordinator of the originating office.

[FR Doc.75-9553 Filed 4-11-75; 8:45 am]

EPA PROCEDURAL UPDATE
VOLUME 47, NUMBER 45, 8 MARCH, 1982

ENVIRONMENTAL PROTECTION AGENCY

40 CFR PART 6
[ER-FRL-1953-2a]

Categorical Exclusion from EPA Procedures Implementing the National Environmental Policy Act

AGENCY: Environmental Protection Agency (EPA).

ACTION: Interim final rule.

SUMMARY: This document amends EPA's procedures implementing the National Environmental Policy Act (NEPA). It includes procedures for granting categorical exclusions from the substantive environmental review requirements of 40 CFR Part 6 for EPA actions subject to NEPA. This action is taken to comply with the Council on Environmental Quality's (CEQ) regulations implementing the procedural provisions of NEPA. The intended effect of this action is to reduce the regulatory requirements on recipients of EPA grants for the planning and construction of wastewater treatment facilities while maintaining the integrity of EPA's environmental review responsibilities. Technical changes will be made during the comment period to make the procedures consistent with Pub. L. 97-117 (Municipal Wastewater Treatment Construction Grant Amendments of 1981) and related changes to 40 CFR Part 35.

DATES: Effective date: Interim final rule effective March 8, 1982.
Comment date: Comments must be received on or before April 7, 1982.

ADDRESS: Comments may be mailed to the Office of Federal Activities, A-104, Environmental Protection Agency, 401 M St. SW., Washington, D.C. 20460; Attention: Richard Otis.

FOR FURTHER INFORMATION CONTACT: John Gustafson, John Gerba, or Richard Otis, Office of Federal Activities, (202) 755-8835.

SUPPLEMENTARY INFORMATION:

Classification

The Office of Federal Activities has determined that this revision is not a "major" rule within the meaning of Executive Order (E.O.) 12291. This is because the revision will not: (1) Have an annual effect on the economy of $100 million or more; (2) cause a major increase in costs or prices for consumers, individual industries, geographic regions, or Federal, State, or local government agencies; or (3) have significant adverse effects on competition, employment, investment productivity, innovation, or on the ability of a United States-based enterprise to compete with Foreign-based enterprises in domestic or export markets.

The purpose and effect of this amendment is to reduce unnecessary regulatory burdens on municipalities and other recipients of wastewater treatment facility construction grants. No increased paperwork burdens are imposed by the amendment.

This amendment (will be) was submitted to the Office of Management and Budget (OMB) for review as required by E.O. 12291. Any comments from OMB to EPA and any EPA response to those comments are available for public inspection in the Document Control Office, EPA, Room 107, 401 M Street, SW., Washington, D.C. 20460.

Because this amendment is not "major," it is effective immediately as an interim final rule. It may be revised before becoming final if substantive comments are received by the Office of Federal Activities (OFA) within thirty (30) calendar days of the effective date.

Regulatory Analysis

Under E.O. 12291, EPA must determine if a regulation is "major" and therefore subject to a Regulatory Impact Analysis. Since EPA believes that this amendment is not "major," it is not subject to such an analysis. However, this amendment will significantly reduce the regulatory burden on recipients of wastewater treatment facility construction grants.

Action Being Undertaken

EPA is amending its procedures for implementing CEQ's regulations (40 CFR 1508.4) authorized by NEPA to (1) make several technical changes; (2) be consistent with the revised regulations for EPA's Construction Grants Program (40 CFR Part 35); and (3) include categorical exclusions as provided in the CEQ regulations. Technical revisions include changing the name of the Office of Environmental Review to the Office of Federal Activities and correcting references in § 6.600 to the New Source National

Pollution Discharge Elimination System consolidated permit program regulations (40 CFR Part 122-125). Alterations to the public participation process are made to be consistent with the revised Construction Grants regulations. Amendments incorporating the categorical exclusion process will exclude specified categories of EPA actions from the substantive environmental review requirements of 40 CFR Part 6.

Definitions

Categorical Exclusion—In part, CEQ's NEPA regulations state that: "Categorical exclusion means a category of actions which does not individually or cumulatively have a significant effect on the human environment and which have been found to have no such effect in procedures adopted by a Federal agency * * * and for which, therefore, neither an environmental assessment nor an environmental impact statement is required." EPA is amending its implementing procedures (40 CFR Part 6) to provide descriptions of the categories of actions within the Construction Grants Program that EPA has determined do not have a significant effect upon the human environment. It also provides the procedures and conditions for granting an exclusion and for identifying additional categories of actions.

It is EPA's intent to apply this exclusion process to actions which, through Agency, State, and grantee experience, have not required an environmental impact statement as defined in 40 CFR 6.506 and which have not generally required the application of procedures described in 40 CFR Subpart C, "Coordination With Other Environmental Review and Consultation Requirements." Categorical exclusions are intended to apply to actions that are small scale, minor, and routine.

Minor Rehabilitation, Minor Expansion, Replacement, and Minor Upgrading—In addition to avoiding primary effects, certain actions eligible for categorical exclusions within the Construction Grants Program are described as being "minor" to avoid the negative secondary environmental effects of induced new development. The intent is to ensure that categorical exclusions assist the Agency's efforts to achieve direct environmental improvements from the Construction Grants Program and not to fund or unintentionally subsidize new development.

Excluded actions involving replacement, minor rehabilitation, minor expansion, or minor upgrading of facilities should not result in increasing the overall design capacity of treatment works, nor the pipe size of interceptors or collector sewers.

Therefore, actions which may result in new community development shall not receive categorical exclusions except for infilling largely built-up areas and where new development is insignificant. Allowing some new development to occur without an environmental review recognizes that there is a trade-off between the potential negative secondary effects of the development and the benefits provided by granting categorical exclusions. It also recognizes that for these actions the possible negative secondary effects are outweighed by the primary benefits. Incidental new development unrelated to the action should not disqualify the action from receiving an exclusion.

Consistency—In granting an exclusion, § 6.506(c) uses a consistency test to determine if action is eligible for exclusion under one of the categories in § 6.506(c)(1). The consistency test requires the characteristics of the likely alternatives (including but not limited to treatment technology, location, capacity, and possible secondary effects) which may be considered in the planning process or the potential action if facilities planning has begun, to be substantially similar to or described by the characteristics of the actions outlined in a category.

Review Required for Application—The review included as part of the request for an exclusion described in § 6.507(a) shall include a description of the characteristics of the likely alternatives which may be considered in the planning process or the potential action if facility planning has begun. It should also state that the action under consideration is consistent with a category of actions eligible for exclusion and should identify the category. If the environmental review process has not begun, information required for this review should come from existing sources. If such information is not available before the environmental review process begins the action shall not be granted an exclusion at that time even if the action appears consistent with actions described by a category eligible for exclusion. But, information generated during the development of an Environmental Information Document (EID) or an environmental assessment can be used in producing this review.

Timing—Categorical exclusions may be granted for actions before the environmental review process has begun as well as during the process. If, in preparing an EID or environmental assessment, the information generated clearly indicates that the project meets the requirements for an exclusion, the responsible official may grant an exclusion. EPA may revoke an exclusion at any time as described in § 6.107(e).

Response to Comments

Comments from EPA headquarters and regional offices generated during two or more review periods have been addressed or incorporated into this amendment. Comments from major constituent groups and delegated states have also been addressed or incorporated. Several significant comments are discussed below.

Public Review Period—A thirty day public review period had been originally proposed where comments on a publicly available proposed notice of categorical exclusion could be addressed to the official responsible for granting an exclusion. Comments on the notice were intended to assist the responsible official in determining if any serious local or environmental issues exist or if Federal, State, or local laws would be violated by granting an exclusion. After receiving comments from several parties on this topic, and with the responsible official's ability to revoke an exclusion after it is granted (as provided in § 6.107(e)) the regulation development work group concluded that the comment period was unnecessary.

Notice of Categorical Exclusion—In order to further reduce paperwork, the preparation and public issuance of a "notice of categorical exclusion" was removed from the amendment. Instead, the responsible official is required to review the request for an exclusion and document the resulting decision.

New Categories—The work group considered various alternative processes for creating additional categories of excluded actions. This interim final rule will use EPA's minor regulation development process to create new categories. New categories will be published in the **Federal Register** with a subsequent 30 day comment period because CEQ's regulations require public notice and a comment period.

Suggestions are invited from EPA Regional Offices, State governments, and interested parties for new categories of minor construction grants actions that may be eligible for exclusion and which are not already identified in this amendment. Comments are also invited on the process for creating new categories.

Anne M. Gorsuch,
Administrator.
March 1, 1982.

PART 6—IMPLEMENTATION OF PROCEDURES ON THE NATIONAL ENVIRONMENTAL POLICY ACT

For the reasons set out in the preamble, 40 CFR Part 6 is amended as follows:

1. The authority for Part 6 reads as follows:

Authority: Sec. 101, 102 and 103 of the National Environmental Policy Act of 1969 (42 U.S.C. 4321 et seq.); also, the Council on Environmental Quality Regulations promulgated November 29, 1978 (40 CFR Part 1500).

§§ ***6.102 and 6.103*** [***Amended***]

2. Sections 6.102(c) and 6.103(a)(1)(A)(5) are amended by removing the words "Office of Environmental Review, (OER)" and inserting, in their place, the words "Office of Federal Activities, (OFA)."

3. Section 6.103(a)(2) is amended by removing the words "Office of Environmental Review" and inserting, in their place, the words "Office of Federal Activities."

§§ ***6.103, 6.104, 6.106, 6.400, 6.401, 6.402, 6.404*** [***Amended***]

4. Amend this part by removing "OER" and inserting in its place "OFA" in the following places:

(a) 40 CFR 6.103(a)(3)(iii) and (iv)
(b) 40 CFR 6.104
(c) 40 CFR 6.106(a)
(d) 40 CFR 6.106(b)(1), (2), and (3)
(e) 40 CFR 6.400(b), (d), and (e)
(f) 40 CFR 6.401(a), (c), (d), and (e)
(g) 40 CFR 6.402(b)
(h) 40 CFR 6.404(b)

5. Section 6.103 is amended by revising paragraph (a)(1)(iii); deleting § 6.103(a)(1)(iii)(A)(1), (2), and (3); and renumbering § 6.103(a)(1)(iii)(A) (4), (5), and (6) as § 6.103(a)(1) (iv), (v), and (vi) to read as follows:

§ ***6.103 Responsibilities.***

(a) * * *

(1) * * *

(iii) When an EIS is not prepared, assuring documentation of the decision to grant a categorical exclusion, or assuring that findings of no significant impact (FNSIs) and environmental assessments are prepared and distributed for those actions requiring them.

(iv) Consulting with appropriate officials responsible for other environmental law set forth in Subpart C.

(v) Consulting with the Office of Federal Activities (OFA) on actions involving unresolved conflicts concerning this part or other Federal agencies.

(vi) When required, assuring that public participation requirements are met.

* * * * *

6. Section 6.105 is amended by revising paragraph (b) to read as follows:

§ ***6.105 Synopsis of EIS procedures.***

* * * * *

(b) *Environmental information documents.* Environmental information documents must be prepared by applicants, grantees, or permittees and submitted to EPA as required in Subparts E, F, G, H, and I. Environmental information documents will be of sufficient scope to enable the responsible official to prepare an environmental assessment as described under § 6.105(d) of this part and Subparts E through I. Environmental information documents will not have to be issued for actions where a categorical exclusion has been granted.

* * * * *

7. Subpart A of this part is amended by adding § 6.107 to read as follows:

§ ***6.107 Categorical exclusions.***

(a) *General.* Categories of actions which do not individually, cumulatively over time, or in conjunction with other Federal, State, local, or private actions have a significant effect on the quality of the human environment and which have been identified as having no such effect in the requirements set forth in Subpart E, may be exempted from the substantive environmental review requirements of this part. Generally, environmental information documents and environmental assessments or environmental impact statements will not be required for excluded actions.

(b) *Determination.* For each excluded action, the responsible official shall determine whether an action is eligible for a categorical exclusion as established in Subpart E. The determination shall be made as early as possible following the receipt of an application or notification that an application will be filed. The responsible official shall document the decision to issue or deny an exclusion. The documentation shall include the application, a brief description of the proposed action, and a brief statement of how the action meets the criteria for a categorical exclusion.

(c) *Consultation.* The documentation outlined in § 6.107(b) of this part shall be made available to the public and a copy be sent to The Office of Federal Activities.

(d) *Extraordinary circumstances.* If undertaking a normally excluded action may violate Federal, State, or local environmental laws or may involve serious local or environmental issues, the full environmental review procedures of this part must be followed. The responsible official shall ensure that actions requiring environmental assessment under this part are not processed as categorical exclusions.

(e) *Revocation.* The responsible official shall revoke a categorical exclusion and shall require a full environmental review if subsequent to the granting of an exclusion, the responsible official determines that: (1) The proposed action no longer meets the requirements for a categorical exclusion due to changes in the proposed action; or (2) determines from new evidence that serious local or environmental issues exist; or (3) that Federal, State, or local laws are being or may be violated.

8. Section 6.400 is amended by adding paragraph (f) to read as follows:

§ ***6.400 [Amended]***

* * * * *

(f) *Categorical Exclusions.* The responsible official shall make the documentation described in § 6.107(b) of this part available to the public and shall send a copy to the Office of Federal Activities.

9. Section 6.502 is amended by revising paragraph (a) introductory text to read as follows:

§ ***6.502 Applicability.***

(a) *Administrative actions covered.* This subpart applies to administrative actions listed below (except as provided in § 6.503 of this part):

* * * * *

10. Section 6.504 is amended by revising paragraph (a), deleting paragraph (b), and renumbering subsequent paragraphs (c) and (d) as (b) and (c) to read as follows:

§ *6.504 Public participation.*

(a) *General.* It is EPA policy that public participation be achieved during the environmental review process as deemed appropriate by the responsible official under 40 CFR Part 25 and implementing provisions of Part 35, Subpart E of this chapter. Compliance with Part 25 and implementing provisions constitutes compliance with public participation requirements under this part.

(b) Public participation activities undertaken in connection with the environmental review process should be coordinated with the facility planning public participation program wherever possible.

(c) The responsible official may institute such additional NEPA-related public participation procedures as he deems necessary during the environmental review process.

11. Section 6.505 is revised to read as follows:

§ *6.505 Limitations on actions during the environmental review process.*

No administrative action under § 6.502(a) shall be taken until the environmental review process has been completed except as provided under § 6.106, § 6.107 and § 6.503 of this part. The responsible official shall ensure compliance in accordance with 40 CFR 1506.1 and Subparts A, C, and D of this part, and all policies, guidance and regulations adopted to implement the requirements under 42 U.S.C. 7616 of the Clean Air Act.

12. Section 6.506 is amended by revising paragraph (a) introductory text, removing paragraph (c), and adding a new paragraph (c) to read as follows:

§ *6.506 Criteria for preparing EISs and granting categorical exclusions.*

(a) *EISs.* The responsible official shall assure that an EIS will be issued when he determines that any of the following conditions exist:

* * * * *

(b) * * *

(c) *Categorical Exclusions.* Except as provided in § 6.107(d), the responsible official shall determine and document as provided in § 6.107(b) and § 6.507(a) that an action is consistent with the categories eligible for exclusion identified in § 6.506(c)(1). The responsible official shall ensure that actions consistent with the categories listed in § 6.506(c)(1) but requiring an environmental assessment under § 6.506(c)(2) are not granted categorical exclusions.

(1) Categories of actions eligible for exclusion. For this Subpart, actions consistent with the following categories are eligible for a categorical exclusion:

(i) Actions for which the facilities planning is solely directed toward minor rehabilitation of existing facilities, functional replacement of equipment, or towards the construction of new ancillary facilities adjacent or appurtenant to existing facilities which do not affect the degree of treatment or capacity of the existing facility. Such actions include but are not limited to infiltration and inflow analyses, sewer system evaluation studies, replacement of existing mechanical equipment or structures, and the construction of new small on-site structures.

(ii) Actions in sewered communities of less than 3,500 persons which are for minor upgrading and minor expansion of existing treatment works. This category does not include actions that directly or indirectly involve the extension of new collection systems funded with Federal or other sources of funds.

(iii) Actions in unsewered communities of less than 3,500 persons where onsite or other alternative technologies are proposed.

(iv) Other actions developed in accordance with § 6.506(c)(3).

(2) Criteria for not granting a categorical exclusion. Notwithstanding the provisions of

§ 6.506(c)(1), the responsible official shall ensure that an adequate environmental information document and environmental assessment are prepared for actions that meet any of the following criteria. An action, therefore, that meets any of the following criteria shall not be granted a categorical exclusion.

(i) The facilities to be provided will create a new discharge to surface or ground waters.

(ii) The facilities will result in substantial increases in the volume of discharge or the loading of pollutants from an existing source or from new facilities to receiving waters.

(iii) The facilities would provide capacity to serve a population 20% greater than the existing population.

(iv) The action is known or expected to have a significant effect on the quality of the human environment, either individually, cumulatively over time, or in conjunction with other Federal, State, local, or private actions.

(v) The action is known or expected to directly or indirectly affect sensitive environmental resources or areas, such as floodplains, wetlands, prime or unique agricultural lands, aquifer recharge zones, archaeological and historic sites, endangered or threatened species, or other areas identified in guidance issued by the OFA.

(vi) The action is known or expected not to be cost-effective or to cause significant public controversy.

(3) Developing new categories of excluded actions. The responsible official or other interested parties may request that a new category of excluded actions be created, or that an existing category be amended or deleted. The request shall be made in writing to the Direçtor, OFA and shall contain adequate information to support the request. Under the direction of OFA, proposed new categories shall be developed through EPA's "non-major" rule-making process (E.O. 12291), including publication as an interim final rule in the **Federal Register** and a subsequent thirty (30) day public comment period. The following shall be considered in evaluating proposals for new categories:

(i) Actions in the proposed category should seldom result in the effects identified in § 6.506(c)(2).

(ii) Based upon previous environmental reviews of similar actions, the proposed category of actions would seldom require the preparation of an EIS.

(iii) Whether information adequate to determine if a potential action is consistent with the proposed category will normally be available when needed.

13. Section 6.507 is amended by revising paragraph (a) to read as follows:

§ *6.507 Environmental review process.*

* * * * *

(a) *Award of a facilities planning grant (Step 1).* Prior to award of Step 1 assistance, or within no more than 30 days thereafter, EPA shall review, or request that the State review, if the facilities plan review is delegated under § 205(g) of the Clean Water Act, the likely project alternatives in the potential action and the existence of environmentally sensitive areas in the facilities planning area, including those identified in § 6.506(c)(2)(v) of this Subpart. The responsible official shall use this review and any additional analysis he may deem necessary to determine whether the action is eligible for a categorical exclusion from the substantive environmental review requirements of this part. This review is intended to be brief and concise, drawing on existing information and knowledge of EPA, State agencies, regional planning agencies, water quality management agencies, and grantees. If a categorical exclusion is granted, the grantee will not be required to prepare a formal environmental information document and environmental assessment during facilities planning. This review should also be used to determine the scope of the environmental information document prepared by the grantee when an action does not meet the requirements for a categorical exclusion. It should also be used to make an early determination of the need for an EIS. Whenever possible, this initial review should be discussed at the first conference held with the potential grantee.

* * * * *

14. Section 6.508 is amended by revising paragraph (a) (1) and (2), and paragraph (b) to read as follows:

§ *6.508 Limits on delegation to States.*

(a) * * *

(1) The determination of whether or not to prepare an EIS shall be solely that of EPA. EPA shall consider a State's recommendations, but the ultimate decision under NEPA cannot be delegated.

(2) Categorical exclusions, findings of no significant impact, and the environmental assessment shall be approved, finalized, and issued by EPA.

(3) * * *

(b) *Elimination of duplication.* The responsible official shall assure that maximum efforts are undertaken to minimize duplication within the limits described under § 6.506 and § 6.507(a) of this part. In carrying out requirements under this subpart, maximum consideration shall be given to eliminating duplication in accordance with 40 CFR 1506.2. Where there are State or local procedures comparable to NEPA, EPA should enter into Memoranda of Understanding with a State concerning workload distribution and responsibilities for implementing the environmental review and facilities planning process.

15. Section 6.600 is amended by revising paragraph (b) to read as follows:

§ *6.600 Purpose.*

* * * * *

(b) *Permit regulations.* All references in this subpart to the "permit regulations" shall mean Parts 122 and 124 of Title 40 of the Code of Federal Regulations relating to the NPDES program.

* * * * *

16. Section 6.601 is amended by revising paragraph (a) to read as follows:

§ *6.601 Definitions.*

(a) The term "administrative action" for the sake of this subpart means the issuance by EPA of an NPDES permit to discharge as a new source, pursuant to 40 CFR 124.15.

* * * * *

17. Section 6.602 is amended by revising paragraph (b) to read as follows:

§ *6.602 Applicability.*

* * * * *

(b) *New Source Determination.* An NPDES permittee must be determined a "new source" before these procedures apply. New source determinations will be undertaken pursuant to the provisions of the permit regulations under 40 CFR 122.66 (a) and (b) and 122.53(h).

* * * * *

18. Section 6.603 is revised to read as follows:

§ *6.603 Limitations on actions during environmental review process.*

The processing and review of an applicant's NPDES permit application shall proceed concurrently with the procedures within this subpart. Actions undertaken by the applicant or EPA shall be performed consistent with the requirements of 40 CFR 122.66(c).

19. Section 6.604 is amended by revising paragraph (h) to read as follows:

§ ***6.604 Environmental review process.***

* * * * *

(h) *Documents for the administrative record.* Pursuant to 40 CFR 124.9(b)(6) and 124.18(b)(5) any environmental assessment, FNSI EIS, or supplement to an EIS shall be made a part of the administrative record related to permit issuance.

20. Section 6.606 is amended by revising paragraph (a) to read as follows:

§ ***6.606 General.***

At the time of permit award, the responsible official shall prepare a record of decision in those cases where a final EIS was issued in accordance with 40 CFR 1505.2 and pursuant to the provisions of the permit regulations under 40 CFR 124.15 and 124.18(b)(5). The record of decision shall list any mitigation measures necessary to make the recommended alternative environmentally acceptable.

* * * * *

21. Section 6.607 is amended to read as follows:

§ ***6.607 Monitoring.***

In accordance with 40 CFR 1505.3 and pursuant to 40 CFR 122.66(c) and 122.10 the responsible official shall ensure that there is adequate monitoring of compliance with all NEPA related requirements contained in the permit.

[FR Doc. 82-6165 Filed 3-5-82; 8:45 am]

---

40 CFR Part 6

[ER-FRL-1953-2b]

Segmentation Consultation Process

AGENCY: Environmental Protection Agency (EPA).

ACTION: Interim final rule.

SUMMARY: This document revises EPA's procedures implementing the National Environmental Policy Act (40 CFR Part 6) by removing the Council on Environmental Quality (CEQ) from the consultation process on requests to segment wastewater treatment facility construction grant projects and by making several other technical changes.

DATES: Interim final rule effective March 8, 1982. Comments must be received on or before April 7, 1982.

ADDRESSES: Comments may be mailed to the Office of Federal Activities, A-104, Environmental Protection Agency, 401 M St. S.W., Washington, D.C. 20460; Attention: Richard Otis.

For Further Information Contact: John Gustafson or Richard Otis, Office of Federal Activities; (202) 755-8835.

Supplementary Information: *Classification:* The Office of Federal Activities has determined that this revision is not a "major" rule within the meaning of Executive Order (E.O.) 12291. This is because the revision will not: (1) Have an annual effect on the economy of $100 million or more; (2) cause a major increase in costs or prices for consumers, individual industries, geographic regions, or Federal, State, or local government agencies; or (3) have significant adverse effects on competition, employment, investment, productivity, innovation, or on the ability of United States-based enterprise to compete with Foreign-based enterprise in domestic or export markets.

The purpose and effect of this revision is to reduce unnecessary regulatory burdens on municipalities and other recipients of EPA grants for planning and constructing wastewater treatment facilities. No paperwork burdens are imposed by the amendment.

This revision (will be) was submitted to the Office of Management and Budget (OMB) for review as required by E.O. 12291. Any comments from OMB to EPA and any EPA response to those comments are available for public inspection in the Document Control Office, EPA, Room 107, 401 M Street, S.W., Washington, D.C. 20460.

Because this revision is not "major," it is effective immediately as an interim final rule. It may be revised before becoming final if substantive comments are received by the Office of Federal Activities (OFA) within thirty (30) calendar days of the effective date.

Regulatory Analysis: Under E.O. 12291, EPA must determine if a regulation is "major" and therefore subject to a Regulatory Impact Analysis. Since EPA believes that this revision is not "major," it is not subject to such an analysis.

Action Being Undertaken

This revision is being made in response to a request from CEQ and after EPA and CEQ determined that the consulting process for approving the segmentation of wastewater treatment facility construction projects now represents a noncontroversial routine action.

The segmentation process as outlined in 40 CFR 6.503 allows EPA to award a Step 2 grant for facilities design or a Step 3 grant for construction of a discrete segment of a project before the environmental review for the project as a whole is complete. EPA is currently required to consult with CEQ before awarding grants under this procedure. This revision removes CEQ from the consultation process and also makes several minor technical changes.

The Office of Federal Activities (OFA) is responsible for handling these segmentation requests. Examples include: Award of a Step 2 grant for preparation of plans and specifications for a large treatment plant when the only unresolved issue is the site for final effluent or sludge disposal; and award of a Step 3 grant for construction of collectors and an interceptor to transport flows to an existing plant proposed for expansion from an area of failing septic tanks affecting a public water supply.

OFA will continue to review requests under the procedures currently identified in § 6.503 of this part. In reviewing a request either through the regional office from a delegated State or directly from a regional office in the case of a nondelegated State, OFA will consult with the Office of Water and return its finding to the originating EPA regional office.

Anne M. Gorsuch,
Administrator.
March 1, 1982.

PART 6—IMPLEMENTATION OF PROCEDURES ON THE NATIONAL ENVIRONMENTAL POLICY ACT

For the reasons set out in the preamble, 40 CFR 6.503 is revised to read as follows:

§ 6.503 Consultation during the environmental review process.

When there are overriding considerations of costs or impaired program effectiveness, the responsible official may award a Step 2 or Step 3 grant for a discrete segment of the project plans or construction before the environmental review is completed. The segmented portion of the project must be noncontroversial, necessary to correct water quality or other immediate environmental problems, and cannot, by its completion, foreclose any reasonable options being considered in the environmental review. If a project is segmented, the entire project shall be evaluated to determine if an EIS is required. In applying the criteria to determine if an EIS is required, regional EIS preparation staff shall be consulted. In no case may these types of grant assistance for Step 2 or Step 3 projects be awarded unless the OFA has been consulted, an FNSI has been issued on the segments permitted to proceed at least 30 days prior to grant award, and the grant award contains a specific agreement prohibiting action on the segment of planning or construction for which the environmental review is not complete. The Director, OFA is responsible for consulting with the Assistant Administrator for Water.

[FR DOC. 82-6168 Filed 3-5-82; 8:45 am]

EPA PROCEDURAL UPDATE
VOLUME 48, NUMBER 5, 7 JANUARY, 1983

ENVIRONMENTAL PROTECTION AGENCY

40 CFR PART 6
[FA-FRL 2097-8a]

Categorical Exclusion from EPA Procedures Implementing the National Environmental Policy Act

AGENCY: Environmental Protection Agency (EPA).

ACTION: Interim final rule revisions and extension to comment period.

SUMMARY: This document: (1) Extends the public comment period on amendments to EPA's Procedures Implementing the National Environmental Policy Act (NEPA) (40 CFR Part 6) as published in the **Federal Register** on March 8, 1982. (These amendments created a process for excluding certain categories of action from the substantive environmental review requirements of EPA's NEPA procedures.); (2) revises the categories of actions eligible for exclusion and the criteria for not granting an exclusion in § 6.506; and (3) corrects a factual error in § 6.507 on the responsibility for preparation of a final environmental assessment. A related action involving additional changes to 40 CFR Part 6 is published elsewhere in this issue of the **Federal Register.**

DATES: Effective Date: Revisions to interim final rule effective January 7, 1983.

Comment Date: Comments on amendments to Categorical Exclusion from EPA

Procedures Implementing the National Environmental Policy Act published in the **Federal Register** March 8, 1982 (47 FR 9827) and on the revisions and correction made herein must be received on or before February 7, 1983.

ADDRESS: Comments may be mailed to the Office of Federal Activities (A-104), U.S. Environmental Protection Agency, 401 M Street S.W., Washington, D.C. 20460. Attention: Paul C. Cahill, Director.

FOR FURTHER INFORMATION CONTACT: John Gerba, Office of Federal Activities, telephone: (202) 382-5910.

SUPPLEMENTARY INFORMATION:

OMB Control Number
2000-0422.

Background
An interim final rule amending EPA's NEPA regulations to include categorical exclusion procedures, authorized by CEQ regulations (40 CFR 1508.4), was published on pp. 9827-9831 in the March 8, 1982 **Federal Register.** These procedures largely apply to the environmental reviews of the Agency's Municipal Wastewater Treatment Construction Grants program and the National Pollution Discharge Elimination System (NPDES). EPA invited public comment on these procedures for 30 days ending April 7, 1982.

Since publication of the categorical exclusion procedures, several related regulatory actions have occurred. The EPA's Office of Water published amendments to its Construction Grants regulations (40 CFR Part 35) in the May 12, 1982 issue of the **Federal Register** which reference NEPA requirements. Also, in a separate document elsewhere in this issue of the **Federal Register,** the OFA has published proposed amendments to Subpart E of its NEPA regulations (40 CFR Part 6) "Environmental Review Procedures for the Wastewater Treatment Construction Grants Program," which incorporate many of the interim final categorical exclusion amendments. In response to these regulatory actions, and to recommendations received from several commentors on the March 8, 1982 NEPA amendments, the Agency has made a decision to extend the comment period on the interim final rule as revised by this document.

Included in this document are revisions necessary to make the categorical exclusion process consistent with the Construction Grants regulations, and to correct a factual error. Comments and suggestions received during the initial public comment period and not discussed in this document, and others received during the extended public comment period, will be considered together during final rule development.

Publication of the final rule on categorical exclusions will be consolidated with the amendments to Subpart E of 40 CFR Part 6 as mentioned above. The consolidation of these two documents for the final rule will permit making all changes to Subpart E needed to make it consistent with changes to the Municipal Construction Grants Amendments of 1981 at one time.

Interested parties wishing to comment on the categorical exclusion amendments to EPA's NEPA procedures are encouraged to review the following **Federal Register** documents for related issues: the amendments establishing the procedures published March 8, 1982; the amendments to the Wastewater Treatment Construction Grants regulations published as interim final in the **Federal Register** May 12, 1982; and the proposed amendments to 40 CFR Part 6 Subpart E published elsewhere in this issue of the **Federal Register.**

Classification

The Office of Federal Activities has determined that these revisions are not considered to be a "major" rule within the meaning of Executive Order (E.O.) 12291.

This amendment was submitted to the OMB for clearance as required by E.O. 12291 as well as for its information collection burden on the public. Any comments from OMB to EPA and any EPA response to those comments are available for public inspection in the Document Control Office, EPA, Room 107, 401 M Street, S.W., Washington, D.C. 20460.

Because these revisions are not "major," it is effective immediately as a corrected interim final rule.

Regulatory Analysis

Since EPA believes that these revisions are not "major," they are not subject to the Regulatory Impact Analysis procedures as outlined under E.O. 12291.

Paperwork Reduction

Information collection requirements contained in this regulation (Section 6.506(c)) have been approved by the Office of Management and Budget under the provisions of the Paperwork Reduction Act of 1980 (44 U.S.C. 3501 *et seq.*) and have been assigned OMB control number 2000-0422.

Discussion of Issues

Though there were other suggestions and issues raised in the comments received in response to the interim final rule published March 8, the following discussion focuses on those issues needed to make the interim final rule consistent with regulations for Municipal Wastewater Treatment Construction Grants.

Categories and Criteria

Past experience with the Municipal Wastewater Treatment Facility Construction Grants (WWTCG) program has indicated that full EISs are rarely undertaken on projects involving service populations under 10,000. Projects of this size, as a general rule, have not involved signficant long term impacts on the human environment; therefore the Office of Water suggested that the population limit in § 6.506(c)(1)(ii) and (iii) be revised upward from 3,500 to 10,000. This would allow more projects to be considered for an exclusion and reduce the regulatory burden on many grant applicants. Adequate protection exists within the regulation to ensure that full environmental reviews will be undertaken for projects that are likely to have significant environmental impacts.

The Office of Water also recommended that the growth capacity in § 6.506(c)(2)(iii) should correspond to the increment used by States in screening generic facilities under WWTCG's small communities program. OFA agrees with this position and has revised the allowable capacities upward from 20% to 30% to be consistent with the screening criterion.

The Office of Water also pointed out that the inclusion of "alternative" technologies in § 6.506(c)(1)(iii) was inappropriate because "alternative" technologies include many off-site as well as on-site technologies. Since the intent of this section is limited to consideration of on-site technologies, deleting the term "alternative" from the paragraph removes this inconsistency without precluding the use of those on-site technologies that are either "alternative" or "innovative" from consideration.

The effect of the suggestions discussed above would be to permit a greater number of projects to be considered for an exclusion.

Responsibilty for Environmental Assessment

In the March 8, 1982 **Federal Register** publication of the interim final rule, the preparation of a formal environmental assessment § 6.507(a) was erroneously identified as a grantee responsibility. Although delegated States are frequently involved in preparing preliminary assessments, the final assessment preparation is an IPA responsibility. This action corrects that error. Changes necessary to make this section consistent with the Municipal Construction Grants Amendments of 1981 are published elsewhere in this issue of the **Federal Register.**

Actions Taken

In response to regulatory developments in the MWTCG program and to several related comments and suggestions provided during the original public comment period, the following generally describes the changes made to the interim final rule published in the **Federal Register** March 8, 1982: 1) Changing the community size criterion used in § 6.506(c)(1)(ii) and (iii) from 3,500 to 10,000 population; 2) removal of the reference to alternative technologies from § 6.506(c)(1)(iii); and 3) increasing the excess capacity in § 6.506(c)(2)(iii) from 20% to 30%. The change to correct the factual error on responsibility for preparing an environmental assessment is found in § 6.507(a).

List of Subjects in 40 CFR Part 6

Environmental impact statements, foreign relations.

Dated: November 5, 1982.

Anne M. Gorsuch,
Administrator.

PART 6—[AMENDED]

For the reasons set out in the preamble, 40 CFR Part 6 is amended as follows:

1. Section 6.506 is correctecd by revising paragraph (c)(1) (ii) and (iii); and revising (c)(2)(iii) to read as follows:

§ ***6.506 Criteria for preparing EISs and granting categorical exclusions.***

* * * * *

(c) * * *

(1) * * *

(ii) Actions in sewered communities of less than 10,000 persons which are for minor upgrading and minor expansion of existing treatment works. This category does not include actions that directly or indirectly involve the extension of new collection systems funded with Federal or other sources of funds.

(iii) Actions in unsewered communities of less than 10,000 persons where onsite technologies are proposed.

(iv) * * *

(2) * * *

(iii) The facilities would provide capacity to serve a population 30% greater than the existing population.

* * * * *

2. Section 6.507 is amended by revising paragraph (a) to read as follows:

§ ***6.507 Environmental review process.***

* * * * *

(a) *Facilities planning (Step 1).* Early in facilities planning, the grantee should evaluate the likely project alternatives and the existence of environmentally sensitive areas in the facilities planning area, including those identified in § 6.506(c)(2)(v) of this Subpart. This evaluation is intended to be brief and concise and should draw on existing information from EPA, State agencies, regional planning agencies, areawide water quality management agencies, and the potential grant applicant. The evaluation and any additional analysis deemed necessary may be used by EPA to determine whether the action is eligible for a categorical exclusion from the substantive environmental review requirements of this Part. It is recommended that the potential applicant submit the information to EPA or a delegated State at the earliest possible time to allow EPA to determine if the action is eligible for a categorical exclusion. If a categorical exclusion is granted, the grantee will not be required to prepare a formal environmental information document nor will EPA need to prepare a formal environmental assessment during facilities planning. If an action has not been granted a categorical exclusion, this evaluation may be used to determine the scope of the environmental information document required of the grantee. It also should be used to make an early determination of the need for an EIS. Whenever possible, the potential grant applicant should discuss this initial evaluation with EPA or a delegated State, whichever is appropriate.

* * * * *

[FR Doc. 83-195 Filed 1-6-83; 8:45 am]

ENVIRONMENTAL PROTECTION AGENCY

40 CFR PART 6
[FA-FRL 2097-8]

National Environmental Policy Act; Environmental Review Procedures for the Wastewater Treatment Construction Grants Program

AGENCY: Environmental Protection Agency (EPA).

ACTION: Proposed rule.

SUMMARY: This document provides procedural and minor substantive amendments to EPA's Procedures Implementing the National Environmental Policy Act (NEPA) for the Wastewater Treatment Construction Grants Program (40 CFR Part 6 Subpart E). The procedural amendments accommodate recent changes in EPA's regulations for the Construction Grants Program (40 CFR Part 35) which have been modified to incorporate the Municipal Wastewater Treatment Construction Grants Amendments of 1981 (Pub. L. 97-117). The modifications in the grant program change the process recipients of EPA grants follow in the planning and construction of wastewater treatment facilities. The minor substantive amendments to Subpart E streamline the criteria for preparing an EIS. Further recommendations for additional substantive changes will be proposed in the near future and will comprehensively apply to all of 40 CFR Part 6.

DATE: Comments on this proposed rule must be received by February 7, 1983.

ADDRESSES: Comments may be mailed to the Office of Federal Activities, A-104, U.S. Environmental Protection Agency, 401 M Street, SW, Washington, D.C. 20460. Attention: Paul Cahill, Director.

For Further Information Contact: John Gerba, Office of Federal Activities, (202) 382-5910.

Supplementary Information:

Classification

The Office of Federal Activities has determined that this revision is not a "major" rule within the meaning of Executive Order (E.O.) 12291. This is because the revision will not: (1) have an annual effect on the economy of $100 million or more; (2) cause a major increase in costs or prices for consumers, individual industries, geographic regions, or Federal, State, or local government agencies; or (3) have significant adverse effects on competition, employment, investment, productivity, innovation, or on the ability of United States-based enterprises to compete with foreign-based enterprises in domestic or export markets.

The purpose and effect of this amendment to the environmental review process for the Construction Grants Program is to accommodate recent changes in the grant program and to make minor substantive changes. No increased paperwork burdens are imposed by the amendments.

This amendment was submitted to the Office of Management and Budget (OMB) for review as required by E.O. 12291. Any OMB comments on its reporting or information collection requirements will be addressed in the Final Rule.

This amendment is being published as a proposed rule to allow for public comment. Comments must be received by the Office of Federal Activities (OFA) before February 7, 1983.

Regulatory Analysis

Under E.O. 12291, EPA must determine if a regulation is "major" and therefore subject to a Regulatory Impact Analysis. Since EPA believes that this amendment is not "major," it is not subject to such an analysis.

Background

On December 29, 1981 President Reagan signed the Municipal Wastewater Treatment Construction Grants Amendments of 1981 (Pub. L. 97-117). The amendments reflect Congressional and Administration objectives to: (1) Reduce the Federal cost and involvement in the construction of municipal wastewater treatment facilities; (2) streamline the construction grants process; and (3) maintain the environmental integrity of the program. They also express the Administration's policy to delegate the operation of Federal programs to the appropriate level of government and to provide both States and municipalities with more flexibility in carrying out this responsibility. Although the amendments do not alter EPA's responsibility to make NEPA determinations, they do substantially affect how NEPA is applied by eliminating Step 1 and Step 2 Federal grant assistance.

Meeting NEPA Requirements

NEPA reviews have been most effective when they addressed environmental issues during the facility planning phase. With the elimination of Step 1 and Step 2 grants, official Federal involvement does not occur until after the completion of facilities planning and design. This effectively postpones the "major Federal action" which would trigger NEPA involvement until much of the planning and design phases are completed. The application of NEPA at this point in the development process could cause unnecessary waste and delay if potential Step 3 grantees propose environmentally unsuitable alternatives for Federal funding. The interim final amendments to 40 CFR Part 35 in the May 12, 1982 **Federal Register** address this issue by requiring that NEPA requirements (40 CFR Part 6) be met before submission of an application for a Step 3 (construction) grant. More specifically, the regulations at Section 35.2113 encourages potential applicants to work with the State and EPA as early as possible in the facility planning process to "ascertain the appropriateness of a categorical exclusion, a finding of no significant impact, or an environmental impact statement." They also allow a potential applicant to request a NEPA review early in the facilities planning or design stages. The amendments proposed here reflect this approach.

Categorical Exclusions

On March 8, 1982, an interim final regulation was printed in the **Federal Register** establishing the process for granting categorical exclusions from NEPA procedures for certain categories of wastewater treatment construction grant projects. This process will likely exclude 20 percent of the EPA funded projects from substantive environmental review. The interim final regulation (as revised by a document published in this issue of the **Federal Register**) will be combined with the proposed amendments to Subpart E, and together they will be published as a final regulation.

Action Being Taken: Subpart E Amendments

EPA is proposing to amend its procedures for implementing the National Environmental Policy Act (NEPA) to: (1) Be consistent with the Municipal Wastewater Treatment Construction Grant Amendments of 1981 (Pub. L. 97-117); (2) be consistent with changes in the Wastewater Treatment Construction Grants Program's regulations (40 CFR Part 35); (3) reorder the sections of Subpart E to more closely reflect the sequence of the steps undertaken in the environmental review process; and (4) make minor substantive changes to the criteria for deciding whether to prepare an EIS. The proposed amendments also provide that a decision by the responsible official to issue a finding of no significant impact or to prepare an EIS shall not be subject to administrative appeal before the EPA Board of Assistance Appeals. This provision is intended to reflect a proposed change in the Agency's general grant regulations (40 CFR Part 30) which excludes NEPA determinations under 40 CFR Part 6 from the Board's jurisdiction.

Pre- and Post-December 29, 1981 Grants

There are approximately 5,000 wastewater treatment facility planning projects at various stages of development that received Step 1 grants from EPA on or before December 29, 1981. Except as noted in the revised § 6.504 (b) and (c), the requirements of these proposed amendments apply to those projects and to projects subject to the Municipal Wastewater Treatment Construction Grant Amendments of 1981 (projects that did not receive a Step 1 grant on or before December 29, 1981). Although these proposed amendments include provisions for projects that received Step 1 grants on or before December 29, 1981, they do not substantively change the environmental review process for such projects and thus avoid the imposition of retroactive requirements.

Reordering and Clarifying of Subpart E Sections

The existing order of the sections and subsections of Subpart E does not follow the sequence of the environmental review process. In order to make the regulation more understandable, the order of the sections has been revised to follow the process. Tables provided below are a guide to the reordering of the text.

Distribution Table—Subpart E

Old Section	New Section
6.500 Purposes	6.500.
6.501 Definitions	6.501.
6.502 Applicability. [interim final*]	Obsolete.
6.503 Consultation during the environment review process. [revised interim final†].	6.512 Segmenting projects.
6.504 Public participation. [interim final].	6.513.
6.505 Limitations. [interim final].	Obsolete.
6.506(c) [interim final] and (b); Criteria for preparing EISs.	6.506 Criteria for initiating EISs.
6.506(c) *Categorical exclusions.* [revised interim final].	6.505.
6.507(a) *Categorical exclusions.* [revised interim final].	6.505.
6.507(a) [interim final] and (b); *Award of a facilities grant (Step 1). And Midcourse review.*	6.504 Consultation during the facility planning process.
6.507(c) and (d) *Review of completed facilities plan. And Environmental review.*	6.506 Environmental review process.
6.507(e) *Finding of No Significant Impact.*	6.507 ... (FNSI) determination.
6.507(f), (g) and (h) *Notice of intent, Scoping, and IES method.*	6.509 Environmental Impact Statement (EIS)—preparation.
6.508 Limits on delegation to States. [interim final].	6.514.
6.509 Identification of mitigation measures.	6.510 Record of decision and ...
6.510 Monitoring	6.511 Monitoring compliance.

*FR, March 8, 1982, pp. 9829-32.
†FR, of this date, following this action.

Derivation Table—Subpart E

New Section	Old Section
6.500 Purposes	6.500.
6.501 Definitions. (a)–(f)	6.501(a)–(f).
6.501(g)	New.
6.502 Applicability and limitations.	New.
6.503 Overview of the environmental review process.	New.
6.504(a) Consultation during the facility planning process.	6.507 Introductory paragraph.
6.504(a)(1) and (2)	6.507(a) [interim final*] and (b); new title.
6.504(c)	New.
6.505(a) Categorical exclusions.	6.506(c) Introductory paragraph [interim final].
6.505(b)	6.506(c)(1) [revised interim final].
6.505(c)	6.506(c)(2) [revised interim final].
6.505(d)	6.506(c)(3) [interim final].
6.506(a) and (b) Environmental review process.	6.507(o).
6.506(c)	6.507(d).
6.507 Finding of No Significant Impact (FNSI) determination.	6.507(e).
6.508 Criteria for initiating EISs.	6.506(a) [interim final] and (b).
6.509 Environmental Impact Statement (EIS) preparation. (introductory paragraph).	New.
6.509(a)	6.507(f).
6.509(b)	6.507(g).
6.509(c)	6.507(h).
6.510 Record of decision and indentification of mitigation measures.	6.509.
6.511 Monitoring compliance.	6.510.
6.512 Segmenting projects	6.503 [revised interim final].
6.513 Public participation	6.504 [interim final].
6.514 Delegation to States	6.508 [interim final].

*FR, March 8, 1982, pp. 9829-32.
†FR, of this date, following this action.

Minor Substantive Changes

EPA's Office of Water suggested a revision to the criteria for preparing an EIS (§ 6.508(a)(1)). The revision removes examples of land use related criteria that are currently recommended as a basis for preparing an EIS. In practice, these criteria have not been used as a basis for preparing EISs and are covered in other paragraphs of the same section. The revised language more succinctly states the land-use-related circumstances which require the preparation of an EIS.

Public and Agency Participation

These amendments were developed by a work group with representatives from EPA headquarters and regional offices. Their efforts followed the extensive public and regional comment process carried out by the Construction Grant program in developing amendments to 40 CFR Part 35 during which NEPA implementation was considered.

List of Subjects on 40 CFR Part 6

Environmental Impact Statements, Foreign relations.

Dated: October 27, 1982.

John W. Hernandez.
Acting Administrator.

For the reasons set out in the preamble, 40 CFR Part 6 is proposed to be amended as follows:

1. The authority citation for Part 6 reads as follows:

Authority: Sections 101, 102, and 103 of the National Environmental Policy Act of 1969 (42 U.S.C. 4321 et seq.); also, the Council on Environmental Quality Regulations dated November 29, 1978 (40 CFR Part 1500).

2. The title to 40 CFR Part 6 is revised to read as follows:

PART 6—PROCEDURES FOR IMPLEMENTING THE NATIONAL ENVIRONMENTAL POLICY ACT

3. Subpart E is revised to read as follows:

SUBPART E—ENVIRONMENTAL REVIEW PROCEDURES FOR THE WASTEWATER TREATMENT CONSTRUCTION GRANTS PROGRAM.

Sec.

Note.—To facilitate the identification of proposed changes for the reader, the text of both proposed Revised and New sections or paragraphs are enclosed by arrows (▶).◀

SUBPART E—ENVIRONMENTAL REVIEW PROCEDURES FOR THE WASTEWATER TREATMENT CONSTRUCTION GRANTS PROGRAM

§ ***6.500 Purpose.***

This subpart amplifies the procedures described in Subparts A through D with detailed environmental review procedures for the wastewater treatment works construction grants program under Title II of the Clean Water Act.

§ ***6.501 Definitions.***

▶(a) "Step 1 facilities planning" means preparation of a plan for facilities as described in 40 CFR Part 35, Subpart E or I.

(b) "Step 2" means preparation of design drawings and specifications as described in 40 CFR Part 35, Subpart E or I.

(c) "Step 3" means building of a publicly owned treatment works as described in 40 CFR Part 35, Subpart E or I.

(d) "Step 2 + 3" means a project which combines preparation of design drawings and specifications as described in § 6.501(b), and building as described in § 6.501(c).◀

(e) "Applicant"▶means any individual, agency, or entity which has filed an application for grant assistance under 40 CFR Part 35, Subpart E or I.◀

(f) "Grantee"▶means any individual, agency, or entity which has been awarded wastewater treatment construction grant assistance under 40 CFR Part 35, Subpart E or I.◀

▶(g) "Responsible official" means the Federal or State decision maker authorized to fulfill the requirements of this subpart. The responsible Federal official is the EPA Regional Administrator and the responsible State official is as defined in a delegation agreement under § 205(g) of the Clean Water Act subject to the limitations in § 6.514 of this subpart.◀

▶§ ***6.502 Applicability and limitations.***

(a) *Applicability.* This Subpart applies to the following actions:

(1) Projects that received Step 1 grant assistance on or before December 29, 1981;

(2) Approval of grant assistance for a project involving Step 3 or Step 2 + 3; and

(3) Award of grant assistance for a project where significant change has occurred in the project or its impact since compliance with this Part.

(b) *Limitations.* Recipients of Step 1 grant assistance must comply with the requirements, steps, and procedures described in this Subpart. As specified in 40 CFR 35.2113, projects that have not received Step 1 grant assistance must comply with the requirements of this subpart prior to submission of an application for Step 3 or Step 2 + 3 grant assistance.◀

▶§ ***6.503 Overview of the environmental review process.***

The process for conducting an environmental review of wastewater treatment construction grant projects includes several steps whose procedures are described in subsequent sections of this subpart. The steps are:

(a) *Consultation.* The Step 1 grantee or the potential Step 3 or Step 2 + 3 applicant is encouraged to consult with EPA early in project formulation or facilities planning stage to determine whether a project is eligible for a categorical exclusion from the remaining substantive environmental review requirements of this part (§ 6.505) and to identify potential environmental issues.

(b) *Determining categorical exclusion eligibility.* At the request of a potential Step 3 or Step 2 + 3 grant applicant or a Step 1 facilities planning grantee, EPA determines the eligibility of the project for a categorical exclusion. A Step 1 facilities planning grantee awarded a Step 1 grant on or before December 29, 1981 may request a categorical exclusion at any time during Step 1 facilities planning or Step 2 design work. A potential Step 3 or Step 2 + 3 grant applicant may request a categorical exclusion at any time before the submission of a Step 3 or Step 2 + 3 grant application.

(c) *Documenting environmental information.* If the project is determined to be ineligible for a categorical exclusion, the potential Step 3 or Step 2 + 3 applicant or the Step 1 grantee subsequently prepares an Environmental Information Document (EID) (§ 6.506) for the project.

(d) *Preparing environmental assessments.* Except as provided in § 6.506(c)(3) and following a review of the EID by EPA or by a State with delegated authority, EPA prepares an environmental assessment (§ 6.506), or a State with delegated authority (§ 6.514) prepares a preliminary environmental assessment. EPA reviews and finalizes any preliminary assessments. EPA subsequently:

(1) prepares and issues a Finding of No Significant Impact (FNSI); or

(2) prepares and issues an Environmental Impact Statement (EIS) (§ 6.509) and record of decision (§ 6.510).

(e) *Monitoring.* The construction and post-construction operation and maintenance of the facilities is monitored (§ 6.511) to ensure the implementation of mitigation measures (§ 6.510) identified in the FNSI, final EIS or record of decision.

▶§ ***6.504 Consultation during the facility planning process.***

(a) *General.* Consistent with 40 CFR 1501.2, EPA shall initiate the environmental review process as early as possible in order to identify environmental effects, avoid delays, and resolve conflicts. The environmental review process should be integrated throughout the facilities planning process (Step 1). Two processes for consultation are described in this section to meet this directive. The first addresses projects which were awarded Step 1 grant assistance on or before December 29, 1981. The second applies to projects which did not receive grant assistance for facilities planning on or before December 29, 1981 and are, therefore, subject to the regulations implementing the Municipal Wastewater Treatment Construction Grant Amendments of 1981 (40 CFR Part 35 Subpart I).

(b) *Projects that received Step 1 grant assistance on or before December 29, 1981.* (1) Early in facilities planning, the grantee should evaluate the likely project alternatives and the existence of environmentally sensitive areas in the facilities planning area, including those identified in § 6.508 of this Subpart. This evaluation is intended to be brief and concise and should draw on existing information from EPA, State agencies, regional planning agencies, areawide water quality management agencies, and the Step 1 grantee. The evaluation and any additional analysis deemed necessary may be used by EPA to determine whether the action is eligible for a categorical exclusion from the substantive environmental review requirements of this Part. It is recommended that the Step 1 grantee submit the information to EPA or a delegated State at the earliest possible time to allow EPA to determine if the action is eligible for a categorical exclusion. If a categorical exclusion is granted, the grantee will not be required to prepare a formal EID nor will EPA need to prepare an environmental assessment. If an action has not been granted a categorical exclusion this evaluation may be used to determine the scope of the EID required of the grantee. It also should be used to make an early determination of the need for an EIS. Whenever possible, the Step 1 grantee should discuss this initial evaluation with EPA or a delegated State, whichever is appropriate.

(2) A review of environmental information developed by the grantee should be conducted to the extent practicable whenever meetings are held to assess the progress of facilities plan development. These meetings should be held after completion of the majority of the EID document and before a preferred alternative is selected. Since any required EIS must be completed before the approval of a facility plan for a project which received a Step 1 grant on or before December 29, 1981, a decision whether to prepare an EIS is encouraged early during the facilities planning process. These meetings may assist in this early determination. EPA should inform interested parties of the following:

(i) The preliminary nature of the Agency's position on preparing an EIS;

(ii) The relationship between the facilities planning and environmental review processes;

(iii) The desirability of public input; and

(iv) A contact person for further information.

(c) *Projects that did not receive grant assistance for Step 1 facility planning on or before December 29, 1981.* Potential Step 3 or Step 2 + 3 grant applicants are encouraged to consult with EPA or the State during the facilities planning process. In accordance with § 35.2030(c), the potential applicant should work with the State and EPA as early as possible in the facilities planning process to determine the appropriateness of a categorical exclusion, the scope of an EID, or the appropriateness of the early preparation of a FNSI or an EIS. The consultation would be most useful if initiated during the evaluation of project alternatives and prior to the selection of a preferred alternative. This consultation may also assist the potential applicant in resolving any identified environmental problems.◀

▶§ *6.505 Categorical exclusions.*

(a) *General.* At the request of an existing Step 1 facilities planning grantee or of a potential Step 3 or Step 2 + 3 grant applicant, the responsible official, as provided for in § 6.107(b) and § 6.504(a), shall determine from existing information whether an action is consistent with the categories eligible for exclusion identified in § 6.505(b). The responsible official shall document this determination as provided for in § 6.107(b).◀

(b) *Categories of actions eligible for exclusion.* For this subpart, actions consistent with the following categories are eligible for a categorical exclusion:

▶(1) Actions for which the facilities planning is solely directed toward minor rehabilitation of existing facilities, functional replacement of equipment, or towards the construction of new ancillary facilities adjacent or appurtenant to existing facilities which do not affect the degree of treatment or capacity of the existing facility. Such actions include but are not limited to infiltration and inflow corrections grant eligible replacement of existing mechanical equipment or structures, and the construction of new small on-site structures.◀

▶(2)◀ Actions in sewered communities of less than 10,000 persons which are for minor upgrading and minor expansion of existing treatment works. This category does not include actions that directly or indirectly involve the extension of new collection systems funded with Federal or other sources of funds.

▶(3)◀ Actions in unsewered communities of less than 10,000 persons where onsite technologies are proposed.

▶(4) Other actions developed in accordance with paragraph (d) of this section.

(c) *Criteria for not granting a categorical exclusion.* (1) The full environmental review procedures of this part must be followed if undertaking an action consistent with the categories described in § 6.505(b) may involve serious local or environmental issues, or meets any of the criteria listed below:◀

(i) The facilities to be provided will create a new discharge to surface or ground waters;

(ii) The facilities will result in substantial increases in the volume of discharge or the loading of pollutants from an existing source or from new facilities to receiving waters;

(iii) The facilities would provide capacity to serve a population 30% greater than the existing population;

(iv) The action is known or expected to have a significant effect on the quality of the human environment, either individually, cumulatively over time, or in conjunction with other Federal, State, local, or private actions;

(v) The action is known or expected to directly or indirectly affect sensitive environmental resources or areas, such as floodplains, wetlands, prime or unique agricultural lands, aquifer recharge zones, archaeological and historic sites, endangered or threatened species, or other areas identified in guidance issued by the OFA; or

(vi) The action is known or expected not to be cost-effective or to cause significant public controversy.

▶(2) Notwithstanding the provisions of § 6.505(b), if any of the above conditions exist, the responsible official shall ensure:

(i) That a categorical exclusion is not granted;

(ii) That an adequate EID and environmental assessment are prepared.

(iii) That either a FNSI or an EIS and record of decision is prepared and issued.◀

▶(d)◀ *Developing new categories of excluded actions.* The responsible official or other interested parties may request that a new category of excluded actions be created, or that an existing category be amended or deleted. The request shall be made in writing to the Director, OFA and shall contain adequate information to support the request. Under the direction of OFA, proposed new categories shall be developed through EPA's "non-major" rulemaking process (E.O. 12291), including publication as an interim final rule in the **Federal Register** and a subsequent thirty (30) day public comment period. The following shall be considered in evaluating proposals for new categories:

▶(1) Actions in the proposed category should seldom result in the effects identified in § 6.505(c);

(2) Based upon previous environmental reviews, actions consistent with the proposed category have not required the preparation of an EIS; and◀

▶(3)◀ Whether information adequate to determine if a potential action is consistent with the proposed category will normally be available when needed.

§ ***6.506 Environmental review process.***

(a) *Review of completed facilities plans.* EPA, or the State where the program is delegated, shall review the completed facilities plan with particular attention to the EID and its utilization in the development of alternatives and the selection of a preferred alternative. An adequate EID shall be an integral part of any facilities plan submitted to EPA or to a State. The EID shall be of sufficient scope to enable the responsible official to prepare an environmental assessment.

(b) *Environmental assessment.* The environmental assessment shall cover all potentially significant environmental impacts. For those States where the review of facilities plans has been delegated, State personnel shall prepare a preliminary environmental assessment which serves as an adequate basis for EPA's decision to issue a FNSI or an EIS. Each of the following subjects shall be critically reviewed to identify potentially significant environmental concerns and shall be addressed in the environmental assessment.

(1) *Description of the existing environment.* For the delineated facilities planning area, the existing environmental conditions relevant to the analysis of alternatives or to determining the environmental impacts of the proposed action shall be considered.

(2) *Description of the future environment without the project.* The relevant future environmental conditions shall be described. The no action alternative should be discussed.◀

(3) *Purpose and need.* This should include a summary discussion and demonstration of the need for wastewater treatment in the facilities planning area, with particular emphasis on existing public health or water quality problems and their severity and extent.

▶(4) *Documentation.* Sources of information used to describe the existing environment and to assess future environmental impacts should be clearly referenced. These sources should include regional, State, and Federal agencies with responsibility or interest in the types of conditions listed in § 6.508 and in Subpart C.

(5) *Evaluation of Alternatives.* This discussion shall include a comparative analysis of feasible alternatives, including the no action alternative, throughout the study area. The alternatives shall be screened with respect to capital and operating costs; significant direct and indirect environmental effects; physical, legal, or institutional constraints; and compliance with regulatory requirements. Special attention should be given to long term, irreversible, and induced impacts. The reasons for rejecting any alternatives shall be presented in addition to any significant environmental benefits precluded by rejection of an alternative. The analysis should consider when relevant to the project:◀

(i) Flow and waste reduction measures, including infiltration/inflow reduction;
(ii) Appropriate water conservation measures;
(iii) Alternative locations, capacities, and construction phasing of facilities;
(iv) Alternative waste management techniques, including treatment and discharge, wastewater reuse, land application, and individual systems;
(v) Alternative methods for management of sludge, other residual materials, including utilization options such as land application, composting, and conversion of sludge for marketing as a soil conditioner or fertilizer;
(vi) Improving effluent quality through more efficient operation and maintenance;
(vii) Appropriate energy reduction measures; and
(viii) Multiple use, including recreation and education.

(6) *Environmental consequences.* Relevant impacts of the proposed action shall be considered, steps to mitigate significant adverse impacts, any irreversible or irretrievable commitments of resources to the project and the relationship between local short term uses of the environment and the maintenance and enhancement of long term productivity. Any specific requirements, including grant conditions and areawide waste treatment management plan requirements, should be identified and referenced. In addition to these items, the responsible official may require that other analyses and data which are needed to satisfy environmental review requirements, be included with the facilities plan. Such requirements should be discussed whenever meetings are held with Step 1 grantees or potential Step 3 or Step 2 + 3 applicants. The responsible official also may require submission of supplementary information before the award of grant assistance if needed for compliance with environmental review requirements. Requests for supplementary information shall be made in writings.◀

(7) *Steps to minimize significant adverse effects.* (i) This section shall describe structural and nonstructural measures, if any in the facilities plan, or additional measures identified during the review, to mitigate or eliminate significant adverse effects on the human and natural environments. Structural provisions include changes in facility design, size, and location; non-structural provisions include staging facilities as well as developing and enforcing land use regulations and environmental protection regulations.

▶(ii) The responsible official shall not award grant assistance if the grantee has not made, or agreed to make, pertinent changes in the project, in accordance with determinations made in a FNSI or EIS. The responsible official shall condition a grant to seek other ways of compliance, to ensure that the grantee will comply with such environmental review determinations.

(c) *FNSI/EIS determination.* The responsible official shall apply the criteria under § 6.508 to any of the following:◀

(1) A complete facilities plan and the EID, whenever review of facilities plan has not been delegated;

(2) A complete facilities plan, the applicant's EID, information document and the preliminary environmental assessment prepared by the State, for a State which has been delegated authority for facilities plan review; or

(3) Other documentation, deemed necessary by the responsible official or submitted by a State with delegated review authority, adequate to make an EIS determination by EPA. Where EPA determines that an EIS is to be prepared, there is no need to prepare a formal environmental assessment.

▶If EPA or the State identifies deficiencies in the EID, preliminary environmental assessment, or other supporting documentation, necessary corrections shall be made before the conditions of the Step 1 grant are considered satisfied or before the Step 3 or

Step 2 + 3 application is considered complete. The responsible official's determination to issue a FNSI or to prepare an EIS shall constitute final Agency action and shall not be subject to administrative appeal to the EPA Board of Assistance Appeals under 40 CFR Part 30.◀

▶§ ***6.507 Findings of No Significant Impact (FNSI) determination.***

If, after completion of the environmental review, EPA determines that an EIS will not be required, the responsible official shall prepare and distribute a FNSI in accordance with § 6.104 and Subpart D of this Chapter. The FNSI will be based on EPA's independent review and the environmental assessment which will either be incorporated into or attached to the FNSI. In accordance with 40 CFR 1508.2, the FNSI shall list any mitigation measures necessary to make the recommended alternative environmentally acceptable. Once an environmental assessment and a FNSI have been prepared for the facilities plan for a certain area, grant awards may proceed without preparation of additional FNSIs, unless the responsible official determines that the project has changed significantly from that which underwent environmental assessment.◀

§ ***6.508 Criteria for initiating Environmental Impact Statements (EISs).***

(a) *Conditions requiring EISs.* The responsible official shall assure that an EIS will be prepared and issued when he determines that any of the following conditions exist:

(1) The treatment works in and of itself will significantly affect the pattern and type of land use (industrial, commercial, agricultural, residential) or the potential effects resulting from the construction or operation of the treatment works will conflict with established land use plans or policies;◀

(2) The treatment works or collector system will have significant adverse effects on wetlands, including indirect effects, or any major part of the treatment works will be located on wetlands;

(3) The treatment works or collector system will significantly affect a habitat identified on the Department of the Interior's or a State's threatened and endangered species lists, or their treatment works will be located on the habitat;

(4) Implementation of the treatment works or plan may directly cause or induce changes that significantly:

(i) Displace population;

(ii) Alter the character of an existing residential area;

(iii) Adversely affect a floodplain; or

(iv) Adversely affect significant amounts of prime or unique agricultural land, or agricultural operations on this land as defined in EPA's Policy to Protect Environmentally Significant Agricultural Land.

(5) The treatment works will have significant adverse direct or indirect effects on parklands, other public lands or areas of recognized scenic, recreational, archaeological, or historic value; or

(6) The treatment works may directly or through induced development have a significant adverse effect upon local ambient air quality, local ambient noise levels, surface or groundwater quality or quantity, fish, wildlife, and their natural habitats.

(7) The treated effluent is being discharged into a body of water where the present classification is too lenient or is being challenged as too low to protect present or recent uses, and the effluent will not be of sufficient quailty or quantity to meet the requirements of these uses.

▶(b) *Other conditions.* The responsible official shall consider preparing an EIS if it is determined that the treatment works may threaten a violation of Federal, State, or local law or requirements imposed for the protection of the environment.◀

▶§ *6.509* ***Environmental Impact Statement (EIS) preparation.***

In addition to the requirements specified in subpart B, C, and D of this part, EPA will conduct the following activities:

(a) *Notice of intent.* If a determination is made that an EIS will be required, the responsible official shall prepare and distribute a notice of intent as required in Subpart D in accordance with § 6.104.

(b) *Scoping.* As soon as possible, after the publication of the notice of intent, the responsible official will convene a meeting of affected Federal, State, and local agencies, the grantee and other interested parties to determine the scope of the EIS. A notice of this scoping meeting will meet the requirements of Subpart D. As part of the scoping meeting EPA will as a minimum:◀

(1) Determine the scope and the significant issues to be analyzed in depth in the EIS;

(2) Identify those issues which are not significant;

(3) Determine what information is needed from cooperating agencies or other parties;

(4) Discuss the method for EIS preparation and the public participation strategy;

(5) Identify consultation requirements of other environmental laws, in accordance with subpart C; and

(6) Determine the relationship between the EIS and the completion of the facilities plan and any necessary coordination arrangements between the preparers of both documents.

▶(c) *Methods for preparing EISs.* EPA shall prepare this EIS by any one of the following means:◀

(1) Directly by its own staff;

(2) By contracting directly with a qualified consulting firm; or

(3) By utilizing a joint EIS process, whereby the grantee contracts directly with a qualified consulting firm. In this case the draft EIS serves the purpose of and satisfies the requirement for an EID. In this instance, the following selection requirements shall be fulfilled:

(i) A Memorandum of Understanding shall be developed between EPA, the grantee, and where possible, the State, outlining the responsibilities of each party and their relationship to the EIS consultant;

(ii) EPA shall approve evaluation criteria to be used in the consultant selection process;

(iii) EPA shall review and approve the selection process; and

(iv) EPA shall approve the consultant selected for EIS preparation.

▶§ *6.510* ***Record of decision and identification of mitigation measures.***

(a) *Record of decision.* When a final EIS has been issued, the responsible official shall prepare a record of decision in accordance with 40 CFR 1505.2 prior to the submission of an application for grant asssistance (40 CFR Part 35.2113). The record of decision shall include identification of mitigation measures derived from the EIS process which are necessary to make the recommended alternative environmentally acceptable.

(b) *Specific mitigation measures.* Prior to the approval of grant assistance, the responsible official must ensure that effective mitigation measures identified in the FNSI, final EIS, or record of decision are implemented by the grantee. This should be done by revising the facilities plan, initiating other steps to mitigate adverse effects, or agreeing to conditions in grants requiring actions to minimize effects. Care should be exercised if a condition is to be imposed in a grant document to assure that the applicant possesses the authority to fulfill the conditions.◀

▶§ *6.511 Monitoring for compliance.*

(a) *General.* The responsible official shall ensure there is adequate monitoring of mitigation measures and other grant conditions identified in the FNSI, final EIS, and record of decision.

(b) *Enforcement.* The responsible official may consider taking the following actions consistent with 40 CFR 35.965 and 30.430 if the grantee fails to comply with grant conditions:◀

(1) Terminating or annulling the grant;

(2) Disallowing project costs related to noncompliance;

(3) Withholding project payments;

▶(4) Finding the grantee to be nonresponsible or ineligible for future Federal assistance or for approval for future contract awards under EPA grants;

(5) Seeking an injunction against the grantee; or

(6) Instituting such other administrative or judicial action as may be legally available and appropriate.◀

▶§ *6.512 Segmenting projects.*

(a) *Criteria for segmenting.* When there are overriding considerations of costs or impaired program effectiveness, a Step 3 grant for the building of a discrete segment of the treatment works may be awarded before the environmental review is completed if the segmented portion of the treatment works:

(1) is noncontroversial;

(2) is necessary to correct water quality or other immediate environmental problems; and

(3) will not, by its completion, foreclose any reasonable options being considered in the environmental review.

(b) *EIS determination.* If a treatment works is to be segmented, the entire treatment works shall be evaluated to determine if an EIS is required. In applying the criteria to determine if an EIS is required, the regional EIS preparation staff shall be consulted.

(c) *Steps in segmenting.* In no case may grant assistance for a segmented Step 3 project be awarded unless:

(1) the OFA has been consulted;

(2) a FNSI on the segment permitted to proceed has been issued at least 30 days prior to grant award; and

(3) the grant award contains a specific agreement prohibiting the building of additional or different segments of the treatment works for which the environmental review is not complete.◀

▶§ *6.513 Public participation.*

(a) *General.* It is EPA policy that optimum public participation be achieved during the environmental review process as deemed appropriate by the responsible official. Compliance with public participation activities require under this part, Part 25, and Part 35 Subpart E or I constitutes compliance with the requirements for public participation under this subpart.

(b) *Coordination.* NEPA related public participation activities undertaken in connection with the environmental review process should be coordinated with any applicable public participation program wherever possible.

(c) *Scope.* Consistent with 40 CFR 1506.6, the responsible official may institute such additional NEPA-related public participation procedures as is deemed necessary during the environmental review process.◀

▶§ ***6.514 Delegation to States.***

(a) *General.* In cases where the authority for facilities plan review has been delegated to the State under section 205(g) of the Clean Water Act, the State may be delegated the responsibility for carrying out all EPA activities under this part except for the following responsibilities:◀

(1) The determination of whether or not to prepare an EIS shall be solely that of EPA. EPA shall consider a State's recommendations, but the ultimate decision under NEPA cannot be delegated;

(2) Categorical exclusions, Findings of No Significant Impact, and the environmental assessment shall be approved, finalized and issued by EPA; and

(3) Notices of intent shall be prepared and issued by EPA.

(b) *Elimination of duplication.* ▶The responsible official shall assure that maximum efforts are undertaken to minimize duplication within the limits described under § 6.508 and under paragraph (a) of this section. In carrying out requirements under this subpart, maximum consideration shall be given to eliminating duplication in accordance with 40 CFR 1506.2. Where there are State or local procedures comparable to NEPA, EPA should enter into memoranda of understanding with a State concerning workload distribution and responsibilities for implementing the environmental review and facilities planning process.◀

[FR Doc. 83-193 Filed 1-6-83; 8:45 am]

APPENDIX I
EPA Sample Consolidated Permit Forms

United States
Environmental Protection
Agency

Office of
Enforcement
Washington, DC 20460

EPA Form 3510-1
June 1980

Permits Division

EPA

Application Form 1 - General Information

Consolidated Permits Program

This form must be completed by all persons applying for a permit under EPA's Consolidated Permits Program. See the general instructions to Form 1 to determine which other application forms you will need.

DESCRIPTION OF CONSOLIDATED PERMIT APPLICATION FORMS

The Consolidated Permit Application Forms are:

Form 1 – General Information *(included in this part)*;

Form 2 – Discharges to Surface Water *(NPDES Permits)*:

2A. Publicly Owned Treatment Works *(Reserved – not included in this package)*,

2B. Concentrated Animal Feeding Operations and Aquatic Animal Production Facilities *(not included in this package)*,

2C. Existing Manufacturing, Commercial, Mining, and Silvicultural Operations *(not included in this package)*, and

2D. New Manufacturing, Commercial, Mining, and Silvicultural Operations *(Reserved – not included in this package)*;

Form 3 – Hazardous Waste Application Form *(RCRA Permits – not included in this package)*;

Form 4 – Underground Injection of Fluids *(UIC Permits – Reserved – not included in this package)*; and

Form 5 – Air Emissions in Attainment Areas *(PSD Permits – Reserved – not included in this package)*.

FORM 1 PACKAGE TABLE OF CONTENTS

SECTION A – GENERAL INSTRUCTIONS

Who Must Apply

With the exceptions described in Section C of these instructions, Federal laws prohibit you from conducting any of the following activities without a permit.

NPDES *(National Pollutant Discharge Elimination System Under the Clean Water Act, 33 U.S.C. 1251)*. Discharge of pollutants into the waters of the United States.

RCRA *(Resource Conservation and Recovery Act, 42 U.S.C. 6901)*. Treatment, storage, or disposal of hazardous wastes.

UIC *(Underground Injection Control Under the Safe Drinking Water Act, 42 U.S.C. 300f)*. Injection of fluids underground by gravity flow or pumping.

PSD *(Prevention of Significant Deterioration Under the Clean Air Act, 72 U.S.C. 7401)*. Emission of an air pollutant by a new or modified facility in or near an area which has attained the National Ambient Air Quality Standards for that pollutant.

Each of the above permit programs is operated in any particular State by either the United States Environmental Protection Agency *(EPA)* or by an approved State agency. You must use this application form to apply for a permit for those programs administered by EPA. For those programs administered by approved States, contact the State environmental agency for the proper forms.

If you have any questions about whether you need a permit under any of the above programs, or if you need information as to whether a particular program is administered by EPA or a State agency, or if you need to obtain application forms, contact your EPA Regional office *(listed in Table 1)*.

Upon your request, and based upon information supplied by you, EPA will determine whether you are required to obtain a permit for a particular facility. Be sure to contact EPA if you have a question, because Federal laws provide that **you may be heavily penalized if you do not apply for a permit when a permit is required.**

Form 1 of the EPA consolidated application forms collects general information applying to all programs. You must fill out Form 1 regardless of which permit you are applying for. In addition, you must fill out one of the supplementary forms *(Forms 2 – 5)* for each permit needed under each of the above programs. Item II of Form 1 will guide you to the appropriate supplementary forms.

You should note that there are certain exclusions to the permit requirements listed above. The exclusions are described in detail in Section C of these instructions. If your activities are excluded from permit requirements then you do not need to complete and return any forms.

NOTE: Certain activities not listed above also are subject to EPA administered environmental permit requirements. These include permits for ocean dumping, dredged or fill material discharging, and certain types of air emissions. Contact your EPA Regional office for further information.

Table 1. Addresses of EPA Regional Contacts and States Within the Regional Office Jurisdictions

REGION I

Permit Contact, Environmental and Economic Impact Office, U.S. Environmental Protection Agency, John F. Kennedy Building, Boston, Massachusetts 02203, (617) 223–4635, FTS 223–4635.
Connecticut, Maine, Massachusetts, New Hampshire, Rhode Island, and Vermont.

REGION II

Permit Contact, Permits Administration Branch, Room 432, U.S. Environmental Protection Agency, 26 Federal Plaza, New York, New York 10007, (212) 264–9880, FTS 264–9880.
New Jersey, New York, Virgin Islands, and Puerto Rico.

REGION III

Permit Contact *(3 EN 23)*, U.S. Environmental Protection Agency, 6th & Walnut Streets, Philadelphia, Pennsylvania 19106, (215) 597–8816, FTS 597–8816.
Delaware, District of Columbia, Maryland, Pennsylvania, Virginia, and West Virginia.

REGION IV

Permit Contact, Permits Section, U.S. Environmental Protection Agency, 345 Courtland Street, N.E., Atlanta, Georgia 30365, (404) 881–2017, FTS 257–2017.
Alabama, Florida, Georgia, Kentucky, Mississippi, North Carolina, South Carolina, and Tennessee.

REGION V

Permit Contact *(5EP)*, U.S. Environmental Protection Agency, 230 South Dearborn Street, Chicago, Illinois 60604, (312) 353–2105, FTS 353–2105.
Illinois, Indiana, Michigan, Minnesota, Ohio, and Wisconsin.

SECTION A – GENERAL INSTRUCTIONS *(continued)*

Table 1 *(continued)*

REGION VI

Permit Contact *(6AEP)*, U.S. Environmental Protection Agency, First International Building, 1201 Elm Street, Dallas, Texas 75270, (214) 767–2765, FTS 729–2765.
Arkansas, Louisiana, New Mexico, Oklahoma, and Texas.

REGION VII

Permit Contact, Permits Branch, U.S. Environmental Protection Agency, 324 East 11th Street, Kansas City, Missouri 64106, (816) 758–5955, FTS 758–5955.
Iowa, Kansas, Missouri, and Nebraska.

REGION VIII

Permit Contact *(8E–WE)*, Suite 103, U.S. Environmental Protection Agency, 1816 Lincoln Street, Denver, Colorado 80203, (303) 837–4901, FTS 837–4901.
Colorado, Montana, North Dakota, South Dakota, Utah, and Wyoming.

REGION IX

Permit Contact, Permits Branch *(E–4)*, U.S. Environmental Protection Agency, 215 Fremont Street, San Francisco, California 94105, (415) 556–3450, FTS 556–3450.
Arizona, California, Hawaii, Nevada, Guam, American Samoa, and Trust Territories.

REGION X

Permit Contact *(M/S 521)*, U.S. Environmental Protection Agency, 1200 6th Avenue, Seattle, Washington 98101, (206) 442–7176, FTS 399–7176.
Alaska, Idaho, Oregon, and Washington.

Where to File

The application forms should be mailed to the EPA Regional office whose Region includes the State in which the facility is located *(see Table 1)*.

If the State in which the facility is located administers a Federal permit program under which you need a permit, you should contact the appropriate State agency for the correct forms. Your EPA Regional office *(Table 1)* can tell you to whom to apply and can provide the appropriate address and phone number.

When to File

Because of statutory requirements, the deadlines for filing applications vary according to the type of facility you operate and the type of permit you need. These deadlines are as follows:[1]

Table 2. Filing Dates for Permits

FORM*(permit)*	WHEN TO FILE
2A*(NPDES)*	180 days before your present NPDES permit expires.
2B*(NPDES)*	180 days before your present NPDES permit expires[2], or 180 days prior to startup if you are a new facility.
2C*(NPDES)*	180 days before your present NPDES permit expires[2].
2D*(NPDES)*	180 days prior to startup.
3*(Hazardous Waste)*	Existing facility: Six months following publication of regulations listing hazardous wastes. New facility: 180 days before commencing physical construction.
4*(UIC)*	A reasonable time prior to construction for new wells; as directed by the Director for existing wells.
5*(PSD)*	Prior to commencement of construction.

[1] Please note that some of these forms are not yet available for use and are listed as "Reserved" at the beginning of these instructions. Contact your EPA Regional office for information on current application requirements and forms.

[2] If your present permit expires on or before November 30, 1980, the filing date is the date on which your permit expires. If your permit expires during the period December 1, 1980 – May 31, 1981, the filing date is 90 days before your permit expires.

Federal regulations provide that you may not begin to construct a new source in the NPDES program, a new hazardous waste management facility, a new injection well, or a facility covered by the PSD program before the issuance of a permit under the applicable program. Please note that if you are required to obtain a permit before beginning construction, as described above, you may need to submit your permit application well in advance of an applicable deadline listed in Table 2.

Fees

The U.S. EPA does not require a fee for applying for any permit under the consolidated permit programs. *(However, some States which administer one or more of these programs require fees for the permits which they issue.)*

Availability of Information to Public

Information contained in these application forms will, upon request, be made available to the public for inspection and copying. However, you may request confidential treatment for certain information which you submit on certain supplementary forms. The specific instructions for each supplementary form state what information on the form, if any, may be claimed as confidential and what procedures govern the claim. No information on Forms 1 and 2A through 2D may be claimed as confidential.

Completion of Forms

Unless otherwise specified in instructions to the forms, each item in each form must be answered. To indicate that each item has been considered, enter "NA," for not applicable, if a particular item does not fit the circumstances or characteristics of your facility or activity.

If you have previously submitted information to EPA or to an approved State agency which answers a question, you may either repeat the information in the space provided or attach a copy of the previous submission. Some items in the form require narrative explanation. If more space is necessary to answer a question, attach a separate sheet entitled "Additional Information."

Financial Assistance for Pollution Control

There are a number of direct loans, loan guarantees, and grants available to firms and communities for pollution control expenditures. These are provided by the Small Business Administration, the Economic Development Administration, the Farmers Home Administration, and the Department of Housing and Urban Development. Each EPA Regional office *(Table 1)* has an economic assistance coordinator who can provide you with additional information.

EPA's construction grants program under Title II of the Clean Water Act is an additional source of assistance to publicly owned treatment works. Contact your EPA Regional office for details.

SECTION B – FORM 1 LINE–BY–LINE INSTRUCTIONS

This form must be completed by all applicants.

Completing This Form

Please type or print in the unshaded areas only. Some items have small graduation marks in the fill–in spaces. These marks indicate the number of characters that may be entered into our data system. The marks are spaced at 1/6" intervals which accommodate elite type *(12 characters per inch)*. If you use another type you may ignore the marks. If you print, place each character between the marks. Abbreviate if necessary to stay within the number of characters allowed for each item. Use one space for breaks between words, but not for punctuation marks unless they are needed to clarify your response.

Item I

Space is provided at the upper right hand corner of Form 1 for insertion of your EPA Identification Number. If you have an existing facility, enter your Identification Number. If you don't know your EPA Identification Number, please contact your EPA Regional office *(Table 1)*, which will provide you with your number. If your facility is new *(not yet constructed)*, leave this item blank.

Item II

Answer each question to determine which supplementary forms you need to fill out. Be sure to check the glossary in Section D of these instructions for the legal definitions of the **bold faced words**. Check Section C of these instructions to determine whether your activity is excluded from permit requirements.

If you answer "no" to every question, then you do not need a permit, and you do not need to complete and return any of these forms.

If you answer "yes" to any question, then you must complete and file the supplementary form by the deadline listed in Table 2 along with this form. *(The applicable form number follows each question and is enclosed in parentheses.)* You need not submit a supplementary form if you already have a permit under the appropriate Federal program, unless your permit is due to expire and you wish to renew your permit.

Questions (I) and (J) of Item II refer to major new or modified sources subject to Prevention of Significant Deterioration *(PSD)* requirements under the Clean Air Act. For the purpose of the PSD program, major sources are defined as: (A) Sources listed in Table 3 which have the potential to emit 100 tons or more per year emissions; and (B) All other sources with the potential to emit 250 tons or more per year. See Section C of these instructions for discussion of exclusions of certain modified sources.

Table 3. 28 Industrial Categories Listed in Section 169(1) of the Clean Air Act of 1977

Fossil fuel–fired steam generators of more than 250 million BTU per hour heat input;
Coal cleaning plants *(with thermal dryers)*;
Kraft pulp mills;
Portland cement plants;
Primary zinc smelters;
Iron and steel mill plants;
Primary aluminum ore reduction plants;
Primary copper smelters;
Municipal incinerators capable of charging more than 250 tons of refuse per day;
Hydrofluoric acid plants;
Nitric acid plants;
Sulfuric acid plants;
Petroleum refineries;
Lime plants;
Phosphate rock processing plants;
Coke oven batteries;
Sulfur recovery plants;
Carbon black plants *(furnace process)*;
Primary lead smelters;
Fuel conversion plants;
Sintering plants;
Secondary metal production plants;
Chemical process plants;
Fossil fuel boilers *(or combination thereof)* totaling more than 250 million BTU per hour heat input;

Table 3 *(continued)*

Petroleum storage and transfer units with a total storage capacity exceeding 300,000 barrels;
Taconite ore processing plants;
Glass fiber processing plants; and
Charcoal production plants.

Item III

Enter the facility's official or legal name. Do not use a colloquial name.

Item IV

Give the name, title, and work telephone number of a person who is thoroughly familiar with the operation of the facility and with the facts reported in this application and who can be contacted by reviewing offices if necessary.

Item V

Give the complete mailing address of the office where correspondence should be sent. This often is not the address used to designate the location of the facility or activity.

Item VI

Give the address or location of the facility identified in Item III of this form. If the facility lacks a street name or route number, give the most accurate alternative geographic information *(e.g., section number or quarter section number from county records or at intersection of Rts. 425 and 22)*.

Item VII

List, in descending order of significance, the four 4–digit standard industrial classification *(SIC)* codes which best describe your facility in terms of the principal products or services you produce or provide. Also, specify each classification in words. These classifications may differ from the SIC codes describing the operation generating the discharge, air emissions, or hazardous wastes.

SIC code numbers are descriptions which may be found in the "Standard Industrial Classification Manual" prepared by the Executive Office of the President, Office of Management and Budget, which is available from the Government Printing Office, Washington, D.C. Use the current edition of the manual. If you have any questions concerning the appropriate SIC code for your facility, contact your EPA Regional office *(see Table 1)*.

Item VIII–A

Give the name, as it is legally referred to, of the person, firm, public organization, or any other entity which operates the facility described in this application. This may or may not be the same name as the facility. The operator of the facility is the legal entity which controls the facility's operation rather than the plant or site manager. Do not use a colloquial name.

Item VIII–B

Indicate whether the entity which operates the facility also owns it by marking the appropriate box.

Item VIII–C

Enter the appropriate letter to indicate the legal status of the operator of the facility. Indicate "public" for a facility solely owned by local government*(s)* such as a city, town, county, parish, etc.

Items VIII–D – H

Enter the telephone number and address of the operator identified in Item VIII–A.

SECTION B – FORM 1 LINE–BY–LINE INSTRUCTIONS *(continued)*

Item IX

Indicate whether the facility is located on Indian Lands.

Item X

Give the number of each presently effective permit issued to the facility for each program or, if you have previously filed an application but have not yet received a permit, give the number of the application, if any. Fill in the unshaded area only. If you have more than one currently effective permit for your facility under a particular permit program, you may list additional permit numbers on a separate sheet of paper. List any relevant environmental Federal *(e.g., permits under the Ocean Dumping Act, Section 404 of the Clean Water Act or the Surface Mining Control and Reclamation Act)*, State *(e.g., State permits for new air emission sources in nonattainment areas under Part D of the Clean Air Act or State permits under Section 404 of the Clean Water Act)*, or local permits or applications under "other."

Item XI

Provide a topographic map or maps of the area extending at least to one mile beyond the property boundaries of the facility which clearly show the following:

The legal boundaries of the facility;

The location and serial number of each of your existing and proposed intake and discharge structures;

All hazardous waste management facilities;

Each well where you inject fluids underground; and

All springs and surface water bodies in the area, plus all drinking water wells within 1/4 mile of the facility which are identified in the public record or otherwise known to you.

If an intake or discharge structure, hazardous waste disposal site, or injection well associated with the facility is located more than one mile from the plant, include it on the map, if possible. If not, attach additional sheets describing the location of the structure, disposal site, or well, and identify the U.S. Geological Survey *(or other)* map corresponding to the location.

On each map, include the map scale, a meridian arrow showing north, and latitude and longitude at the nearest whole second. On all maps of rivers, show the direction of the current, and in tidal waters, show the directions of the ebb and flow tides. Use a 7-1/2 minute series map published by the U.S. Geological Survey, which may be obtained through the U.S. Geological Survey Offices listed below. If a 7-1/2 minute series map has not been published for your facility site, then you may use a 15 minute series map from the U.S. Geological Survey. If neither a 7-1/2 nor 15 minute series map has been published for your facility site, use a plat map or other appropriate map, including all the requested information; in this case, briefly describe land uses in the map area *(e.g., residential, commercial)*.

You may trace your map from a geological survey chart, or other map meeting the above specifications. If you do, your map should bear a note showing the number or title of the map or chart it was traced from. Include the names of nearby towns, water bodies, and other prominent points. An example of an acceptable location map is shown in Figure 1–1 of these instructions. *(NOTE: Figure 1–1 is provided for purposes of illustration only, and does not represent any actual facility.)*

U.S.G.S. OFFICES	AREA SERVED
Eastern Mapping Center National Cartographic Information Center U.S.G.S. 536 National Center Reston, Va. 22092 Phone No. (703) 860–6336	Ala., Conn., Del., D.C., Fla., Ga., Ind., Ky., Maine, Md., Mass., N.H., N.J., N.Y., N.C., S.C., Ohio, Pa., Puerto Rico, R.I., Tenn., Vt., Va., W. Va., and Virgin Islands.
Mid Continent Mapping Center National Cartographic Information Center U.S.G.S. 1400 Independance Road Rolla, Mo. 65401 Phone No. (314) 341–0851	Ark., Ill., Iowa, Kans., La., Mich., Minn., Miss., Mo., N. Dak., Nebr., Okla., S. Dak., and Wis.
Rocky Mountain Mapping Center National Cartographic Infomation Center U.S.G.S. Stop 504, Box 25046 Federal Center Denver, Co. 80225 Phone No. (303) 234–2326	Alaska, Colo., Mont., N. Mex., Tex., Utah, and Wyo.
Western Mapping Center National Cartographic Information Center U.S.G.S. 345 Middlefield Road Menlo Park, Ca. 94025 Phone No. (415) 323–8111	Ariz., Calif., Hawaii, Idaho, Nev., Oreg., Wash., American Samoa, Guam, and Trust Territories

Item XII

Briefly describe the nature of your business *(e.g., products produced or services provided)*.

Item XIII

Federal statues provide for severe penalties for submitting false information on this application form.

18 U.S.C. Section 1001 provides that "Whoever, in any matter within the jurisdiction of any department or agency of the United States knowingly and willfully falsifies, conceals or covers up by any trick, scheme, or device a material fact, or makes or uses any false writing or document knowing same to contain any false, fictitious or fraudulent statement or entry, shall be fined not more than $10,000 or imprisoned not more than five years, or both."

Section 309(c)(2) of the Clean Water Act and Section 113(c)(2) of the Clean Air Act each provide that "Any person who knowingly makes any false statement, representation, or certification in any application, . . . shall upon conviction, be punished by a fine of no more than $10,000 or by imprisonment for not more than six months, or both."

In addition, Section 3008(d)(3) of the Resource Conservation and Recovery Act provides for a fine up to $25,000 per day or imprisonment up to one year, or both, for a first conviction for making a false statement in any application under the Act, and for double these penalties upon subsequent convictions.

FEDERAL REGULATIONS REQUIRE THIS APPLICATION TO BE SIGNED AS FOLLOWS:

A. For a corporation, by a principal executive officer of at least the level of vice president. However, if the only activity in Item II which is marked "yes" is Question G, the officer may authorize a person having responsibility for the overall operations of the well or well field to sign the certification. In that case, the authorization must be written and submitted to the permitting authority.

B. For partnership or sole proprietorship, by a general partner or the proprietor, respectively; or

C. For a municipality, State, Federal, or other public facility, by either a principal executive officer or ranking elected official.

SECTION C – ACTIVITIES WHICH DO NOT REQUIRE PERMITS

I. National Pollutant Discharge Elimination System Permits Under the Clean Water Act. You are not required to obtain an NPDES permit if your discharge is in one of the following categories, as provided by the Clean Water Act *(CWA)* and by the NPDES regulations *(40 CFR Parts 122–125)*. However, under Section 510 of CWA a discharge exempted from the federal NPDES requirements may still be regulated by a State authority; contact your State environmental agency to determine whether you need a State permit.

A. DISCHARGES FROM VESSELS. Discharges of sewage from vessels, effluent from properly functioning marine engines, laundry, shower, and galley sink wastes, and any other discharge incidental to the normal operation of a vessel do not require NPDES permits. However, discharges of rubbish, trash, garbage, or other such materials discharged overboard require permits, and so do other discharges when the vessel is operating in a capacity other than as a means of transportation, such as when the vessel is being used as an energy or mining facility, a storage facility, or a seafood processing facility, or is secured to the bed of the ocean, contiguous zone, or waters of the United States for the purpose of mineral or oil exploration or development.

B. DREDGED OR FILL MATERIAL. Discharges of dredged or fill material into waters of the United States do not need NPDES permits if the dredging or filling is authorized by a permit issued by the U.S. Army Corps of Engineers or an EPA approved State under Section 404 of CWA.

C. DISCHARGES INTO PUBLICLY OWNED TREATMENT WORKS *(POTW)*. The introduction of sewage, industrial wastes, or other pollutants into a POTW does not need an NPDES permit. You must comply with all applicable pretreatment standards promulgated under Section 307(b) of CWA, which may be included in the permit issued to the POTW. If you have a plan or an agreement to switch to a POTW in the future, this does not relieve you of the obligation to apply for and receive an NPDES permit until you have stopped discharging pollutants into waters of the United States.

(NOTE: Dischargers into privately owned treatment works do not have to apply for or obtain NPDES permits except as otherwise required by the EPA Regional Administrator. The owner or operator of the treatment works itself, however, must apply for a permit and identify all users in its application. Users so identified will receive public notice of actions taken on the permit for the treatment works.)

D. DISCHARGES FROM AGRICULTURAL AND SILVICULTURAL ACTIVITIES. Most discharges from agricultural and silvicultural activities to waters of the United States do not require NPDES permits. These include runoff from orchards, cultivated crops, pastures, range lands, and forest lands. However, the discharges listed below do require NPDES permits. Definitions of the terms listed below are contained in the Glossary section of these instructions.

1. Discharges from Concentrated Animal Feeding Operations. *(See Glossary for definitions of "animal feeding operations" and "concentrated animal feeding operations." Only the latter require permits.)*

2. Discharges from Concentrated Aquatic Animal Production Facilities. *(See Glossary for size cutoffs.)*

3. Discharges associated with approved Aquaculture Projects.

4. Discharges from Silvicultural Point Sources. *(See Glossary for the definition of "silvicultural point source.")* Nonpoint source silvicultural activities are excluded from NPDES permit requirements. However, some of these activities, such as stream crossings for roads, may involve point source discharges of dredged or fill material which may require a Section 404 permit. See 33 CFR 209.120.

E. DISCHARGES IN COMPLIANCE WITH AN ON–SCENE COORDINATOR'S INSTRUCTIONS.

II. Hazardous Waste Permits Under the Resource Conservation and Recovery Act. You may be excluded from the requirement to obtain a permit under this program if you fall into one of the following categories:

Generators who accumulate their own hazardous waste on–site for less than 90 days as provided in 40 CFR 262.34;

Farmers who dispose of hazardous waste pesticide from their own use as provided in 40 CFR 262.51;

Certain persons treating, storing, or disposing of small quantities of hazardous waste as provided in 40 CFR 261.4 or 261.5; and

Owners and operators of totally enclosed treatment facilities as defined in 40 CFR 260.10.

Check with your Regional office for details. Please note that even if you are excluded from permit requirements, you may be required by Federal regulations to handle your waste in a particular manner.

III. Underground Injection Control Permits Under the Safe Drinking Water Act. You are not required to obtain a permit under this program if you:

Inject into existing wells used to enhance recovery of oil and gas or to store hydrocarbons *(note, however, that these underground injections are regulated by Federal rules)*; or

Inject into or above a stratum which contains, within 1/4 mile of the well bore, an underground source of drinking water *(unless your injection is the type identified in Item II-H, for which you do need a permit)*. However, you must notify EPA of your injection and submit certain required information on forms supplied by the Agency, and your operation may be phased out if you are a generator of hazardous wastes or a hazardous waste management facility which uses wells or septic tanks to dispose of hazardous waste.

IV. Prevention of Significant Deterioration Permits Under the Clean Air Act. The PSD program applies to newly constructed or modified facilities *(both of which are referred to as "new sources")* which increase air emissions. The Clean Air Act Amendments of 1977 exclude small new sources of air emissions from the PSD review program. Any new source in an industrial category listed in Table 3 of these instructions whose potential to emit is less than 100 tons per year is not required to get a PSD permit. In addition, any new source in an industrial category not listed in Table 3 whose potential to emit is less than 250 tons per year is exempted from the PSD requirements.

Modified sources which increase their net emissions *(the difference between the total emission increases and total emission decreases at the source)* less than the significant amount set forth in EPA regulations are also exempt from PSD requirements. Contact your EPA Regional office *(Table 1)* for further information.

SECTION D – GLOSSARY

NOTE: This Glossary includes terms used in the instructions and in Forms 1, 2B, 2C, and 3. Additional terms will be included in the future when other forms are developed to reflect the requirements of other parts of the Consolidated Permits Program. If you have any questions concerning the meaning of any of these terms, please contact your EPA Regional office *(Table 1)*.

ALIQUOT means a sample of specified volume used to make up a total composite sample.

ANIMAL FEEDING OPERATION means a lot or facility *(other than an aquatic animal production facility)* where the following conditions are met:

A. Animals *(other than aquatic animals)* have been, are, or will be stabled or confined and fed or maintained for a total of 45 days or more in any 12 month period; and

B. Crops, vegetation, forage growth, or post–harvest residues are not sustained in the normal growing season over any portion of the lot or facility.

Two or more animal feeding operations under common ownership are a single animal feeding operation if they adjoin each other or if they use a common area or system for the disposal of wastes.

ANIMAL UNIT means a unit of measurement for any animal feeding operation calculated by adding the following numbers: The number of slaughter and feeder cattle multiplied by 1.0; Plus the number of mature dairy cattle multiplied by 1.4; Plus the number of swine weighing over 25 kilograms *(approximately 55 pounds)* multiplied by 0.4; Plus the number of sheep multiplied by 0.1; Plus the number of horses multiplied by 2.0.

APPLICATION means the EPA standard national forms for applying for a permit, including any additions, revisions, or modifications to the forms; or forms approved by EPA for use in approved States, including any approved modifications or revisions. For RCRA, "application" also means "Application, Part B."

APPLICATION, PART A means that part of the Consolidated Permit Application forms which a RCRA permit applicant must complete to qualify for interim status under Section 3005(e) of RCRA and for consideration for a permit. Part A consists of Form 1 *(General Information)* and Form 3 *(Hazardous Waste Application Form)*.

APPLICATION, PART B means that part of the application which a RCRA permit applicant must complete to be issued a permit. *(NOTE: EPA is not developing a specific form for Part B of the permit application, but an instruction booklet explaining what information must be supplied is available from the EPA Regional office.)*

APPROVED PROGRAM or APPROVED STATE means a State program which has been approved or authorized by EPA under 40 CFR Part 123.

AQUACULTURE PROJECT means a defined managed water area which uses discharges of pollutants into that designated area for the maintenance or production of harvestable freshwater, estuarine, or marine plants or animals. "Designated area" means the portions of the waters of the United States within which the applicant plans to confine the cultivated species, using a method of plan or operation *(including, but not limited to, physical confinement)* which, on the basis of reliable scientific evidence, is expected to ensure the specific individual organisms comprising an aquaculture crop will enjoy increased growth attributable to the discharge of pollutants and be harvested within a defined geographic area.

AQUIFER means a geological formation, group of formations, or part of a formation that is capable of yielding a significant amount of water to a well or spring.

AREA OF REVIEW means the area surrounding an injection well which is described according to the criteria set forth in 40 CFR Section 146.06.

AREA PERMIT means a UIC permit applicable to all or certain wells within a geographic area, rather than to a specified well, under 40 CFR Section 122.37.

ATTAINMENT AREA means, for any air pollutant, an area which has been designated under Section 107 of the Clean Air Act as having ambient air quality levels better than any national primary or secondary ambient air quality standard for that pollutant. Standards have been set for sulfur oxides, particulate matter, nitrogen dioxide, carbon monoxide, ozone, lead, and hydrocarbons. For purposes of the Glossary, "attainment area" also refers to "unclassifiable area," which means, for any pollutants, an area designated under Section 107 as unclassifiable with respect to that pollutant due to insufficient information.

BEST MANAGEMENT PRACTICES *(BMP)* means schedules of activities, prohibitions of practices, maintenance procedures, and other management practices to prevent or reduce the pollution of waters of the United States. BMP's include treatment requirements, operation procedures, and practices to control plant site runoff, spillage or leaks, sludge or waste disposal, or drainage from raw material storage.

BIOLOGICAL MONITORING TEST means any test which includes the use of aquatic algal, invertebrate, or vertebrate species to measure acute or chronic toxicity, and any biological or chemical measure of bioaccumulation.

BYPASS means the intentional diversion of wastes from any any portion of a treatment facility.

CONCENTRATED ANIMAL FEEDING OPERATION means an animal feeding operation which meets the criteria set forth in either (A) or (B) below or which the Director designates as such on a case–by–case basis:

A. More than the numbers of animals specified in any of the following categories are confined:

1. 1,000 slaughter or feeder cattle,
2. 700 mature dairy cattle *(whether milked or dry cows)*,
3. 2,500 swine each weighing over 25 kilograms *(approximately 55 pounds)*,
4. 500 horses,
5, 10,000 sheep or lambs,
6. 55,000 turkeys,
7. 100,000 laying hens or broilers *(if the facility has a continuous overflow watering)*,
8. 30,000 laying hens or broilers *(if the facility has a liquid manure handling system)*,
9. 5,000 ducks, or
10. 1,000 animal units; or

B. More than the following numbers and types of animals are confined:

1. 300 slaughter or feeder cattle,
2. 200 mature dairy cattle *(whether milked or dry cows)*,
3. 750 swine each weighing over 25 kilograms *(approximately 55 pounds)*,
4. 150 horses,

SECTION D – GLOSSARY *(continued)*

CONCENTRATED ANIMAL FEEDING OPERATION *(continued)*

5. 3,000 sheep or lambs,

6. 16,500 turkeys,

7. 30,000 laying hens or broilers *(if the facility has continuous overflow watering)*,

8. 9,000 laying hens or broilers *(if the facility has a liquid manure handling system)*,

9. 1,500 ducks, or

10. 300 animal units; AND

Either one of the following conditions are met: Pollutants are discharged into waters of the United States through a manmade ditch, flushing system or other similar manmade device *("manmade" means constructed by man and used for the purpose of transporting wastes)*; or Pollutants are discharged directly into waters of the Unites States which originate outside of and pass over, across, or through the facility or otherwise come into direct contact with the animals confined in the operation.

Provided, however, that no animal feeding operation is a concentrated animal feeding operation as defined above if such animal feeding operation discharges only in the event of a 25 year, 24 hour storm event.

CONCENTRATED AQUATIC ANIMAL PRODUCTION FACILITY means a hatchery, fish farm, or other facility which contains, grows or holds aquatic animals in either of the following categories, or which the Director designates as such on a case–by–case basis:

A. Cold water fish species or other cold water aquatic animals including, but not limited to, the Salmonidae family of fish *(e.g., trout and salmon)* in ponds, raceways or other similar structures which discharge at least 30 days per year but does not include:

1. Facilities which produce less than 9,090 harvest weight kilograms *(approximately 20,000 pounds)* of aquatic animals per year; and

2. Facilities which feed less than 2,272 kilograms *(approximately 5,000 pounds)* of food during the calendar month of maximum feeding.

B. Warm water fish species or other warm water aquatic animals including, but not limited to, the Ameiuridae, Cetrarchidae, and Cyprinidae families of fish *(e.g., respectively, catfish, sunfish, and minnows)* in ponds, raceways, or other similar structures which discharge at least 30 days per year, but does not include:

1. Closed ponds which discharge only during periods of excess runoff; or

2. Facilities which produce less than 45,454 harvest weight kilograms *(approximately 100,000 pounds)* of aquatic animals per year.

CONTACT COOLING WATER means water used to reduce temperature which comes into contact with a raw material, intermediate product, waste product other than heat, or finished product.

CONTAINER means any portable device in which a material is stored, transported, treated, disposed of, or otherwise handled.

CONTIGUOUS ZONE means the entire zone established by the United States under article 24 of the convention of the Territorial Sea and the Contiguous Zone.

CWA means the Clean Water Act *(formerly referred to the Federal Water Pollution Control Act)* Pub. L. 92–500, as amended by Pub. L. 95–217 and Pub. L. 95–576, 33 U.S.C. 1251 *et seq.*

DIKE means any embankment or ridge of either natural or manmade materials used to prevent the movement of liquids, sludges, solids, or other materials.

DIRECT DISCHARGE means the discharge of a pollutant as defined below.

DIRECTOR means the EPA Regional Administrator or the State Director as the context requires.

DISCHARGE *(OF A POLLUTANT)* means:

A. Any addition of any pollutant or combination of pollutants to waters of the United States from any point source; or

B. Any addition of any pollutant or combination of pollutants to the waters of the contiguous zone or the ocean from any point source other than a vessel or other floating craft which is being used as a means of transportation.

This definition includes discharges into waters of the United States from: Surface runoff which is collected or channelled by man; Discharges through pipes, sewers, or other conveyances owned by a State, municipality, or other person which do not lead to POTW's; and Discharges through pipes, sewers, or other conveyances, leading into privately owned treatment works. This term does not include an addition of pollutants by any indirect discharger.

DISPOSAL *(in the RCRA program)* means the discharge, deposit, injection, dumping, spilling, leaking, or placing of any hazardous waste into or on any land or water so that the hazardous waste or any constituent of it may enter the environment or be emitted into the air or discharged into any waters, including ground water.

DISPOSAL FACILITY means a facility or part of a facility at which hazardous waste is intentionally placed into or on land or water, and at which hazardous waste will remain after closure.

EFFLUENT LIMITATION means any restriction imposed by the Director on quantities, discharge rates, and concentrations of pollutants which are discharged from point sources into waters of the United States, the waters of the continguous zone, or the ocean.

EFFLUENT LIMITATION GUIDELINE means a regulation published by the Administrator under Section 304(b) of the Clean Water Act to adopt or revise effluent limitations.

ENVIRONMENTAL PROTECTION AGENCY *(EPA)* means the United States Environmental Protection Agency.

EPA IDENTIFICATION NUMBER means the number assigned by EPA to each generator, transporter, and facility.

EXEMPTED AQUIFER means an aquifer or its portion that meets the criteria in the definition of USDW, but which has been exempted according to the procedures in 40 CFR Section 122.35(b).

EXISTING HWM FACILITY means a Hazardous Waste Management facility which was in operation, or for which construction had commenced, on or before October 21, 1976. Construction had commenced if (A) the owner or operator had obtained all necessary Federal, State, and local preconstruction approvals or permits, and either (B1) a continuous on–site, physical construction program had begun, or (B2) the owner or operator had entered into contractual obligations, which could not be cancelled or modified without substantial loss, for construction of the facility to be completed within a reasonable time.

(NOTE: This definition reflects the literal language of the statute. However, EPA believes that amendments to RCRA now in conference will shortly be enacted and will change the date for determining when a facility is an "existing facility" to one no earlier than May of 1980, indications are the conferees are considering October 30, 1980. Accordingly, EPA encourages every owner or operator of a facility which was built or under construction as of the promulgation date of the RCRA program regulations to file Part A of its permit application so that it can be quickly processed for interim status when the change in the law takes effect. When those amendments are enacted, EPA will amend this definition.)

EXISTING SOURCE or **EXISTING DISCHARGER** *(in the NPDES program)* means any source which is not a new source or a new discharger.

SECTION D – GLOSSARY *(continued)*

EXISTING INJECTION WELL means an injection well other than a new injection well.

FACILITY means any HWM facility, UIC underground injection well, NPDES point source, PSD stationary source, or any other facility or activity *(including land or appurtenances thereto)* that is subject to regulation under the RCRA, UIC, NPDES, or PSD programs.

FLUID means material or substance which flows or moves whether in a semisolid, liquid, sludge, gas, or any other form or state.

GENERATOR means any person by site, whose act or process produces hazardous waste identified or listed in 40 CFR Part 261.

GROUNDWATER means water below the land surface in a zone of saturation.

HAZARDOUS SUBSTANCE means any of the substances designated under 40 CFR Part 116 pursuant to Section 311 of CWA. *(NOTE: These substances are listed in Table 2c–4 of the instructions to Form 2C.)*

HAZARDOUS WASTE means a hazardous waste as defined in 40 CFR Section 261.3 published May 19, 1980.

HAZARDOUS WASTE MANAGEMENT FACILITY *(HWM facility)* means all contiguous land, structures, appurtenances, and improvements on the land, used for treating, storing, or disposing of hazardous wastes. A facility may consist of several treatment, storage, or disposal operational units *(for example, one or more landfills, surface impoundments, or combinations of them)*.

IN OPERATION means a facility which is treating, storing, or disposing of hazardous waste.

INCINERATOR *(in the RCRA program)* means an enclosed device using controlled flame combustion, the primary purpose of which is to thermally break down hazardous waste. Examples of incinerators are rotary kiln, fluidized bed, and liquid injection incinerators.

INDIRECT DISCHARGER means a nondomestic discharger introducing pollutants to a publicly owned treatment works.

INJECTION WELL means a well into which fluids are being injected.

INTERIM AUTHORIZATION means approval by EPA of a State hazardous waste program which has met the requirements of Section 3006(c) of RCRA and applicable requirements of 40 CFR Part 123, Subparts A, B, and F.

LANDFILL means a disposal facility or part of a facility where hazardous waste is placed in or on land and which is not a land treatment facility, a surface impoundment, or an injection well.

LAND TREATMENT FACILITY *(in the RCRA program)* means a facility or part of a facility at which hazardous waste is applied onto or incorporated into the soil surface; such facilities are disposal facilities if the waste will remain after closure.

LISTED STATE means a State listed by the Administrator under Section 1422 of SDWA as needing a State UIC program.

MGD means millions of gallons per day.

MUNICIPALITY means a city, village, town, borough, county, parish, district, association, or other public body created by or under State law and having jurisdiction over disposal of sewage, industrial wastes, or other wastes, or an Indian tribe or an authorized Indian tribal organization, or a designated and approved management agency under Section 208 of CWA.

NATIONAL POLLUTANT DISCHARGE ELIMINATION SYSTEM *(NPDES)* means the national program for issuing modifying, revoking and reissuing, terminating, monitoring, and enforcing permits and imposing and enforcing pretreatment requirements, under Sections 307, 318, 402, and 405 of CWA. The term includes an approved program.

NEW DISCHARGER means any building, structure, facility, or installation: (A) From which there is or may be a new or additional discharge of pollutants at a site at which on October 18, 1972, it had never discharged pollutants; (B) Which has never received a finally effective NPDES permit for discharges at that site; and (C) Which is not a "new source." This definition includes an indirect discharger which commences discharging into waters of the United States. It also includes any existing mobile point source, such as an offshore oil drilling rig, seafood processing vessel, or aggregate plant that begins discharging at a location for which it does not have an existing permit.

NEW HWM FACILITY means a Hazardous Waste Management facility which began operation or for which construction commenced after October 21, 1976.

NEW INJECTION WELL means a well which begins injection after a UIC program for the State in which the well is located is approved.

NEW SOURCE *(in the NPDES program)* means any building, structure, facility, or installation from which there is or may be a discharge of pollutants, the construction of which commenced:

A. After promulgation of standards of performance under Section 306 of CWA which are applicable to such source; or

B. After proposal of standards of performance in accordance with Section 306 of CWA which are applicable to such source, but only if the standards are promulgated in accordance with Section 306 within 120 days of their proposal.

NON–CONTACT COOLING WATER means water used to reduce temperature which does not come into direct contact with any raw material, intermediate product, waste product *(other than heat)*, or finished product.

OFF–SITE means any site which is not "on–site."

ON–SITE means on the same or geographically contiguous property which may be divided by public or private right*(s)*–of–way, provided the entrance and exit between the properties is at a cross–roads intersection, and access is by crossing as opposed to going along, the right*(s)*–of–way. Non–contiguous properties owned by the same person, but connected by a right–of–way which the person controls and to which the public does not have access, is also considered on–site property.

OPEN BURNING means the combustion of any material without the following characteristics:

A. Control of combustion air to maintain adequate temperature for efficient combustion;

B. Containment of the combustion–reaction in an enclosed device to provide sufficient residence time and mixing for complete combustion; and

C. Control of emission of the gaseous combustion products.

(See also "incinerator" and "thermal treatment").

OPERATOR means the person responsible for the overall operation of a facility.

OUTFALL means a point source.

OWNER means the person who owns a facility or part of a facility.

SECTION D – GLOSSARY *(continued)*

PERMIT means an authorization, license, or equivalent control document issued by EPA or an approved State to implement the requirements of 40 CFR Parts 122, 123, and 124.

PHYSICAL CONSTRUCTION *(in the RCRA program)* means excavation, movement of earth, erection of forms or structures, or similar activity to prepare a HWM facility to accept hazardous waste.

PILE means any noncontainerized accumulation of solid, nonflowing hazardous waste that is used for treatment or storage.

POINT SOURCE means any discernible, confined, and discrete conveyance, including but not limited to any pipe, ditch, channel, tunnel, conduit, well, discrete fissure, container, rolling stock, concentrated animal feeding operation, vessel or other floating craft from which pollutants are or may be discharged. This term does not include return flows from irrigated agriculture.

POLLUTANT means dredged spoil, solid waste, incinerator residue, filter backwash, sewage, garbage, sewage sludge, munitions, chemical waste, biological materials, radioactive materials *(except those regulated under the Atomic Energy Act of 1954, as amended [42 U.S.C. Section 2011 et seq.])*, heat, wrecked or discarded equipment, rocks, sand, cellar dirt and industrial, municipal, and agriculture waste discharged into water. It does not mean:

A. Sewage from vessels; or

B. Water, gas, or other material which is injected into a well to facilitate production of oil or gas, or water derived in association with oil and gas production and disposed of in a well, if the well used either to facilitate production or for disposal purposes is approved by authority of the State in which the well is located, and if the State determines that the injection or disposal will not result in the degradation of ground or surface water resources.

(NOTE: Radioactive materials covered by the Atomic Energy Act are those encompassed in its definition of source, byproduct, or special nuclear materials. Examples of materials not covered include radium and accelerator produced isotopes. See Train v. Colorado Public Interest Research Group, Inc., 426 U.S. 1 [1976].)

PREVENTION OF SIGNIFICANT DETERIORATION *(PSD)* means the national permitting program under 40 CFR 52.21 to prevent emissions of certain pollutants regulated under the Clean Air Act from significantly deteriorating air quality in attainment areas.

PRIMARY INDUSTRY CATEGORY means any industry category listed in the NRDC Settlement Agreement *(Natural Resources Defense Council v. Train, 8 ERC 2120 [D.D.C. 1976], modified 12 ERC 1833 [D.D.C. 1979])*.

PRIVATELY OWNED TREATMENT WORKS means any device or system which is: (A) Used to treat wastes from any facility whose operator is not the operator of the treatment works; and (B) Not a POTW.

PROCESS WASTEWATER means any water which, during manufacturing or processing, comes into direct contact with or results from the production or use of any raw material, intermediate product, finished product, byproduct, or waste product.

PUBLICLY OWNED TREATMENT WORKS or POTW means any device or system used in the treatment *(including recycling and reclamation)* of municipal sewage or industrial wastes of a liquid nature which is owned by a State or municipality. This definition includes any sewers, pipes, or other conveyances only if they convey wastewater to a POTW providing treatment.

RENT means use of another's property in return for regular payment.

RCRA means the Solid Waste Disposal Act as amended by the Resource Conservation and Recovery Act of 1976 *(Pub. L. 94–580, as amended by Pub. L. 95–609, 42 U.S.C. Section 6901 et seq.)*.

ROCK CRUSHING AND GRAVEL WASHING FACILITIES are facilities which process crushed and broken stone, gravel, and riprap *(see 40 CFR Part 436, Subpart B, and the effluent limitations guidelines for these facilities)*.

SDWA means the Safe Drinking Water Act *(Pub. L. 95–523, as amended by Pub. L. 95–1900, 42 U.S.C. Section 300[f] et seq.)*.

SECONDARY INDUSTRY CATEGORY means any industry category which is not a primary industry category.

SEWAGE FROM VESSELS means human body wastes and the wastes from tiolets and other receptacles intended to receive or retain body wastes that are discharged from vessels and regulated under Section 312 of CWA, except that with respect to commercial vessels on the Great Lakes this term includes graywater. For the purposes of this definition "graywater" means galley, bath, and shower water.

SEWAGE SLUDGE means the solids, residues, and precipitate separated from or created in sewage by the unit processes of a POTW. "Sewage" as used in this definition means any wastes, including wastes from humans, households, commercial establishments, industries, and storm water runoff, that are discharged to or otherwise enter a publicly owned treatment works.

SILVICULTURAL POINT SOURCE means any discernable, confined and discrete conveyance related to rock crushing, gravel washing, lo sorting, or log storage facilities which are operated in connection wit silvicultural activities and from which pollutants are discharged int waters of the United States. This term does not include nonpoir source silvicultural activities such as nursery operations, site prepara tion, reforestation and subsequent cultural treatment, thinning, pre scribed burning, pest and fire control, harvesting operations, surfac drainage, or road construction and maintenance from which there natural runoff. However, some of these activities *(such as stream cross ing for roads)* may involve point source discharges of dredged or fi material which may require a CWA Section 404 permit. "Log sortin and log storage facilities" are facilities whose discharges result from th holding of unprocessed wood, e.g., logs or roundwood with bark c after removal of bark in self–contained bodies of water *(mill ponds c log ponds)* or stored on land where water is applied intentionally o the logs *(wet decking). (See 40 CFR Part 429, Subpart J, and the efflu ent limitations guidelines for these facilities.)*

STATE means any of the 50 States, the District of Columbia, Guam the Commonwealth of Puerto Rico, the Virgin Islands, America Samoa, the Trust Territory of the Pacific Islands *(except in the case c RCRA)*, and the Commonwealth of the Northern Mariana Islanc *(except in the case of CWA)*.

STATIONARY SOURCE *(in the PSD program)* means any buildin structure, facility, or installation which emits or may emit any air pc lutant regulated under the Clean Air Act. "Building, structure, facilit or installation" means any grouping of pollutant–emitting activiti which are located on one or more contiguous or adjacent properti and which are owned or operated by the same person *(or by perso under common control)*.

STORAGE *(in the RCRA program)* means the holding of hazardo waste for a temporary period at the end of which the hazardous was is treated, disposed, or stored elsewhere.

STORM WATER RUNOFF means water discharged as a result of ra snow, or other precipitation.

SURFACE IMPOUNDMENT or IMPOUNDMENT means a facility part of a facility which is a natural topographic depression, manma excavation, or diked area formed primarily of earthen materials *(a though it may be lined with manmade materials)*, which is designed hold an accumulation of liquid wastes or wastes containing free liquic and which is not an injection well. Examples of surface impoundmer are holding, storage, settling, and aeration pits, ponds, and lagoor

TANK *(in the RCRA program)* means a stationary device, designed contain an accumulation of hazardous waste which is constructed pr marily of non–earthen materials *(e.g., wood, concrete, steel, plasti* which provide structural support.

SECTION D – GLOSSARY *(continued)*

THERMAL TREATMENT *(in the RCRA program)* means the treatment of hazardous waste in a device which uses elevated temperature as the primary means to change the chemical, physical, or biological character or composition of the hazardous waste. Examples of thermal treatment processes are incineration, molten salt, pyrolysis, calcination, wet air oxidation, and microwave discharge. *(See also "incinerator" and "open burning").*

TOTALLY ENCLOSED TREATMENT FACILITY *(in the RCRA program)* means a facility for the treatment of hazardous waste which is directly connected to an industrial production process and which is constructed and operated in a manner which prevents the release of any hazardous waste or any constituent thereof into the environment during treatment. An example is a pipe in which waste acid is neutralized.

TOXIC POLLUTANT means any pollutant listed as toxic under Section 307(a)(1) of CWA.

TRANSPORTER *(in the RCRA program)* means a person engaged in the off–site transportation of hazardous waste by air, rail, highway, or water.

TREATMENT *(in the RCRA program)* means any method, technique, or process, including neutralization, designed to change the physical, chemical, or biological character or composition of any hazardous waste so as to neutralize such waste, or so as to recover energy or material resources from the waste, or so as to render such waste non–hazardous, or less hazardous; safer to transport, store, or dispose of; or amenable for recovery, amenable for storage, or reduced in volume.

UNDERGROUND INJECTION means well injection.

UNDERGROUND SOURCE OF DRINKING WATER or USDW means an aquifer or its portion which is not an exempted aquifer and:

A. Which supplies drinking water for human consumption; or

B. In which the ground water contains fewer than 10,000 mg/l total dissolved solids.

UPSET means an exceptional incident in which there is unintentional and temporary noncompliance with technology–based permit effluent limitations because of factors beyond the reasonable control of the permittee. An upset does not include noncompliance to the extent caused by operational error, improperly designed treatment facilities, inadequate treatment facilities, lack of preventive maintenance, or careless or improper operation.

WATERS OF THE UNITED STATES means:

A. All waters which are currently used, were used in the past, or may be susceptible to use in interstate or foreign commerce, including all waters which are subject to the ebb and flow of the tide;

B. All interstate waters, including interstate wetlands;

C. All other waters such as intrastate lakes, rivers, streams *(including intermittent streams)*, mudflats, sandflats, wetlands, sloughs, prairie potholes, wet meadows, playa lakes, and natural ponds, the use, degradation, or destruction of which would or could affect interstate or foreign commerce including any such waters:

1. Which are or could be used by interstate or foreign travelers for recreational or other purposes,

2. From which fish or shellfish are or could be taken and sold in interstate or foreign commerce,

3. Which are used or could be used for industrial purposes by industries in interstate commerce;

D. All impoundments of waters otherwise defined as waters of the United States under this definition;

E. Tributaries of waters identified in paragraphs (A) – (D) above;

F. The territorial sea; and

G. Wetlands adjacent to waters *(other than waters that are themselves wetlands)* identified in paragraphs (A) – (F) of this definition.

Waste treatment systems, including treatment ponds or lagoons designed to meet requirement of CWA *(other than cooling ponds as defined in 40 CFR Section 423.11(m) which also meet the criteria of this definition)* are not waters of the United States. This exclusion applies only to manmade bodies of water which neither were originally created in waters of the United States *(such as a disposal area in wetlands)* nor resulted from the impoundments of waters of the United States.

WELL INJECTION or UNDERGROUND INJECTION means the subsurface emplacement of fluids through a bored, drilled, or driven well; or through a dug well, where the depth of the dug well is greater than the largest surface dimension.

WETLANDS means those areas that are inundated or saturated by surface or groundwater at a frequency and duration sufficient to support, and that under normal circumstances do support, a prevalence of vegetation typically adapted for life in saturated soil conditions. Wetlands generally include swamps, marshes, bogs, and similar areas.

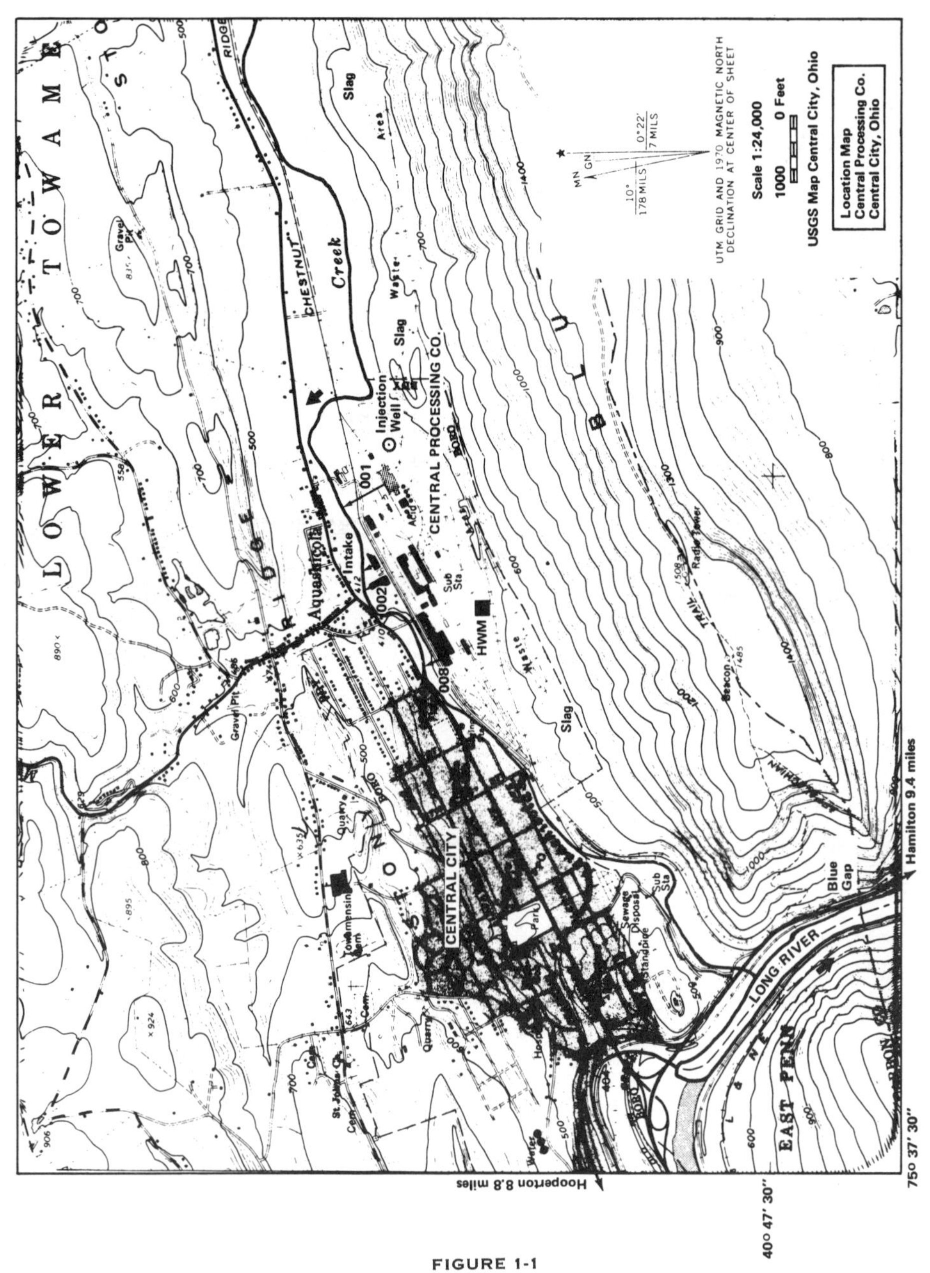
LOWER TOWAMENSING
CENTRAL CITY
CENTRAL PROCESSING CO.
Injection Well
Intake
Aquashicola
Chestnut
Creek
Slag
Waste
HWM
001
002
008
LONG RIVER
Blue Gap
Radio Tower
Beacon
Sewage Disposal
Gravel Pit
Quarry
UTM GRID AND 1970 MAGNETIC NORTH DECLINATION AT CENTER OF SHEET
Scale 1:24,000
USGS Map Central City, Ohio
Location Map
Central Processing Co.
Central City, Ohio
Hamilton 9.4 miles
Hooperton 8.8 miles
40° 47' 30"
75° 37' 30"

FIGURE 1-1

Please print or type in the unshaded areas only
(fill–in areas are spaced for elite type, i.e., 12 characters/inch).

Form Approved OMB No. 158-R0175

FORM 1 GENERAL — EPA — U.S. ENVIRONMENTAL PROTECTION AGENCY

GENERAL INFORMATION

Consolidated Permits Program

(Read the "General Instructions" before starting.)

I. EPA I.D. NUMBER

F D

LABEL ITEMS

I. EPA I.D. NUMBER

III. FACILITY NAME

V. FACILITY MAILING ADDRESS

VI. FACILITY LOCATION

PLEASE PLACE LABEL IN THIS SPACE

GENERAL INSTRUCTIONS

If a preprinted label has been provided, affix it in the designated space. Review the information carefully; if any of it is incorrect, cross through it and enter the correct data in the appropriate fill–in area below. Also, if any of the preprinted data is absent *(the area to the left of the label space lists the information that should appear)*, please provide it in the proper fill–in area(s) below. If the label is complete and correct, you need not complete Items I, III, V, and VI *(except VI-B which must be completed regardless)*. Complete all items if no label has been provided. Refer to the instructions for detailed item descriptions and for the legal authorizations under which this data is collected.

II. POLLUTANT CHARACTERISTICS

INSTRUCTIONS: Complete A through J to determine whether you need to submit any permit application forms to the EPA. If you answer "yes" to any questions, you must submit this form and the supplemental form listed in the parenthesis following the question. Mark "X" in the box in the third column if the supplemental form is attached. If you answer "no" to each question, you need not submit any of these forms. You may answer "no" if your activity is excluded from permit requirements; see Section C of the instructions. See also, Section D of the instructions for definitions of bold–faced terms.

SPECIFIC QUESTIONS	YES	NO	FORM ATTACHED	SPECIFIC QUESTIONS	YES	NO	FORM ATTACHED
A. Is this facility a **publicly owned treatment works** which results in a **discharge** to **waters of the U.S.**? (FORM 2A)	16	17	18	B. Does or will this facility *(either existing or proposed)* include a **concentrated animal feeding operation** or **aquatic animal production facility** which results in a **discharge** to **waters of the U.S.**? (FORM 2B)	19	20	21
C. Is this a facility which currently results in **discharges** to **waters of the U.S.** other than those described in A or B above? (FORM 2C)	22	23	24	D. Is this a proposed facility *(other than those described in A or B above)* which will result in a **discharge** to **waters of the U.S.**? (FORM 2D)	25	26	27
E. Does or will this facility treat, store, or dispose of **hazardous wastes**? (FORM 3)	28	29	30	F. Do you or will you inject at this facility industrial or municipal effluent below the lowermost stratum containing, within one quarter mile of the well bore, underground sources of drinking water? (FORM 4)	31	32	33
G. Do you or will you inject at this facility any produced water or other fluids which are brought to the surface in connection with conventional oil or natural gas production, inject fluids used for enhanced recovery of oil or natural gas, or inject fluids for storage of liquid hydrocarbons? (FORM 4)	34	35	36	H. Do you or will you inject at this facility fluids for special processes such as mining of sulfur by the Frasch process, solution mining of minerals, in situ combustion of fossil fuel, or recovery of geothermal energy? (FORM 4)	37	38	39
I. Is this facility a proposed **stationary source** which is one of the 28 industrial categories listed in the instructions and which will potentially emit 100 tons per year of any air pollutant regulated under the Clean Air Act and may affect or be located in an attainment area? (FORM 5)	40	41	42	J. Is this facility a proposed **stationary source** which is NOT one of the 28 industrial categories listed in the instructions and which will potentially emit 250 tons per year of any air pollutant regulated under the Clean Air Act and may affect or be located in an **attainment area**? (FORM 5)	43	44	45

III. NAME OF FACILITY

1 SKIP

IV. FACILITY CONTACT

A. NAME & TITLE *(last, first, & title)*

B. PHONE *(area code & no.)*

V. FACILITY MAILING ADDRESS

A. STREET OR P.O. BOX

B. CITY OR TOWN | C. STATE | D. ZIP CODE

VI. FACILITY LOCATION

A. STREET, ROUTE NO. OR OTHER SPECIFIC IDENTIFIER

B. COUNTY NAME

C. CITY OR TOWN | D. STATE | E. ZIP CODE | F. COUNTY CODE *(if known)*

EPA Form 3510-1 (6-80)

CONTINUE ON REVERSE

CONTINUED FROM THE FRONT

VII. SIC CODES *(4-digit, in order of priority)*

A. FIRST — 7 (specify)

B. SECOND — 7 (specify)

C. THIRD — 7 (specify)

D. FOURTH — 7 (specify)

VIII. OPERATOR INFORMATION

A. NAME — 8

B. Is the name listed in Item VIII-A also the owner? ☐ YES ☐ NO

C. STATUS OF OPERATOR *(Enter the appropriate letter into the answer box; if "Other", specify.)*

F = FEDERAL
S = STATE
P = PRIVATE
M = PUBLIC *(other than federal or state)*
O = OTHER *(specify)*

(specify)

D. PHONE *(area code & no.)* — A

E. STREET OR P.O. BOX

F. CITY OR TOWN — B

G. STATE

H. ZIP CODE

IX. INDIAN LAND

Is the facility located on Indian lands?

☐ YES ☐ NO

X. EXISTING ENVIRONMENTAL PERMITS

A. NPDES *(Discharges to Surface Water)* — 9 N

D. PSD *(Air Emissions from Proposed Sources)* — 9 P

B. UIC *(Underground Injection of Fluids)* — 9 U

E. OTHER *(specify)* — 9 (specify)

C. RCRA *(Hazardous Wastes)* — 9 R

E. OTHER *(specify)* — 9 (specify)

XI. MAP

Attach to this application a topographic map of the area extending to at least one mile beyond property bounderies. The map must show the outline of the facility, the location of each of its existing and proposed intake and discharge structures, each of its hazardous waste treatment, storage, or disposal facilities, and each well where it injects fluids underground. Include all springs, rivers and other surface water bodies in the map area. See instructions for precise requirements.

XII. NATURE OF BUSINESS *(provide a brief description)*

XIII. CERTIFICATION *(see instructions)*

I certify under penalty of law that I have personally examined and am familiar with the information submitted in this application and all attachments and that, based on my inquiry of those persons immediately responsible for obtaining the information contained in the application, I believe that the information is true, accurate and complete. I am aware that there are significant penalties for submitting false information, including the possibility of fine and imprisonment.

A. NAME & OFFICIAL TITLE *(type or print)*	B. SIGNATURE	C. DATE SIGNED

COMMENTS FOR OFFICIAL USE ONLY

C

EPA Form 3510-1 (6-80) REVERSE

Please print or type in the unshaded areas only
(fill–in areas are spaced for elite type, i.e., 12 characters/inch).

Form Approved OMB No. 158-R0175

FORM 1 GENERAL — EPA

U.S. ENVIRONMENTAL PROTECTION AGENCY

GENERAL INFORMATION

Consolidated Permits Program

(Read the "General Instructions" before starting.)

I. EPA I.D. NUMBER

F | T/A | C | D

LABEL ITEMS

I. EPA I.D. NUMBER

III. FACILITY NAME

V. FACILITY MAILING ADDRESS

VI. FACILITY LOCATION

PLEASE PLACE LABEL IN THIS SPACE

GENERAL INSTRUCTIONS

If a preprinted label has been provided, affix it in the designated space. Review the information carefully; if any of it is incorrect, cross through it and enter the correct data in the appropriate fill–in area below. Also, if any of the preprinted data is absent *(the area to the left of the label space lists the information that should appear)*, please provide it in the proper fill–in area*(s)* below. If the label is complete and correct, you need not complete Items I, III, V, and VI *(except VI-B which must be completed regardless)*. Complete all items if no label has been provided. Refer to the instructions for detailed item descriptions and for the legal authorizations under which this data is collected.

II. POLLUTANT CHARACTERISTICS

INSTRUCTIONS: Complete A through J to determine whether you need to submit any permit application forms to the EPA. If you answer "yes" to any questions, you must submit this form and the supplemental form listed in the parenthesis following the question. Mark "X" in the box in the third column if the supplemental form is attached. If you answer "no" to each question, you need not submit any of these forms. You may answer "no" if your activity is excluded from permit requirements; see Section C of the instructions. See also, Section D of the instructions for definitions of **bold–faced** terms.

SPECIFIC QUESTIONS	MARK 'X' YES	NO	FORM ATTACHED	SPECIFIC QUESTIONS	MARK 'X' YES	NO	FORM ATTACHED
A. Is this facility a **publicly owned treatment works** which results in a **discharge** to **waters of the U.S.**? (FORM 2A)	16	17	18	B. Does or will this facility *(either existing or proposed)* include a **concentrated animal feeding operation** or **aquatic animal production facility** which results in a **discharge** to **waters of the U.S.**? (FORM 2B)	19	20	21
C. Is this a facility which currently results in **discharges** to **waters of the U.S.** other than those described in A or B above? (FORM 2C)	22	23	24	D. Is this a proposed facility *(other than those described in A or B above)* which will result in a **discharge** to **waters of the U.S.**? (FORM 2D)	25	26	27
E. Does or will this facility treat, store, or dispose of **hazardous wastes**? (FORM 3)	28	29	30	F. Do you or will you inject at this facility industrial or municipal effluent below the lowermost stratum containing, within one quarter mile of the well bore, underground sources of drinking water? (FORM 4)	31	32	33
G. Do you or will you inject at this facility any produced water or other fluids which are brought to the surface in connection with conventional oil or natural gas production, inject fluids used for enhanced recovery of oil or natural gas, or inject fluids for storage of liquid hydrocarbons? (FORM 4)	34	35	36	H. Do you or will you inject at this facility fluids for special processes such as mining of sulfur by the Frasch process, solution mining of minerals, in situ combustion of fossil fuel, or recovery of geothermal energy? (FORM 4)	37	38	39
I. Is this facility a proposed **stationary source** which is one of the 28 industrial categories listed in the instructions and which will potentially emit 100 tons per year of any air pollutant regulated under the Clean Air Act and may affect or be located in an **attainment area**? (FORM 5)	40	41	42	J. Is this facility a proposed **stationary source** which is NOT one of the 28 industrial categories listed in the instructions and which will potentially emit 250 tons per year of any air pollutant regulated under the Clean Air Act and may affect or be located in an **attainment area**? (FORM 5)	43	44	45

III. NAME OF FACILITY

SKIP 16 - 29 | 30 ... 69

IV. FACILITY CONTACT

A. NAME & TITLE *(last, first, & title)* 15 ... 45

B. PHONE *(area code & no.)* 46 - 48 | 49 - 51 | 52 - 55

V. FACILITY MAILING ADDRESS

A. STREET OR P.O. BOX 15 ... 45

B. CITY OR TOWN 15 ... 40 | C. STATE 41 42 | D. ZIP CODE 47 - 51

VI. FACILITY LOCATION

A. STREET, ROUTE NO. OR OTHER SPECIFIC IDENTIFIER 15 ... 45

B. COUNTY NAME ... 70

C. CITY OR TOWN 15 ... 40 | D. STATE 41 42 | E. ZIP CODE 47 - 51 | F. COUNTY CODE *(if known)* 52 - 54

EPA Form 3510-1 (6-80)

CONTINUE ON REVERSE

CONTINUED FROM THE FRONT

VII. SIC CODES *(4-digit, in order of priority)*

A. FIRST — C 7 (specify)

B. SECOND — C 7 (specify)

C. THIRD — C 7 (specify)

D. FOURTH — C 7 (specify)

VIII. OPERATOR INFORMATION

A. NAME — C 8

B. Is the name listed in Item VIII-A also the owner? ☐ YES ☐ NO

C. STATUS OF OPERATOR *(Enter the appropriate letter into the answer box; if "Other", specify.)*

F = FEDERAL
S = STATE
P = PRIVATE
M = PUBLIC *(other than federal or state)*
O = OTHER *(specify)*

(specify)

D. PHONE *(area code & no.)* — C A

E. STREET OR P.O. BOX

F. CITY OR TOWN — C B

G. STATE

H. ZIP CODE

IX. INDIAN LAND

Is the facility located on Indian lands?

☐ YES ☐ NO

X. EXISTING ENVIRONMENTAL PERMITS

A. NPDES *(Discharges to Surface Water)* — 9 N

D. PSD *(Air Emissions from Proposed Sources)* — 9 P

B. UIC *(Underground Injection of Fluids)* — 9 U

E. OTHER *(specify)* — 9 (specify)

C. RCRA *(Hazardous Wastes)* — 9 R

E. OTHER *(specify)* — 9 (specify)

XI. MAP

Attach to this application a topographic map of the area extending to at least one mile beyond property bounderies. The map must show the outline of the facility, the location of each of its existing and proposed intake and discharge structures, each of its hazardous waste treatment, storage, or disposal facilities, and each well where it injects fluids underground. Include all springs, rivers and other surface water bodies in the map area. See instructions for precise requirements.

XII. NATURE OF BUSINESS *(provide a brief description)*

XIII. CERTIFICATION *(see instructions)*

I certify under penalty of law that I have personally examined and am familiar with the information submitted in this application and all attachments and that, based on my inquiry of those persons immediately responsible for obtaining the information contained in the application, I believe that the information is true, accurate and complete. I am aware that there are significant penalties for submitting false information, including the possibility of fine and imprisonment.

A. NAME & OFFICIAL TITLE *(type or print)*

B. SIGNATURE

C. DATE SIGNED

COMMENTS FOR OFFICIAL USE ONLY

C

EPA Form 3510-1 (6-80) REVERSE

☆U.S. GOVERNMENT PRINTING OFFICE: 1980—323-829:635

United States
Environmental Protection
Agency

Office of
Enforcement
Washington, DC 20460

EPA Form 3510-2C
June 1980

Permits Division

Application Form 2C - Wastewater Discharge Information

Consolidated Permits Program

This form must be completed by all persons applying for an EPA permit to discharge wastewater *(existing manufacturing, commercial, mining, and silvicultural operations)*.

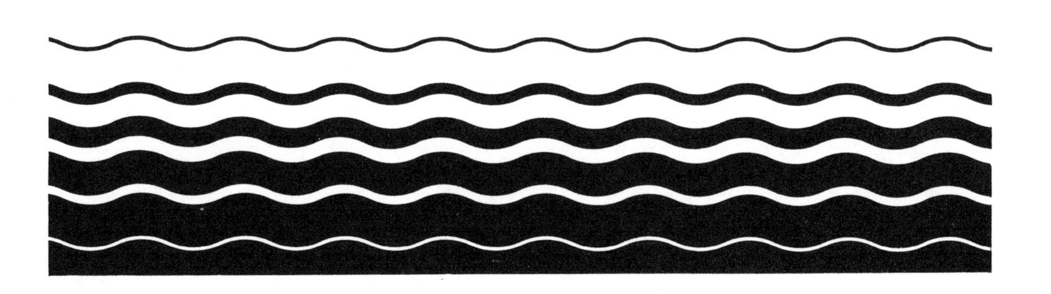

FORM 2C – INSTRUCTIONS

This form must be completed by all applicants who check "yes" to Item II–C in Form 1.

Public Availability of Submitted Information

Your application will not be considered complete unless you answer every question on this form and on Form 1. If an item does not apply to you, enter "NA" *(for not applicable)* to show that you considered the question.

You may not claim as confidential any information required by this form or Form 1, whether the information is reported on the forms or in an attachment. This information will be made available to the public upon request.

Any information you submit to EPA which goes beyond that required by this form and Form 1 you may claim as confidential, but claims for information which is effluent data will be denied. If you do not assert a claim of confidentiality at the time of submitting the information, EPA may make the information public without further notice to you. Claims of confidentiality will be handled in accordance with EPA's business confidentiality regulations at 40 CFR Part 2.

Definitions

All significant terms used in these instructions and in the form are defined in the glossary found in the General Instructions which accompany Form 1.

EPA I.D. Number

Fill in your EPA Identification Number at the top of each page of Form 2C. You may copy this number directly from Item I of Form 1.

Item I

You may use the map you provided for Item XI of Form 1 to determine the latitude and longitude of each of your outfalls and the name of the receiving water.

Item II–A

The line drawing should show generally the route taken by water in your facility from intake to discharge. Show all operations contributing wastewater, including process and production areas, sanitary flows, cooling water, and stormwater runoff. You may group similar operations into a single unit, labeled to correspond to the more detailed listing in Item II–B. The water balance should show average flows. Show all significant losses of water to products, atmosphere, and discharge. You should use actual measurements whenever available; otherwise use your best estimate. An example of any acceptable line drawing appears in Figure 2c–1 to these instructions.

Item II–B

List all sources of wastewater to each outfall. Operations may be described in general terms *(for example, "dye–making reactor" or a "distillation tower")*. You may estimate the flow contributed by each source if no data is available, and for stormwater, you may use any reasonable measure of duration, volume, or frequency. For each treatment unit, indicate its size, flow rate, and retention time, and describe the ultimate disposal of any solid or liquid wastes not discharged. Treatment units should be listed in order and you should select the proper code from Table 2c–1 to fill in column 3–b for each treatment unit. Insert "XX" into column 3–b if no code corresponds to a treatment unit you list.

If you are applying for a permit for a privately owned treatment works, you must also identify all of your contributors in an attached listing.

Item II–C

A discharge is intermittent unless it occurs without interruption during the operating hours of the facility, except for infrequent shutdowns for maintenance, process changes, or other similar activities. A discharge is seasonal if it occurs only during certain parts of the year. Fill in every applicable column in this item for each source of intermittent or seasonal discharges. Base your answers on actual data whenever available; otherwise, provide your best estimate. Report the highest daily value for flow rate and total volume in the "Maximum Daily" columns *(columns 4–a–2 and 4–b–2)*. Report the average of all daily values measured during days when discharge occurred within the last year in the "Long Term Average" columns *(columns 4–a–1 and 4–b–1)*.

Item III–A

All effluent guidelines promulgated by EPA appear in the Federal Register and are published annually in 40 CFR Subchapter N. A guideline applies to you if you have any operations contributing process wastewater in any subcategory covered by a BPT, BCT, or BAT guideline. If you are unsure whether you are covered by a promulgated effluent guideline, check with your EPA Regional office *(Table 1 in the Form 1 instructions)*. You must check "yes" if an applicable effluent guideline has been promulgated, even if the guideline limitations are being contested in court. If you believe that a promulgated effluent guideline has been remanded for reconsideration by a court and does not apply to your operations, you may check "no."

Item III–B

An effluent guideline is expressed in terms of production *(or other measure of operation)* if the limitations are expressed as mass of pollutant per operational parameter; for example, "pounds of BOD per cubic foot of logs from which bark is removed," or "pounds of TSS per megawatt hour of electrical energy consumed by smelting furnace." An example of a guideline not expressed in terms of a measure of operation is one which limits the concentration of pollutants.

Item III–C

This item must be completed only if you checked "yes" to Item III–B. The production information requested here is necessary to apply effluent guidelines to your facility and you may not claim it as confidential. However, you do not have to indicate how the reported information was calculated.

Report quantities in the units of measurement used in the applicable effluent guideline. The figures provided must be a measure of actual operation over a one month period, such as the production for the highest month during the last twelve months, or the monthly average production for the highest year of the last five years, or other reasonable measure of actual operation, but may not be based on design capacity or on predictions of future increases in operation.

Item IV–A

If you check "yes" to this question, complete all parts of the chart, or attach a copy of any previous submission you have made to EPA containing the same information.

Item IV–B

You are not required to submit a description of future pollution control projects if you do not wish to or if none is planned.

Item V–A, B, C, and D

These items require you to collect and report data on the pollutants discharged from each of your outfalls. Each part of this item addresses a different set of pollutants and must be completed in accordance with the specific instructions for that part. The following general instructions apply to the entire item.

FORM 2C – INSTRUCTIONS *(continued)*

ITEM V–A,B,C, and D *(continued)*

GENERAL INSTRUCTIONS. Part A requires you to report at least one analysis for each pollutant listed..Parts B and C require you to report analytical data in two ways. For some pollutants, you may be required to mark "X" in the "Testing Required" column *(column 2–a, Part C)*, and test *(sample and analyze)* and report the levels of the pollutants in your discharge whether or not you expect them to be present in your discharge. For all others, you must mark "X" in either the "Believe Present" column or the "Believe Absent" column *(column 2–a or 2–b, Part B, and column 2–b or 2–c, Part C)* based on your best estimate, and test for those which you believe to be present. Part D requires you to list any of a group of pollutants which you believe to be present, with a brief explanation of why you believe it to be present. *(See specific instructions on the form and below for Parts A through D.)*

Base your determination that a pollutant is present in or absent from your discharge on your knowledge of your raw materials, maintenance chemicals, intermediate and final products and byproducts, and any previous analyses known to you of your effluent or of any similar effluent. *(For example, if you manufacture pesticides, you should expect those pesticides to be present in contaminated stormwater runoff.)* If you would expect a pollutant to be present solely as a result of its presence in your intake water, you must mark "Believe Present" but you are not required to analyze for that pollutant. Instead, mark an "X" in the "Intake" column.

A. REPORTING. All levels must be reported as concentration and as total mass. You may report some or all of the required data by attaching separate sheets of paper instead of filling out pages V–1 thru V–9 if the separate sheets contain all the required information in a format which is consistent with pages V–1 thru V–9 in spacing and in identification of pollutants and columns. *(For example, the data system used in your GC/MS analysis may be able to print data in the proper format.)* Use the following abbreviations in the columns headed "Units" *(column 3, Part A, and column 4, Parts B and C)*.

CONCENTRATION	MASS
ppm. . . . .parts per million	lbs. . . . .pounds
mg/l. . . . .milligrams per liter	ton . . . .tons *(English tons)*
ppb . . . . .parts per billion	mg . . . .milligrams
μg/l. . . . .micrograms per liter	g. . . . . .grams
	kg . . . . .kilograms
	T . . . . .tonnes *(metric tons)*

If you measure only one daily value, complete only the "Maximum Daily Values" columns and insert "1" into the "Number of Analyses" columns *(columns 2–a and 2–d, Part A, and columns 3–a and 3–d, Parts B and C)*. The permitting authority may require you to conduct additional analyses to further characterize your discharges.

For composite samples, the daily value is the total mass or average concentration found in a composite sample taken over the operating hours of the facility during a 24 hour period; for grab samples, the daily value is the arithmetic or flow–weighted total mass or average concentration found in a series of at least four grab samples taken over the operating hours of the facility during a 24 hour period.

If you measure more than one daily value for a pollutant, determine the average of all values within the last year and report the concentration and mass under the "Long Term Average Values" columns *(column 2–c, Part A, and column 3–c, Parts B and C)*, and the total number of daily values under the "Number of Analyses" columns *(column 2–d, Part A, and column 3–d, Parts B and C)*. Also, determine the average of all daily values taken during each calendar month, and report the highest average under the "Maximum 30 Day Values" columns *(column 2–b, Part A, and column 3–b, Parts B and C)*.

B. SAMPLING. The collection of the samples for the reported analyses should be supervised by a person experienced in performing sampling of industrial wastewater. You may contact your EPA or State permitting authority for detailed guidance on sampling techniques and for answers to specific questions. Any specific requirements contained in the applicable analytical methods should be followed for sample containers, sample preservation, holding times, the collection of duplicate samples, etc. The time when you sample should be representative of your normal operation, to the extent feasible, with all processes which contribute wastewater in normal operation, and with your treatment system operating properly with no system upsets. Samples should be collected from the center of the flow channel, where turbulence is at a maximum, at a site specified in your present permit, or at any site adequate for the collection of a representative sample.

ITEM V–A,B,C, and D *(continued)*

Grab and composite samples are defined as follows:

1. GRAB SAMPLE. An individual sample of at least 100 milliliters collected at a randomly–selected time over a period not exceeding 15 minutes.

2. COMPOSITE SAMPLE. A combination of at least 8 sample aliquots of at least 100 milliliters, collected at periodic intervals during the operating hours of a facility over a 24 hour period. For volatile pollutants, aliquots must be combined in the laboratory immediately before analysis. The compostie must be flow proportional; either the time interval between each aliquot or the volume of each aliquot must be proportional to either the stream flow at the time of sampling or the total stream flow since the collection of the previous aliquot. Aliquots may be collected manually or automatically.

C. ANALYSIS. You must use test methods promulgated in 40 CFR Part 136; however, if none has been promulgated for a particular pollutant, you may use any suitable method for measuring the level of the pollutant in your discharge provided that you submit a description of the method or a reference to a published method. Your description should include the sample holding times, preservation techniques, and the quality control measures which you used.

If you have two or more substantially indentical outfalls, you may request permission from your permitting authority to sample and analyze only one outfall and submit the results of the analysis for other substantially identical outfalls. If your request is granted by the permitting authority, on a separate sheet attached to the application form identify which outfall you did test, and describe why the outfalls which you did not test are substantially identical to the outfall which you did test.

D. REPORTING OF INTAKE DATA. You are not required to report data under the "Intake" columns unless you wish to demonstrate your eligibility for a "net" effluent limitation for one or more pollutants, that is, an effluent limitation adjusted by subtracting the average level of the pollutant*(s)* present in your intake water. NPDES regulations allow net limitations only in certain circumstances. To demonstrate your eligibility, under the "Intake" columns report the average of the results of analyses on your intake water *(if your water is treated before use, test the water after it is treated)*, and attach a separate sheet containing the following for each pollutant:

1. A statement that the intake water is drawn from the body of water into which the discharge is made. *(Otherwise, you are not eligible for net limitations.)*

2. A statement of the extent to which the level of the pollutant is reduced by treatment of your wastewater. *(Your limitations will be adjusted only to the extent that the pollutant is not removed.)*

3. When applicable *(for example, when the pollutant represents a class of compounds)*, a demonstration of the extent to which the pollutants in the intake vary physically, chemically, or biologically from the pollutants contained in your discharge. *(Your limitations will be adjusted only to the extent that the intake pollutants do not vary from the discharged pollutants.)*

PART V–A. Part V–A must be completed by all applicants for all outfalls, including outfalls containing only noncontact cooling water or storm runoff. However, at your request, the permitting authority may waive the requirements to test for one or more of these pollutants, upon a determination that testing for the pollutant*(s)* is not appropriate for your effluents.

Use composite samples for all pollutants in this Part, except use grab samples for pH and temperature. See discussion in General Instructions to Item V for definitions of the columns in Part A. The "Long Term Average Values" column *(column 2–c)* and "Maximum 30 Day Values" column *(column 2–b)* are not compulsory but should be filled out if data is available.

PART V–B. Part V–B must be completed by all applicants for all outfalls, including outfalls containing only noncontact cooling water or storm runoff.

Use composite samples for all pollutants you analyze for in this Part, except use grab samples for residual chlorine, oil and grease, and fecal coliform. The "Long Term Average Values" column *(column 3–c)* and "Maximum 30 Day Values" column *(column 3–b)* are not compulsory but should be filled out if data is available.

ITEM V–A,B,C, and D *(continued)*

PART V–C. Table 2c–2 lists the 34 "primary" industry categories in the left–hand column. For each outfall, if any of your processes which contribute wastewater falls into one of those categories, you must mark "X" in "Testing Required" column *(column 2–a)* and test for: (A) All of the toxic metals, cyanide, and total phenols; and (B) The organic toxic pollutants contained in the gas chromotography/mass spectrometry *(GC/MS)* fractions indicated in Table 2c–2 as applicable to your category, unless you qualify as a small business *(see below)*. The organic toxic pollutants are listed by GC/MS fractions on pages V–4 through V–9 in Part V–C. For example, the Organic Chemicals Industry has an "X" in all four fractions; therefore, applicants in this category must test for all organic toxic pollutants in Part V–C. If you are applying for a permit for a privately owned treatment works, determine your testing requirements on the basis of the industry categories of your contributors. When you determine which industry category you are in to find your testing requirements, you are not determining your category for any other purpose and you are not giving up your right to challenge your inclusion in that category *(for example, for deciding whether an effluent guideline is applicable)* before your permit is issued.

For all other cases *(secondary industries, non–process wastewater outfalls, and non–required GC/MS fractions)*, you must mark "X" in either the "Believed Present" column *(column 2–b)* or the "Believed Absent" column *(column 2–c)* for each pollutant, and test for those you believe present *(those marked "X" in column 2–b)*. If you qualify as a small business *(see below)* you are exempt from testing for the organic toxic pollutants, listed on pages V–4 through V–9 in Part C. For pollutants in intake water, see discussion in General Instructions to this item. The "Long Term Average Values" column *(column 3–c)* and "Maximum 30 Day Values" column *(column 3–b)* are not compulsory but should be filled out if data is available.

Use composite samples for all pollutants in this Part, except use grab samples for total phenols and cyanide.

You are required to mark "Testing Required" for dioxin if you use or manufacture one of the following compounds:
A. 2,4,5–trichlorophenoxy acetic acid (2,4,5–T);
B. 2–(2,4,5–trichlorophenoxy) propanoic acid (Silvex, 2,4,5–TP);
C. 2–(2,4,5–trichlorophenoxy) ethyl 2,2–dichloropropionate (Erbon);
D. O,O–dimethyl O–(2,4,5–trichlorophenyl) phosphorothioate (Ronnel);
E. 2,4,5–trichlorophenol (TCP); or
F. Hexachlorophene (HCP).
If you mark "Testing Required" or "Believe Present," you must perform a screening analysis for dioxins, using gas chromatography with an electron capture detector. A TCDD standard for quantitation is not required. Describe the results of this analysis in the space provided; for example, "no measurable baseline deflection at the retention time of TCDD" or "a measurable peak within the tolerances of the retention time of TCDD." The permitting authority may require you to perform a quantitative analysis if you report a positive result.

The Effluent Guidelines Division of EPA has collected and analyzed samples from some plants for the pollutants listed in Part C in the course of its BAT guidelines development program. If your effluents were sampled and analyzed as part of this program in the last three years, you may use this data to answer Part C provided that the permitting authority approves, and provided that no process change or change in raw materials or operating practices has occurred since the samples were taken that would make the analyses unrepresentative of your current discharge.

SMALL BUSINESS EXEMPTION. If you qualify as a "small business," you are exempt from the reporting requirements for the organic toxic pollutants, listed on pages V–4 through V–9 in Part C. If your facility is a coal mine, and if your probable total annual production is less than 100,000 tons per year, you may submit past production data or estimated future production *(such as a schedule of estimated total production under 30 CFR Section 795.14(c))* instead of conducting analyses for the organic toxic pollutants. If your facility is not a coal mine, and if your gross total annual sales for the most recent three years average less than $100,000 per year *(in second quarter 1980 dollars)*, you may submit sales data for those years instead of conducting analyses for the organic toxic pollutants.

ITEM V–A,B,C, and D *(continued)*

The production or sales data must be for the facility which is the source of the discharge. The data should not be limited to production or sales for the process or processes which contribute to the discharge, unless those are the only processes at your facility. For sales data, in situations involving intra–corporate transfers of goods and services, the transfer price per unit should approximate market prices for those goods and services as closely as possible. Sales figures for years after 1980 should be indexed to the second quarter of 1980 by using the gross national product price deflator *(second quarter of 1980 = 100)*. This index is available in "National Income and Product Accounts of the United States" *(Department of Commerce, Bureau of Economic Analysis)*.

PART V–D. List any pollutants in Table 2c–3 that you believe to be present and explain why you believe them to be present. No analysis is required, but if you have analytical data, you must report it.

NOTE: Under 40 CFR 117.12(a)(2), certain discharges of hazardous substances *(listed in Table 2c–4 of these instructions)* may be exempted from the requirements of Section 311 of CWA, which establishes reporting requirements, civil penalties, and liability for clean-up costs for spills of oil and hazardous substances. A discharge of a particular substance may be exempted if the origin, source, and amount of the discharged substance are indentified in the NPDES permit application or in the permit, if the permit contains a requirement for treatment of the discharge, and if the treatment is in place. To apply for an exclusion of the discharge of any hazardous substance from the requirements of Section 311, attach additional sheets of paper to your form, setting forth the following information:

A. The substance and the amount of each substance which may be discharged;

B. The origin and source of the discharge of the substance;

C. The treatment which is to be provided for the discharge by:

1. An on–site treatment system separate from any treatment system treating your normal discharge,

2. A treatment system designed to treat your normal discharge and which is additionally capable of treating the amount of the substance identified under paragraph 1 above, or

3. Any combination of the above.

See 40 CFR Section 117.12(a)(2) and (c), published on August 29, 1979, in 44 FR 50766, or contact your Regional office *(Table 1 in the Form 1 instructions)*, for further information on exclusions from Section 311.

Item VI–A

You may not claim this information as confidential; however, you do not have to distinguish between use or production of the pollutants or list the amounts. Under NPDES regulations your permit will contain limits to control all pollutants you report in answer to this question, as well as all pollutants reported in Item V or VI–B at levels exceeding the technology–based limits appropriate to your facility. Your permit will also require you to report to EPA if you in the future begin or expect that you will begin to use or manufacture as an intermediate or final product or byproduct any toxic pollutant which you did not report here, and your permit may be modified at that time if necessary to control that pollutant.

Item VI–B

For this item, consider only those variations which may result in concentrations of pollutants in effluents which may exceed two times the maximum values you reported in Item V. These variations may be part of your routine operations, or part of your regular cleaning cycles.

Under NPDES regulations your permit will contain limits to control any pollutant you report in answer to this question at levels exceeding the technology–based limits appropriate to your facility. Your permit will also require you to report to EPA if you know or have reason to believe that any activity has occurred or will occur which would make your discharge of any toxic pollutant five times the maximum values reported in Item V–C or in this item, and your permit may be modified at that time if necessary to control the pollutant.

Do not consider variations which are the result of bypasses or upsets. Increased levels of pollutants which are discharged as a result of bypasses or upsets are regulated separately under NPDES regulations.

FORM 2C – INSTRUCTIONS *(continued)*

Item VI–C

Examples of the types of variations to be described here include:
Changes in raw or intermediate materials;
Changes in process equipment or materials;
Changes in product lines;
Significant chemical reactions between pollutants in waste streams; and
Significant variation in removal efficiencies of pollution control equipment.

You may indicate other types of variations as well, except those which are the result of bypasses or upsets. The permitting authority may require you to further investigate or document variations you report here.

Base your prediction of expected levels of these pollutants upon your knowledge of your processes, raw materials, past and projected product ranges, etc., or upon any testing conducted upon your effluents which indicates the range of variability that can be expected in your effluent over the next five years.

EXAMPLE. Outfall 001 discharges water used to clean six 500 gallon tanks. These tanks are used for formulation of dispersions of synthetic resins in water *(adhesives)*. Use of toxic pollutants which can be expected in the next 5 years is:

1. Copper acetate inhibitor, 1/2 lb. per tank;
2. Dibutyl phthalate, 50 lbs. per tank;
3. Toulene, 5 lbs. per tank; and
4. Antimony oxide, 1 lb. per tank.

Based on normal cleaning an average of 1% and a maximum of 3% of the contents of each tank is collected and discharged once every two weeks in the 150 gallons of water used for cleaning. Treatment *(pH adjustment, flocculation, filtration)* removes 85% of metals and 50% of organic compounds.

Item VII

Self explanatory. The permitting authority may ask you to provide additional details after your application is received.

Item VIII

Self explanatory.

Item IX

The Clean Water Act provides for severe penalties for submitting false information on this application form.

Section 309(c)(2) of the Clean Water Act provides that "Any person who knowingly makes any false statement, representation, or certification in any application, . . . shall upon conviction, be punished by a fine of no more than $10,000 or by imprisonment for not more than six months, or both."

FEDERAL REGULATIONS REQUIRE THE CERTIFICATION TO BE SIGNED AS FOLLOWS:

A. For a corporation, by a principal executive officer of at least the level of vice president;

B. For a partnership or sole proprietorship, by a general partner or the proprietor, respectively; or

C. For a municipality, State, Federal, or other public facility, by either a principal executive officer or ranking elected official.

CODES FOR TREATMENT UNITS

PHYSICAL TREATMENT PROCESSES

1–A .... Ammonia Stripping
1–B. .... Dialysis
1–C. .... Diatomaceous Earth Filtration
1–D .... Distillation
1–E. .... Electrodialysis
1–F. .... Evaporation
1–G .... Flocculation
1–H .... Flotation
1–I .... Foam Fractionation
1–J. .... Freezing
1–K .... Gas–Phase Separation
1–L. .... Grinding *(Comminutors)*
1–M .... Grit Removal
1–N .... Microstraining
1–O .... Mixing
1–P. .... Moving Bed Filters
1–Q .... Multimedia Filtration
1–R .... Rapid Sand Filtration
1–S. .... Reverse Osmosis *(Hyperfiltration)*
1–T. .... Screening
1–U .... Sedimentation *(Settling)*
1–V .... Slow Sand Filtration
1–W .... Solvent Extraction
1–X .... Sorption

CHEMICAL TREATMENT PROCESSES

2–A .... Carbon Adsorption
2–B. .... Chemical Oxidation
2–C. .... Chemical Precipitation
2–D .... Coagulation
2–E. .... Dechlorination
2–F. .... Disinfection *(Chlorine)*
2–G .... Disinfection *(Ozone)*
2–H .... Disinfection *(Other)*
2–I .... Electrochemical Treatment
2–J. .... Ion Exchange
2–K .... Neutralization
2–L. .... Reduction

BIOLOGICAL TREATMENT PROCESSES

3–A .... Activated Sludge
3–B. .... Aerated Lagoons
3–C. .... Anaerobic Treatment
3–D .... Nitrification–Denitrification
3–E. .... Pre–Aeration
3–F. .... Spray Irrigation/Land Application
3–G .... Stabilization Ponds
3–H .... Trickling Filtration

OTHER PROCESSES

4–A .... Discharge to Surface Water
4–B. .... Ocean Discharge Through Outfall
4–C. .... Reuse/Recycle of Treated Effluent
4–D .... Underground Injection

SLUDGE TREATMENT AND DISPOSAL PROCESSES

5–A .... Aerobic Digestion
5–B. .... Anaerobic Digestion
5–C. .... Belt Filtration
5–D .... Centrifugation
5–E. .... Chemical Conditioning
5–F. .... Chlorine Treatment
5–G .... Composting
5–H .... Drying Beds
5–I .... Elutriation
5–J. .... Flotation Thickening
5–K .... Freezing
5–L. .... Gravity Thickening
5–M .... Heat Drying
5–N .... Heat Treatment
5–O .... Incineration
5–P. .... Land Application
5–Q .... Landfill
5–R .... Pressure Filtration
5–S. .... Pyrolysis
5–T. .... Sludge Lagoons
5–U .... Vacuum Filtration
5–V .... Vibration
5–W .... Wet Oxidation

TABLE 2C-1

TESTING REQUIREMENTS FOR ORGANIC TOXIC POLLUTANTS INDUSTRY CATEGORY

INDUSTRY CATEGORY	GC/MS FRACTION[1]			
	Volatile	Acid	Base/Neutral	Pesticide
Adhesives and sealants	X	X	X	–
Aluminum forming	X	X	X	–
Auto and other laundries	X	X	X	X
Battery manufacturing	X	–	X	–
Coal mining	X	X	X	X
Coil coating	X	X	X	–
Copper forming	X	X	X	–
Electric and electronic compounds	X	X	X	X
Electroplating	X	X	X	–
Explosives manufacturing	X	X	X	–
Foundries	X	X	X	–
Gum and wood chemicals	X	X	X	X
Inorganic chemicals manufacturing	X	X	X	–
Iron and steel manufacturing	X	X	X	–
Leather tanning and finishing	X	X	X	X
Mechanical products manufacturing	X	X	X	–
Nonferrous metals manufacturing	X	X	X	X
Ore mining	X	X	X	X
Organic chemicals manufacturing	X	X	X	X
Paint and ink formulation	X	X	X	X
Pesticides	X	X	X	X
Petroleum refining	X	X	X	X
Pharmaceutical preparations	X	X	X	–
Photographic equipment and supplies	X	X	X	X
Plastic and synthetic materials manufacturing	X	X	X	X
Plastic processing	X	–	–	–
Porcelain enameling	X	–	X	X
Printing and publishing	X	X	X	X
Pulp and paperboard mills	X	X	X	X
Rubber processing	X	X	X	–
Soap and detergent manufacturing	X	X	X	–
Steam electric power plants	X	X	X	–
Textile mills	X	X	X	X
Timber products processing	X	X	X	X

[1] The pollutants in each fraction are listed in Item V–C.
X = Testing required.
– = Testing not required.

TABLE 2C-2

TOXIC POLLUTANTS AND HAZARDOUS SUBSTANCES REQUIRED TO BE IDENTIFIED BY APPLICANTS IF EXPECTED TO BE PRESENT

TOXIC POLLUTANT

Asbestos

HAZARDOUS SUBSTANCES

Acetaldehyde
Allyl alcohol
Allyl chloride
Amyl acetate
Aniline
Benzonitrile
Benzyl chloride
Butyl acetate
Butylamine
Captan
Carbaryl
Carbofuran
Carbon disulfide
Chlorpyrifos
Coumaphos
Cresol
Crotonaldehyde
Cyclohexane
2,4-D (2,4-Dichlorophenoxyacetic acid)
Diazinon
Dicamba
Dichlobenil
Dichlone
2,2-Dichloropropionic acid

HAZARDOUS SUBSTANCES

Dichlorvos
Diethyl amine
Dimethyl amine
Dintrobenzene
Diquat
Disulfoton
Diuron
Epichlorohydrin
Ethion
Ethylene diamine
Ethylene dibromide
Formaldehyde
Furfural
Guthion
Isoprene
Isopropanolamine
Kelthane
Kepone
Malathion
Mercaptodimethur
Methoxychlor
Methyl mercaptan
Methyl methacrylate
Methyl parathion
Mevinphos
Mexacarbate
Monoethyl amine
Monomethyl amine

HAZARDOUS SUBSTANCES

Naled
Napthenic acid
Nitrotoluene
Parathion
Phenolsulfonate
Phosgene
Propargite
Propylene oxide
Pyrethrins
Quinoline
Resorcinol
Strontium
Strychnine
Styrene
2,4,5-T (2,4,5-Trichlorophenoxyacetic acid)
TDE (Tetrachlorodiphenyl ethane)
2,4,5-TP [2-(2,4,5-Trichlorophenoxy) propanoic acid]
Trichlorofon
Triethanolamine
Triethylamine
Trimethylamine
Uranium
Vanadium
Vinyl acetate
Xylene
Xylenol
Zirconium

TABLE 2C-3

HAZARDOUS SUBSTANCES

1. Acetaldehyde
2. Acetic acid
3. Acetic anhydride
4. Acetone cyanohydrin
5. Acetyl bromide
6. Acetyl chloride
7. Acrolein
8. Acrylonitrile
9. Adipic acid
10. Aldrin
11. Allyl alcohol
12. Allyl chloride
13. Aluminum sulfate
14. Ammonia
15. Ammonium acetate
16. Ammonium benzoate
17. Ammonium bicarbonate
18. Ammonium bichromate
19. Ammonium bifluoride
20. Ammonium bisulfite
21. Ammonium carbamate
22. Ammonium carbonate
23. Ammonium chloride
24. Ammonium chromate
25. Ammonium citrate
26. Ammonium fluoroborate
27. Ammonium fluoride
28. Ammonium hydroxide
29. Ammonium oxalate
30. Ammonium silicofluoride
31. Ammonium sulfamate
32. Ammonium sulfide
33. Ammonium sulfite
34. Ammonium tartrate
35. Ammonium thiocyanate
36. Ammonium thiosulfate
37. Amyl acetate
38. Aniline
39. Antimony pentachloride
40. Antimony potassium tartrate
41. Antimony tribromide
42. Antimony trichloride
43. Antimony trifluoride
44. Antimony trioxide
45. Arsenic disulfide
46. Arsenic pentoxide
47. Arsenic trichloride
48. Arsenic trioxide
49. Arsenic trisulfide
50. Barium cyanide
51. Benzene
52. Benzoic acid
53. Benzonitrile
54. Benzoyl chloride
55. Benzyl chloride
56. Beryllium chloride
57. Beryllium fluoride
58. Beryllium nitrate
59. Butylacetate
60. n-Butylphthalate
61. Butylamine
62. Butyric acid
63. Cadmium acetate
64. Cadmium bromide
65. Cadmium chloride
66. Calcium arsenate
67. Calcium arsenite
68. Calcium carbide
69. Calcium chromate
70. Calcium cyanide
71. Calcium dodecylbenzenesulfonate
72. Calcium hypochlorite
73. Captan
74. Carbaryl
75. Carbofuran
76. Carbon disulfide
77. Carbon tetrachloride
78. Chlordane
79. Chlorine
80. Chlorobenzene
81. Chloroform
82. Chloropyrifos
83. Chlorosulfonic acid
84. Chromic acetate
85. Chromic acid
86. Chromic sulfate
87. Chromous chloride
88. Cobaltous bromide
89. Cobaltous formate
90. Cobaltous sulfamate
91. Coumaphos
92. Cresol
93. Crotonaldehyde
94. Cupric acetate
95. Cupric acetoarsenite
96. Cupric chloride
97. Cupric nitrate
98. Cupric oxalate
99. Cupric sulfate
100. Cupric sulfate ammoniated
101. Cupric tartrate
102. Cyanogen chloride
103. Cyclohexane
104. 2,4-D acid (2,4-Dichlorophenoxyacetic acid)
105. 2,4-D esters (2,4-Dichlorophenoxyacetic acid esters)
106. DDT
107. Diazinon
108. Dicamba
109. Dichlobenil
110. Dichlone
111. Dichlorobenzene
112. Dichloropropane
113. Dichloropropene
114. Dichloropropene-dichlopropropane mix
115. 2,2-Dichloropropionic acid
116. Dichlorvos
117. Dieldrin
118. Diethylamine
119. Dimethylamine
120. Dinitrobenzene
121. Dinitrophenol
122. Dinitrotoluene
123. Diquat
124. Disulfoton
125. Diuron
126. Dodecylbenzesulfonic acid
127. Endosulfan
128. Endrin
129. Epichlorohydrin
130. Ethion
131. Ethylbenzene
132. Ethylenediamine
133. Ethylene dibromide
134. Ethylene dichloride
135. Ethylene diaminetetracetic acid (EDTA)
136. Ferric ammonium citrate
137. Ferric ammonium oxalate
138. Ferric chloride
139. Ferric fluoride
140. Ferric nitrate
141. Ferric sulfate
142. Ferrous ammonium sulfate
143. Ferrous chloride
144. Ferrous sulfate
145. Formaldehyde
146. Formic acid
147. Fumaric acid
148. Furfural
149. Guthion
150. Heptachlor
151. Hexachlorocyclopentadiene
152. Hydrochloric acid
153. Hydrofluoric acid
154. Hydrogen cyanide
155. Hydrogen sulfite
156. Isoprene
157. Isopropanolamine dodecylbenzenesulfonate
158. Kelthane
159. Kepone
160. Lead acetate
161. Lead arsenate
162. Lead chloride
163. Lead fluoborate
164. Lead flourite
165. Lead iodide
166. Lead nitrate
167. Lead stearate
168. Lead sulfate
169. Lead sulfide
170. Lead thiocyanate
171. Lindane
172. Lithium chromate
173. Malathion
174. Maleic acid
175. Maleic anhydride
176. Mercaptodimethur
177. Mercuric cyanide
178. Mercuric nitrate
179. Mercuric sulfate
180. Mercuric thiocyanate
181. Mercurous nitrate
182. Methoxychlor
183. Methyl mercaptan
184. Methyl methacrylate
185. Methyl parathion
186. Mevinphos
187. Mexacarbate
188. Monoethylamine
189. Monomethylamine
190. Naled
191. Napthalene
192. Napthenic acid
193. Nickel ammonium sulfate
194. Nickel chloride
195. Nickel hydroxide
196. Nickel nitrate
197. Nickel sulfate
198. Nitric acid
199. Nitrobenzene
200. Nitrogen dioxide
201. Nitrophenol
202. Nitrotoluene
203. Paraformaldehyde

TABLE 2C-4

HAZARDOUS SUBSTANCES *(continued)*

204. Parathion
205. Pentachlorophenol
206. Phenol
207. Phosgene
208. Phosphoric acid
209. Phosphorus
210. Phosphorus oxychloride
211. Phosphorus pentasulfide
212. Phosphorus trichloride
213. Polychlorinated biphenyls (PCB)
214. Potassium arsenate
215. Potassium arsenite
216. Potassium bichromate
217. Potassium chromate
218. Potassium cyanide
219. Potassium hydroxide
220. Potassium permanganate
221. Propargite
222. Propionic acid
223. Propionic anhydride
224. Propylene oxide
225. Pyrethrins
226. Quinoline
227. Resorcinol
228. Selenium oxide
229. Silver nitrate
230. Sodium
231. Sodium arsenate
232. Sodium arsenite
233. Sodium bichromate
234. Sodium bifluoride
235. Sodium bisulfite
236. Sodium chromate
237. Sodium cyanide
238. Sodium dodecylbenzenesulfonate
239. Sodium fluoride
240. Sodium hydrosulfide
241. Sodium hydroxide
242. Sodium hypochlorite
243. Sodium methylate
244. Sodium nitrite
245. Sodum phosphate (dibasic)
246. Sodium phosphate (tribasic)
247. Sodium selenite
248. Strontium chromate
249. Strychnine
250. Styrene
251. Sulfuric acid
252. Sulfur monochloride
253. 2,4,5-T acid (2,4,5-Trichlorophenoxyacetic acid)
254. 2,4,5-T amines (2,4,5-Trichlorophenoxy acetic acid amines)
255. 2,4,5-T esters (2,4,5-Trichlorophenoxy acetic acid esters)
256. 2,4,5-T salts (2,4,5-Trichlorophenoxy acetic acid salts)
257. 2,4,5-TP acid (2,4,5-Trichlorophenoxy propanoic acid)
258. 2,4,5-TP acid esters (2,4,5-Trichlorophenoxy propanoic acid esters)
259. TDE (Tetrachlorodiphenyl ethane)
260. Tetraethyl lead
261. Tetraethyl pyrophosphate
262. Thallium sulfate
263. Toluene
264. Toxaphene
265. Trichlorofon
266. Trichloroethylene
267. Trichlorophenol
268. Triethanolamine dodecylbenzenesulfonate
269. Triethylamine
270. Trimethylamine
271. Uranyl acetate
272. Uranyl nitrate
273. Vanadium pentoxide
274. Vanadyl sulfate
275. Vinyl acetate
276. Vinylidene chloride
277. Xylene
278. Xylenol
279. Zinc acetate
280. Zinc ammonium chloride
281. Zinc borate
282. Zinc bromide
283. Zinc carbonate
284. Zinc chloride
285. Zinc cyanide
286. Zinc fluoride
287. Zinc formate
288. Zinc hydrosulfonate
289. Zinc nitrate
290. Zinc phenolsulfonate
291. Zinc phosphide
292. Zinc silicofluoride
293. Zinc sulfate
294. Zirconium nitrate
295. Zirconium potassium flouride
296. Zirconium sulfate
297. Zirconium tetrachloride

TABLE 2C-4 *(continued)*

LINE DRAWING

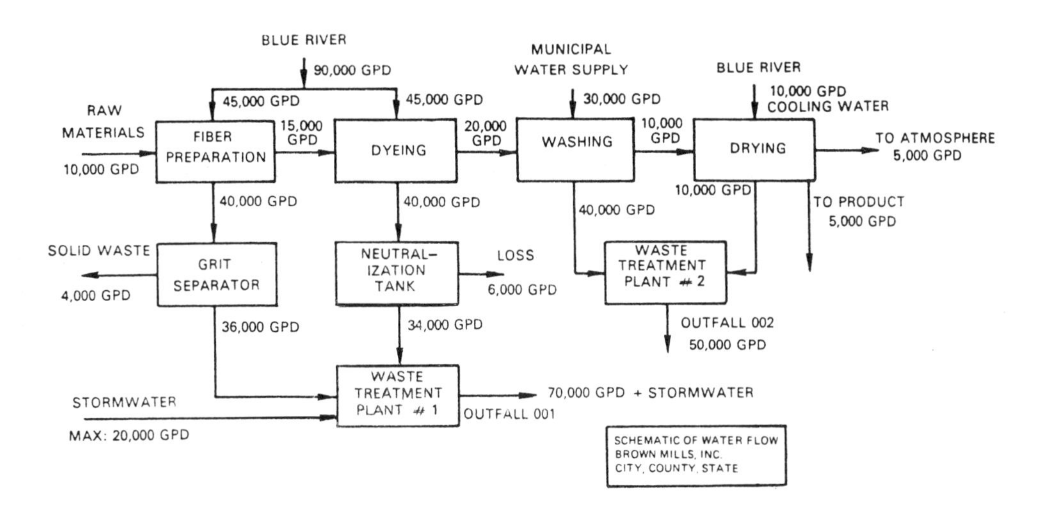
BLUE RIVER
90,000 GPD
MUNICIPAL
WATER SUPPLY
BLUE RIVER
45,000 GPD
45,000 GPD
30,000 GPD
10,000 GPD
COOLING WATER
RAW
MATERIALS
10,000 GPD
FIBER
PREPARATION
15,000
GPD
DYEING
20,000
GPD
WASHING
10,000
GPD
DRYING
TO ATMOSPHERE
5,000 GPD
40,000 GPD
40,000 GPD
40,000 GPD
10,000 GPD
TO PRODUCT
5,000 GPD
SOLID WASTE
4,000 GPD
GRIT
SEPARATOR
NEUTRAL-
IZATION
TANK
LOSS
6,000 GPD
WASTE
TREATMENT
PLANT # 2
36,000 GPD
34,000 GPD
OUTFALL 002
50,000 GPD
STORMWATER
MAX: 20,000 GPD
WASTE
TREATMENT
PLANT # 1
70,000 GPD + STORMWATER
OUTFALL 001
SCHEMATIC OF WATER FLOW
BROWN MILLS, INC.
CITY, COUNTY, STATE

FIGURE 2C-1

EPA I.D. NUMBER *(copy from Item 1 of Form 1)*

Please print or type in the unshaded areas only.

Form Approved OMB No. 158-R0173

FORM 2C NPDES

EPA

U.S. ENVIRONMENTAL PROTECTION AGENCY
APPLICATION FOR PERMIT TO DISCHARGE WASTEWATER
EXISTING MANUFACTURING, COMMERCIAL, MINING AND SILVICULTURAL OPERATIONS
Consolidated Permits Program

I. OUTFALL LOCATION

For each outfall, list the latitude and longitude of its location to the nearest 15 seconds and the name of the receiving water.

A. OUTFALL NUMBER *(list)*	B. LATITUDE			C. LONGITUDE			D. RECEIVING WATER *(name)*
	1. DEG.	2. MIN.	3. SEC.	1. DEG.	2. MIN.	3. SEC.	

II. FLOWS, SOURCES OF POLLUTION, AND TREATMENT TECHNOLOGIES

A. Attach a line drawing showing the water flow through the facility. Indicate sources of intake water, operations contributing wastewater to the effluent, and treatment units labeled to correspond to the more detailed descriptions in Item B. Construct a water balance on the line drawing by showing average flows between intakes, operations, treatment units, and outfalls. If a water balance cannot be determined *(e.g., for certain mining activities)*, provide a pictorial description of the nature and amount of any sources of water and any collection or treatment measures.

B. For each outfall, provide a description of: (1) All operations contributing wastewater to the effluent, including process wastewater, sanitary wastewater, cooling water, and storm water runoff; (2) The average flow contributed by each operation; and (3) The treatment received by the wastewater. Continue on additional sheets if necessary.

1. OUT-FALL NO *(list)*	2. OPERATION(S) CONTRIBUTING FLOW		3. TREATMENT		
	a. OPERATION *(list)*	b. AVERAGE FLOW *(include units)*	a. DESCRIPTION	b. LIST CODES FROM TABLE 2C-1	

OFFICIAL USE ONLY *(effluent guidelines sub-categories)*

CONTINUED FROM THE FRONT

C. Except for storm runoff, leaks, or spills, are any of the discharges described in Items II-A or B intermittent or seasonal?

☐ YES *(complete the following table)* ☐ NO *(go to Section III)*

1. OUTFALL NUMBER *(list)*	2. OPERATION*(s)* CONTRIBUTING FLOW *(list)*	3. FREQUENCY		4. FLOW				
		a. DAYS PER WEEK *(specify average)*	b. MONTHS PER YEAR *(specify average)*	a. FLOW RATE *(in mgd)*		b. TOTAL VOLUME *(specify with units)*		c. DURATION *(in days)*
				1. LONG TERM AVERAGE	2. MAXIMUM DAILY	1. LONG TERM AVERAGE	2. MAXIMUM DAILY	

III. MAXIMUM PRODUCTION

A. Does an effluent guideline limitation promulgated by EPA under Section 304 of the Clean Water Act apply to your facility?

☐ YES *(complete Item III-B)* ☐ NO *(to to Section IV)*

B. Are the limitations in the applicable effluent guideline expressed in terms of production *(or other measure of operation)*?

☐ YES *(complete Item III-C)* ☐ NO *(go to Section IV)*

C. If you answered "Yes" to Item III-B, list the quantity which represents an actual measurement of your maximum level of production, expressed in the terms and units used in the applicable effluent guideline, and indicate the affected outfalls.

1. MAXIMUM QUANTITY			2. AFFECTED OUTFALLS *(list outfall numbers)*
a. QUANTITY PER DAY	b. UNITS OF MEASURE	c. OPERATION, PRODUCT, MATERIAL, ETC. *(specify)*	

IV. IMPROVEMENTS

A. Are you now required by any Federal, State or local authority to meet any implementation schedule for the construction, upgrading or operation of wastewater treatment equipment or practices or any other environmental programs which may affect the discharges described in this application? This includes, but is not limited to, permit conditions, administrative or enforcement orders, enforcement compliance schedule letters, stipulations, court orders, and grant or loan conditions.

☐ YES *(complete the following table)* ☐ NO *(go to Item IV-B)*

1. IDENTIFICATION OF CONDITION, AGREEMENT, ETC.	2. AFFECTED OUTFALLS		3. BRIEF DESCRIPTION OF PROJECT	4. FINAL COMPLIANCE DATE	
	a. NO.	b. SOURCE OF DISCHARGE		a. REQUIRED	b. PROJECTED

B. OPTIONAL: You may attach additional sheets describing any additional water pollution control programs *(or other environmental projects which may affect your discharges)* you now have underway or which you plan. Indicate whether each program is now underway or planned, and indicate your actual or planned schedules for construction. ☐ MARK "X" IF DESCRIPTION OF ADDITIONAL CONTROL PROGRAMS IS ATTACHED

 CONTINUE ON PAGE 3

CONTINUED FROM PAGE 2

EPA I.D. NUMBER *(copy from Item 1 of Form 1)*

Form Approved OMB No. 158-R0173

V. INTAKE AND EFFLUENT CHARACTERISTICS

A, B, & C: See instructions before proceeding — Complete one set of tables for each outfall — Annotate the outfall number in the space provided.
NOTE: Tables V-A, V-B, and V-C are included on separate sheets numbered V-1 through V-9.

D. Use the space below to list any of the pollutants listed in Table 2c-3 of the instructions, which you know or have reason to believe is discharged or may be discharged from any outfall. For every pollutant you list, briefly describe the reasons you believe it to be present and report any analytical data in your possession.

1. POLLUTANT	2. SOURCE	1. POLLUTANT	2. SOURCE

VI. POTENTIAL DISCHARGES NOT COVERED BY ANALYSIS

A. Is any pollutant listed in Item V-C a substance or a component of a substance which you do or expect that you will over the next 5 years use or manufacture as an intermediate or final product or byproduct?

☐ YES *(list all such pollutants below)* ☐ NO *(go to Item VI-B)*

B. Are your operations such that your raw materials, processes, or products can reasonably be expected to vary so that your discharges of pollutants may during the next 5 years exceed two times the maximum values reported in Item V?

☐ YES *(complete Item VI-C below)* ☐ NO *(go to Section VII)*

C. If you answered "Yes" to Item VI-B, explain below and describe in detail the sources and expected levels of such pollutants which you anticipate will be discharged from each outfall over the next 5 years, to the best of your ability at this time. Continue on additional sheets if you need more space.

CONTINUED FROM THE FRONT

VII. BIOLOGICAL TOXICITY TESTING DATA

Do you have any knowledge or reason to believe that any biological test for acute or chronic toxicity has been made on any of your discharges or on a receiving water in relation to your discharge within the last 3 years?

☐ YES *(identify the test(s) and describe their purposes below)* ☐ NO *(go to Section VIII)*

VIII. CONTRACT ANALYSIS INFORMATION

Were any of the analyses reported in Item V performed by a contract laboratory or consulting firm?

☐ YES *(list the name, address, and telephone number of, and pollutants analyzed by, each such laboratory or firm below)* ☐ NO *(go to Section IX)*

A. NAME	B. ADDRESS	C. TELEPHONE *(area code & no.)*	D. POLLUTANTS ANALYZED *(list)*

IX. CERTIFICATION

I certify under penalty of law that I have personally examined and am familiar with the information submitted in this application and all attachments and that, based on my inquiry of those individuals immediately responsible for obtaining the information, I believe that the information is true, accurate and complete. I am aware that there are significant penalties for submitting false information, including the possibility of fine and imprisonment.

A. NAME & OFFICIAL TITLE *(type or print)*	B. PHONE NO. *(area code & no.)*
C. SIGNATURE	D. DATE SIGNED

PLEASE PRINT OR TYPE IN THE UNSHADED AREAS ONLY. You may report some or all of this information on separate sheets *(use the same format)* instead of completing these pages.
SEE INSTRUCTIONS.

EPA I.D. NUMBER *(copy from Item 1 of Form 1)*

Form Approved OMB No. 158-R0173

OUTFALL NO.

V. INTAKE AND EFFLUENT CHARACTERISTICS *(continued from page 3 of Form 2-C)*

PART A - You must provide the results of at least one analysis for every pollutant in this table. Complete one table for each outfall. See instructions for additional details.

1. POLLUTANT	2. EFFLUENT							3. UNITS *(specify if blank)*		4. INTAKE *(optional)*		
	a. MAXIMUM DAILY VALUE		b. MAXIMUM 30 DAY VALUE *(if available)*		c. LONG TERM AVRG. VALUE *(if available)*		d. NO. OF ANALYSES	a. CONCENTRATION	b. MASS	a. LONG TERM AVERAGE VALUE		b. NO. OF ANALYSES
	(1) CONCENTRATION	(2) MASS	(1) CONCENTRATION	(2) MASS	(1) CONCENTRATION	(2) MASS				(1) CONCENTRATION	(2) MASS	
a. Biochemical Oxygen Demand *(BOD)*												
b. Chemical Oxygen Demand *(COD)*												
c. Total Organic Carbon *(TOC)*												
d. Total Suspended Solids *(TSS)*												
e. Ammonia *(as N)*												
f. Flow	VALUE		VALUE		VALUE					VALUE		
g. Temperature *(winter)*	VALUE		VALUE		VALUE			°C		VALUE		
h. Temperature *(summer)*	VALUE		VALUE		VALUE			°C		VALUE		
i. pH	MINIMUM	MAXIMUM	MINIMUM	MAXIMUM				STANDARD UNITS				

PART B - Mark "X" in column 2-a for each pollutant you know or have reason to believe is present. Mark "X" in column 2-b for each pollutant you believe to be absent. If you mark column 2-a for any pollutant, you must provide the results of at least one analysis for that pollutant. Complete one table for each outfall. See the instructions for additional details and requirements.

1. POLLUTANT AND CAS NO. *(if available)*	2. MARK 'X'		3. EFFLUENT							4. UNITS		5. INTAKE *(optional)*		
	a. BELIEVED PRESENT	b. BELIEVED ABSENT	a. MAXIMUM DAILY VALUE		b. MAXIMUM 30 DAY VALUE *(if available)*		c. LONG TERM AVRG. VALUE *(if available)*		d. NO. OF ANALYSES	a. CONCENTRATION	b. MASS	a. LONG TERM AVERAGE VALUE		b. NO. OF ANALYSES
			(1) CONCENTRATION	(2) MASS	(1) CONCENTRATION	(2) MASS	(1) CONCENTRATION	(2) MASS				(1) CONCENTRATION	(2) MASS	
a. Bromide (24959-67-9)														
b. Chlorine, Total Residual														
c. Color														
d. Fecal Coliform														
e. Fluoride (16984-48-8)														
f. Nitrate–Nitrite *(as N)*														

ITEM V-B CONTINUED FROM FRONT

1. POLLUTANT AND CAS NO. *(if available)*	2. MARK 'X'		3. EFFLUENT							4. UNITS		5. INTAKE *(optional)*		
	a. BELIEVED PRESENT	b. BELIEVED ABSENT	a. MAXIMUM DAILY VALUE		b. MAXIMUM 30 DAY VALUE *(if available)*		c. LONG TERM AVRG. VALUE *(if available)*		d. NO. OF ANALYSES	a. CONCENTRATION	b. MASS	a. LONG TERM AVERAGE VALUE		b. NO. OF ANALYSES
			(1) CONCENTRATION	(2) MASS	(1) CONCENTRATION	(2) MASS	(1) CONCENTRATION	(2) MASS				(1) CONCENTRATION	(2) MASS	
g. Nitrogen, Total Organic *(as N)*														
h. Oil and Grease														
i. Phosphorus *(as P)*, Total (7723-14-0)														
j. Radioactivity														
(1) Alpha, Total														
(2) Beta, Total														
(3) Radium, Total														
(4) Radium 226, Total														
k. Sulfate *(as SO_4)* (14808-79-8)														
l. Sulfide *(as S)*														
m. Sulfite *(as SO_3)* (14265-45-3)														
n. Surfactants														
o. Aluminum, Total (7429-90-5)														
p. Barium, Total (7440-39-3)														
q. Boron, Total (7440-42-8)														
r. Cobalt, Total (7440-48-4)														
s. Iron, Total (7439-89-6)														
t. Magnesium, Total (7439-95-4)														
u. Molybdenum, Total (7439-98-7)														
v. Manganese, Total (7439-96-5)														
w. Tin, Total (7440-31-5)														
x. Titanium, Total (7440-32-6)														

EPA Form 3510-2C (6-80) PAGE V-2 CONTINUE ON PAGE V-3

EPA I.D. NUMBER *(copy from Item 1 of Form 1)*	OUTFALL NUMBER

CONTINUED FROM PAGE 3 OF FORM 2-C

Form Approved OMB No. 158-R0173

PART C - If you are a primary industry and this outfall contains process wastewater, refer to Table 2c-2 in the instructions to determine which of the GC/MS fractions you must test for. Mark "X" in column 2-a for all such GC/MS fractions that apply to your industry and for ALL toxic metals, cyanides, and total phenols. If you are not required to mark column 2-a ***(secondary industries, non–process wastewater outfalls, and non–required GC/MS fractions)***, mark "X" in column 2-b for each pollutant you know or have reason to believe is present. Mark "X" in column 2-c for each pollutant you believe to be absent. If you mark either columns 2-a or 2-b for any pollutant, you must provide the results of at least one analysis for that pollutant. Note that there are seven pages to this part; please review each carefully. Complete one table ***(all seven pages)*** for each outfall. See instructions for additional details and requirements.

1. POLLUTANT AND CAS NUMBER *(if available)*	2. MARK 'X'			3. EFFLUENT							4. UNITS		5. INTAKE *(optional)*		
	a. TESTING REQUIRED	b. BELIEVED PRESENT	c. BELIEVED ABSENT	a. MAXIMUM DAILY VALUE		b. MAXIMUM 30 DAY VALUE *(if available)*		c. LONG TERM AVRG. VALUE *(if available)*		d. NO. OF ANALYSES	a. CONCENTRATION	b. MASS	a. LONG TERM AVERAGE VALUE		b. NO. OF ANALYSES
				(1) CONCENTRATION	(2) MASS	(1) CONCENTRATION	(2) MASS	(1) CONCENTRATION	(2) MASS				(1) CONCENTRATION	(2) MASS	
METALS, CYANIDE, AND TOTAL PHENOLS															
1M. Antimony, Total (7440-36-0)															
2M. Arsenic, Total (7440-38-2)															
3M. Beryllium, Total, 7440-41-7)															
4M. Cadmium, Total (7440-43-9)															
5M. Chromium, Total (7440-47-3)															
6M. Copper, Total (7550-50-8)															
7M. Lead, Total (7439-97-6)															
8M. Mercury, Total (7439-97-6)															
9M. Nickel, Total (7440-02-0)															
10M. Selenium, Total (7782-49-2)															
11M. Silver, Total (7440-22-4)															
12M. Thallium, Total (7440-28-0)															
13M. Zinc, Total (7440-66-6)															
14M. Cyanide, Total (57-12-5)															
15M. Phenols, Total															
DIOXIN															
2,3,7,8-Tetrachlorodibenzo-P-Dioxin (1764-01-6)				DESCRIBE RESULTS											

CONTINUE ON REVERSE

CONTINUED FROM THE FRONT

1. POLLUTANT AND CAS NUMBER *(if available)*	2. MARK 'X'			3. EFFLUENT							4. UNITS		5. INTAKE *(optional)*		
	a. TESTING REQUIRED	b. BELIEVED PRESENT	c. BELIEVED ABSENT	a. MAXIMUM DAILY VALUE		b. MAXIMUM 30 DAY VALUE *(if available)*		c. LONG TERM AVRG. VALUE *(if available)*		d. NO. OF ANALYSES	a. CONCENTRATION	b. MASS	a. LONG TERM AVERAGE VALUE		b. NO. OF ANALYSES
				(1) CONCENTRATION	(2) MASS	(1) CONCENTRATION	(2) MASS	(1) CONCENTRATION	(2) MASS				(1) CONCENTRATION	(2) MASS	
GC/MS FRACTION – VOLATILE COMPOUNDS															
1V. Acrolein (107-02-8)															
2V. Acrylonitrile (107-13-1)															
3V. Benzene (71-43-2)															
4V. Bis *(Chloromethyl)* Ether (542-88-1)															
5V. Bromoform (75-25-2)															
6V. Carbon Tetrachloride (56-23-5)															
7V. Chlorobenzene (108-90-7)															
8V. Chlorodibromomethane (124-48-1)															
9V. Chloroethane (75-00-3)															
10V. 2-Chloroethylvinyl Ether (110-75-8)															
11V. Chloroform (67-66-3)															
12V. Dichlorobromomethane (75-27-4)															
13V. Dichlorodifluoromethane (75-71-8)															
14V. 1,1-Dichloroethane (75-34-3)															
15V. 1,2-Dichloroethane (107-06-2)															
16V. 1,1-Dichloroethylene (75-35-4)															
17V. 1,2-Dichloropropane (78-87-5)															
18V. 1,2-Dichloropropylene (542-75-6)															
19V. Ethylbenzene (100-41-4)															
20V. Methyl Bromide (74-83-9)															
21V. Methyl Chloride (74-87-3)															

EPA Form 3510-2C (6-80) PAGE V-4 CONTINUE ON PAGE V-5

CONTINUED FROM PAGE V-4

EPA I.D. NUMBER *(copy from Item 1 of Form 1)*	OUTFALL NUMBER

Form Approved OMB No. 158-R0173

1. POLLUTANT AND CAS NUMBER *(if available)*	2. MARK 'X'			3. EFFLUENT							4. UNITS		5. INTAKE *(optional)*		
	a. TESTING REQUIRED	b. BELIEVED PRESENT	c. BELIEVED ABSENT	a. MAXIMUM DAILY VALUE		b. MAXIMUM 30 DAY VALUE *(if available)*		c. LONG TERM AVRG. VALUE *(if available)*		d. NO. OF ANALYSES	a. CONCENTRATION	b. MASS	a. LONG TERM AVERAGE VALUE		b. NO. OF ANALYSES
				(1) CONCENTRATION	(2) MASS	(1) CONCENTRATION	(2) MASS	(1) CONCENTRATION	(2) MASS				(1) CONCENTRATION	(2) MASS	
GC/MS FRACTION – VOLATILE COMPOUNDS *(continued)*															
22V. Methylene Chloride (75-09-2)															
23V. 1,1,2,2-Tetrachloroethane (79-34-5)															
24V. Tetrachloroethylene (127-18-4)															
25V. Toluene (108-88-3)															
26V. 1,2-Trans-Dichloroethylene (156-60-5)															
27V. 1,1,1-Trichloroethane (71-55-6)															
28V. 1,1,2-Trichloroethane (79-00-5)															
29V. Trichloroethylene (79-01-6)															
30V. Trichlorofluoromethane (75-69-4)															
31V. Vinyl Chloride (75-01-4)															
GC/MS FRACTION – ACID COMPOUNDS															
1A. 2-Chlorophenol (95-57-8)															
2A. 2,4-Dichlorophenol (120-83-2)															
3A. 2,4-Dimethylphenol (105-67-9)															
4A. 4,6-Dinitro-O-Cresol (534-52-1)															
5A. 2,4-Dinitrophenol (51-28-5)															
6A. 2-Nitrophenol (88-75-5)															
7A. 4-Nitrophenol (100-02-7)															
8A. P-Chloro-M-Cresol (59-50-7)															
9A. Pentachlorophenol (87-86-5)															
10A. Phenol (108-95-2)															
11A. 2,4,6-Trichlorophenol															

CONTINUED FROM THE FRONT

1. POLLUTANT AND CAS NUMBER *(if available)*	2. MARK 'X'			3. EFFLUENT							4. UNITS		5. INTAKE *(optional)*		
	a. TESTING REQUIRED	b. BELIEVED PRESENT	c. BELIEVED ABSENT	a. MAXIMUM DAILY VALUE		b. MAXIMUM 30 DAY VALUE *(if available)*		c. LONG TERM AVRG. VALUE *(if available)*		d. NO. OF ANALYSES	a. CONCENTRATION	b. MASS	a. LONG TERM AVERAGE VALUE		b. NO. OF ANALYSES
				(1) CONCENTRATION	(2) MASS	(1) CONCENTRATION	(2) MASS	(1) CONCENTRATION	(2) MASS				(1) CONCENTRATION	(2) MASS	
GC/MS FRACTION — BASE/NEUTRAL COMPOUNDS															
1B. Acenaphthene (83-32-9)															
2B. Acenaphtylene (208-96-8)															
3B. Anthracene (120-12-7)															
4B. Benzidine (92-87-5)															
5B. Benzo *(a)* Anthracene (56-55-3)															
6B. Benzo *(a)* Pyrene (50-32-8)															
7B. 3,4-Benzo-fluoranthene (205-99-2)															
8B. Benzo *(ghi)* Perylene (191-24-2)															
9B. Benzo *(k)* Fluoranthene (207-08-9)															
10B. Bis *(2-Chloro-ethoxy)* Methane (111-91-1)															
11B. Bis *(2-Chloro-ethyl)* Ether (111-44-4)															
12B. Bis *(2-Chloro-isopropyl)* Ether (39638-32-9)															
13B. Bis *(2-Ethyl-hexyl)* Phthalate (117-81-7)															
14B. 4-Bromo-phenyl Phenyl Ether (101-55-3)															
15B. Butyl Benzyl Phthalate (85-68-7)															
16B. 2-Chloro-naphthalene (91-58-7)															
17B. 4-Chloro-phenyl Phenyl Ether (7005-72-3)															
18B. Chrysene (218-01-9)															
19B. Dibenzo *(a,h)* Anthracene (53-70-3)															
20B. 1,2-Dichloro-benzene (95-50-1)															
21B. 1,3-Dichloro-benzene (541-73-1)															

CONTINUE ON PAGE V-7

EPA I.D. NUMBER *(copy from Item 1 of Form 1)* | OUTFALL NUMBER

CONTINUED FROM PAGE V-6

Form Approved OMB No. 158-R0173

1. POLLUTANT AND CAS NUMBER *(if available)*	2. MARK 'X'			3. EFFLUENT							4. UNITS		5. INTAKE *(optional)*		
	a. TESTING REQUIRED	b. BELIEVED PRESENT	c. BELIEVED ABSENT	a. MAXIMUM DAILY VALUE		b. MAXIMUM 30 DAY VALUE *(if available)*		c. LONG TERM AVRG. VALUE *(if available)*		d. NO. OF ANALYSES	a. CONCENTRATION	b. MASS	a. LONG TERM AVERAGE VALUE		b. NO. OF ANALYSES
				(1) CONCENTRATION	(2) MASS	(1) CONCENTRATION	(2) MASS	(1) CONCENTRATION	(2) MASS				(1) CONCENTRATION	(2) MASS	
GC/MS FRACTION – BASE/NEUTRAL COMPOUNDS *(continued)*															
22B. 1,4-Dichlorobenzene (106-46-7)															
23B. 3,3'-Dichlorobenzidine (91-94-1)															
24B. Diethyl Phthalate (84-66-2)															
25B. Dimethyl Phthalate (131-11-3)															
26B. Di-N-Butyl Phthalate (84-74-2)															
27B. 2,4-Dinitrotoluene (121-14-2)															
28B. 2,6-Dinitrotoluene (606-20-2)															
29B. Di-N-Octyl Phthalate (117-84-0)															
30B. 1,2-Diphenylhydrazine *(as Azobenzene)* (122-66-7)															
31B. Fluoranthene (206-44-0)															
32B. Fluorene (86-73-7)															
33B. Hexachlorobenzene (118-71-1)															
34B. Hexachlorobutadiene (87-68-3)															
35B. Hexachlorocyclopentadiene (77-47-4)															
36B. Hexachloroethane (67-72-1)															
37B. Indeno *(1,2,3-cd)* Pyrene (193-39-5)															
38B. Isophorone (78-59-1)															
39B. Naphthalene (91-20-3)															
40B. Nitrobenzene (98-95-3)															
41B. N-Nitrosodimethylamine (62-75-9)															
42B. N-Nitrosodi-N-Propylamine (621-64-7)															

CONTINUE ON REVERSE

CONTINUED FROM THE FRONT

1. POLLUTANT AND CAS NUMBER (*if available*)	2. MARK 'X'			3. EFFLUENT							4. UNITS		5. INTAKE (*optional*)		
	a. TESTING REQUIRED	b. BELIEVED PRESENT	c. BELIEVED ABSENT	a. MAXIMUM DAILY VALUE		b. MAXIMUM 30 DAY VALUE (*if available*)		c. LONG TERM AVRG. VALUE (*if available*)		d. NO. OF ANALYSES	a. CONCENTRATION	b. MASS	a. LONG TERM AVERAGE VALUE		b. NO. OF ANALYSES
				(1) CONCENTRATION	(2) MASS	(1) CONCENTRATION	(2) MASS	(1) CONCENTRATION	(2) MASS				(1) CONCENTRATION	(2) MASS	
GC/MS FRACTION — BASE/NEUTRAL COMPOUNDS (*continued*)															
43B. N-Nitrosodiphenylamine (86-30-6)															
44B. Phenanthrene (85-01-8)															
45B. Pyrene (129-00-0)															
46B. 1,2,4 - Trichlorobenzene (120-82-1)															
GC/MS FRACTION — PESTICIDES															
1P. Aldrin (309-00-2)															
2P. α-BHC (319-84-6)															
3P. β-BHC (319-85-7)															
4P. γ-BHC (58-89-9)															
5P. δ-BHC (319-86-8)															
6P. Chlordane (57-74-9)															
7P. 4,4'-DDT (50-29-3)															
8P. 4,4'-DDE (72-55-9)															
9P. 4,4'-DDD (72-54-8)															
10P. Dieldrin (60-57-1)															
11P. α-Endosulfan (115-29-7)															
12P. β-Endosulfan (115-29-7)															
13P. Endosulfan Sulfate (1031-07-8)															
14P. Endrin (72-20-8)															
15P. Endrin Aldehyde (7421-93-4)															
16P. Heptachlor (76-44-8)															

 CONTINUE ON PAGE V-9

EPA I.D. NUMBER *(copy from Item 1 of Form 1)* | OUTFALL NUMBER

CONTINUED FROM PAGE V-8

Form Approved OMB No. 158-R0173

1. POLLUTANT AND CAS NUMBER *(if available)*	2. MARK 'X'			3. EFFLUENT							4. UNITS		5. INTAKE *(optional)*		
	a. TESTING REQUIRED	b. BELIEVED PRESENT	c. BELIEVED ABSENT	a. MAXIMUM DAILY VALUE		b. MAXIMUM 30 DAY VALUE *(if available)*		c. LONG TERM AVRG. VALUE *(if available)*		d. NO. OF ANALYSES	a. CONCENTRATION	b. MASS	a. LONG TERM AVERAGE VALUE		b. NO. OF ANALYSES
				(1) CONCENTRATION	(2) MASS	(1) CONCENTRATION	(2) MASS	(1) CONCENTRATION	(2) MASS				(1) CONCENTRATION	(2) MASS	
GC/MS FRACTION — PESTICIDES *(continued)*															
17P. Heptachlor Epoxide (1024-57-3)															
18P. PCB-1242 (53469-21-9)															
19P. PCB-1254 (11097-69-1)															
20P. PCB-1221 (11104-28-2)															
21P. PCB-1232 (11141-16-5)															
22P. PCB-1248 (12672-29-6)															
23P. PCB-1260 (11096-82-5)															
24P. PCB-1016 (12674-11-2)															
25P. Toxaphene (8001-35-2)															

EPA Form 3510-2C (6-80)

PAGE V-9

lease print or type in the unshaded areas only.

EPA I.D. NUMBER *(copy from Item 1 of Form 1)*

Form Approved OMB No. 158-R0173

FORM 2C NPDES EPA

U.S. ENVIRONMENTAL PROTECTION AGENCY
APPLICATION FOR PERMIT TO DISCHARGE WASTEWATER
EXISTING MANUFACTURING, COMMERCIAL, MINING AND SILVICULTURAL OPERATIONS
Consolidated Permits Program

I. OUTFALL LOCATION

For each outfall, list the latitude and longitude of its location to the nearest 15 seconds and the name of the receiving water.

A. OUTFALL NUMBER *(list)*	B. LATITUDE			C. LONGITUDE			D. RECEIVING WATER *(name)*
	1. DEG.	2. MIN.	3. SEC.	1. DEG.	2. MIN.	3. SEC.	

II. FLOWS, SOURCES OF POLLUTION, AND TREATMENT TECHNOLOGIES

A. Attach a line drawing showing the water flow through the facility. Indicate sources of intake water, operations contributing wastewater to the effluent, and treatment units labeled to correspond to the more detailed descriptions in Item B. Construct a water balance on the line drawing by showing average flows between intakes, operations, treatment units, and outfalls. If a water balance cannot be determined *(e.g., for certain mining activities)*, provide a pictorial description of the nature and amount of any sources of water and any collection or treatment measures.

B. For each outfall, provide a description of: (1) All operations contributing wastewater to the effluent, including process wastewater, sanitary wastewater, cooling water, and storm water runoff; (2) The average flow contributed by each operation; and (3) The treatment received by the wastewater. Continue on additional sheets if necessary.

1. OUT-FALL NO *(list)*	2. OPERATION(S) CONTRIBUTING FLOW		3. TREATMENT		
	a. OPERATION *(list)*	b. AVERAGE FLOW *(include units)*	a. DESCRIPTION	b. LIST CODES FROM TABLE 2C-1	

FFICIAL USE ONLY *(effluent guidelines sub-categories)*

CONTINUED FROM THE FRONT

C. Except for storm runoff, leaks, or spills, are any of the discharges described in Items II-A or B intermittent or seasonal?

☐ YES *(complete the following table)* ☐ NO *(go to Section III)*

1. OUTFALL NUMBER *(list)*	2. OPERATION*(s)* CONTRIBUTING FLOW *(list)*	3. FREQUENCY		4. FLOW				
		a. DAYS PER WEEK *(specify average)*	b. MONTHS PER YEAR *(specify average)*	a. FLOW RATE *(in mgd)*		b. TOTAL VOLUME *(specify with units)*		c. DURATION *(in days)*
				1. LONG TERM AVERAGE	2. MAXIMUM DAILY	1. LONG TERM AVERAGE	2. MAXIMUM DAILY	

III. MAXIMUM PRODUCTION

A. Does an effluent guideline limitation promulgated by EPA under Section 304 of the Clean Water Act apply to your facility?

☐ YES *(complete Item III-B)* ☐ NO *(to to Section IV)*

B. Are the limitations in the applicable effluent guideline expressed in terms of production *(or other measure of operation)*?

☐ YES *(complete Item III-C)* ☐ NO *(go to Section IV)*

C. If you answered "Yes" to Item III-B, list the quantity which represents an actual measurement of your maximum level of production, expressed in the terms and units used in the applicable effluent guideline, and indicate the affected outfalls.

1. MAXIMUM QUANTITY			2. AFFECTED OUTFALLS *(list outfall numbers)*
a. QUANTITY PER DAY	b. UNITS OF MEASURE	c. OPERATION, PRODUCT, MATERIAL, ETC. *(specify)*	

IV. IMPROVEMENTS

A. Are you now required by any Federal, State or local authority to meet any implementation schedule for the construction, upgrading or operation of wastewater treatment equipment or practices or any other environmental programs which may affect the discharges described in this application? This includes, but is not limited to, permit conditions, administrative or enforcement orders, enforcement compliance schedule letters; stipulations, court orders, and grant or loan conditions.

☐ YES *(complete the following table)* ☐ NO *(go to Item IV-B)*

1. IDENTIFICATION OF CONDITION, AGREEMENT, ETC.	2. AFFECTED OUTFALLS		3. BRIEF DESCRIPTION OF PROJECT	4. FINAL COMPLIANCE DATE	
	a. NO.	b. SOURCE OF DISCHARGE		a. REQUIRED	b. PROJECTED

B. OPTIONAL: You may attach additional sheets describing any additional water pollution control programs *(or other environmental projects which may affect your discharges)* you now have underway or which you plan. Indicate whether each program is now underway or planned, and indicate your actual or planned schedules for construction. ☐ MARK "X" IF DESCRIPTION OF ADDITIONAL CONTROL PROGRAMS IS ATTACHED

 CONTINUE ON PAGE

CONTINUED FROM PAGE 2

EPA I.D. NUMBER *(copy from Item 1 of Form 1)*

Form Approved OMB No. 158-R0173

V. INTAKE AND EFFLUENT CHARACTERISTICS

A, B, & C: See instructions before proceeding — Complete one set of tables for each outfall — Annotate the outfall number in the space provided.
NOTE: Tables V-A, V-B, and V-C are included on separate sheets numbered V-1 through V-9.

D. Use the space below to list any of the pollutants listed in Table 2c-3 of the instructions, which you know or have reason to believe is discharged or may be discharged from any outfall. For every pollutant you list, briefly describe the reasons you believe it to be present and report any analytical data in your possession.

1. POLLUTANT	2. SOURCE	1. POLLUTANT	2. SOURCE

VI. POTENTIAL DISCHARGES NOT COVERED BY ANALYSIS

A. Is any pollutant listed in Item V-C a substance or a component of a substance which you do or expect that you will over the next 5 years use or manufacture as an intermediate or final product or byproduct?

☐ YES *(list all such pollutants below)* ☐ NO *(go to Item VI-B)*

B. Are your operations such that your raw materials, processes, or products can reasonably be expected to vary so that your discharges of pollutants may during the next 5 years exceed two times the maximum values reported in Item V?

☐ YES *(complete Item VI-C below)* ☐ NO *(go to Section VII)*

C. If you answered "Yes" to Item VI-B, explain below and describe in detail the sources and expected levels of such pollutants which you anticipate will be discharged from each outfall over the next 5 years, to the best of your ability at this time. Continue on additional sheets if you need more space.

CONTINUED FROM THE FRONT

VII. BIOLOGICAL TOXICITY TESTING DATA

Do you have any knowledge or reason to believe that any biological test for acute or chronic toxicity has been made on any of your discharges or on a receiving water in relation to your discharge within the last 3 years?

☐ YES *(identify the test(s) and describe their purposes below)* ☐ NO *(go to Section VIII)*

VIII. CONTRACT ANALYSIS INFORMATION

Were any of the analyses reported in Item V performed by a contract laboratory or consulting firm?

☐ YES *(list the name, address, and telephone number of, and pollutants analyzed by, each such laboratory or firm below)* ☐ NO *(go to Section IX)*

A. NAME	B. ADDRESS	C. TELEPHONE *(area code & no.)*	D. POLLUTANTS ANALYZE[D] *(list)*

IX. CERTIFICATION

I certify under penalty of law that I have personally examined and am familiar with the information submitted in this application and a[ll] attachments and that, based on my inquiry of those individuals immediately responsible for obtaining the information, I believe that the in[-] formation is true, accurate and complete. I am aware that there are significant penalties for submitting false information, including th[e] possibility of fine and imprisonment.

A. NAME & OFFICIAL TITLE *(type or print)*	B. PHONE NO. *(area code & no.)*
C. SIGNATURE	D. DATE SIGNED

PLEASE PRINT OR TYPE IN THE UNSHADED AREAS ONLY. You may report some or all of this information on separate sheets *(use the same format)* instead of completing these pages.
SEE INSTRUCTIONS.

EPA I.D. NUMBER *(copy from Item 1 of Form 1)*

Form Approved OMB No. 158-R0173

OUTFALL NO.

V. INTAKE AND EFFLUENT CHARACTERISTICS *(continued from page 3 of Form 2-C)*

PART A - You must provide the results of at least one analysis for every pollutant in this table. Complete one table for each outfall. **See instructions for additional details.**

1. POLLUTANT	2. EFFLUENT							3. UNITS *(specify if blank)*		4. INTAKE *(optional)*		
	a. MAXIMUM DAILY VALUE		b. MAXIMUM 30 DAY VALUE *(if available)*		c. LONG TERM AVRG. VALUE *(if available)*		d. NO. OF ANALYSES	a. CONCENTRATION	b. MASS	a. LONG TERM AVERAGE VALUE		b. NO. OF ANALYSES
	(1) CONCENTRATION	(2) MASS	(1) CONCENTRATION	(2) MASS	(1) CONCENTRATION	(2) MASS				(1) CONCENTRATION	(2) MASS	
a. Biochemical Oxygen Demand *(BOD)*												
b. Chemical Oxygen Demand *(COD)*												
c. Total Organic Carbon *(TOC)*												
d. Total Suspended Solids *(TSS)*												
e. Ammonia *(as N)*												
f. Flow	VALUE		VALUE		VALUE					VALUE		
g. Temperature *(winter)*	VALUE		VALUE		VALUE			°C		VALUE		
h. Temperature *(summer)*	VALUE		VALUE		VALUE			°C		VALUE		
i. pH	MINIMUM	MAXIMUM	MINIMUM	MAXIMUM				STANDARD UNITS				

PART B - Mark "X" in column 2-a for each pollutant you know or have reason to believe is present. Mark "X" in column 2-b for each pollutant you believe to be absent. If you mark column 2-a for any pollutant, you must provide the results of at least one analysis for that pollutant. Complete one table for each outfall. See the instructions for additional details and requirements.

1. POLLUTANT AND CAS NO. *(if available)*	2. MARK 'X'		3. EFFLUENT							4. UNITS		5. INTAKE *(optional)*		
	a. BELIEVED PRESENT	b. BELIEVED ABSENT	a. MAXIMUM DAILY VALUE		b. MAXIMUM 30 DAY VALUE *(if available)*		c. LONG TERM AVRG. VALUE *(if available)*		d. NO. OF ANALYSES	a. CONCENTRATION	b. MASS	a. LONG TERM AVERAGE VALUE		b. NO. OF ANALYSES
			(1) CONCENTRATION	(2) MASS	(1) CONCENTRATION	(2) MASS	(1) CONCENTRATION	(2) MASS				(1) CONCENTRATION	(2) MASS	
a. Bromide (24959-67-9)														
b. Chlorine, Total Residual														
c. Color														
d. Fecal Coliform														
e. Fluoride (16984-48-8)														
f. Nitrate—Nitrite *(as N)*														

CONTINUE ON REVERSE

ITEM V-B CONTINUED FROM FRONT

1. POLLUTANT AND CAS NO. (if available)	2. MARK 'X'		3. EFFLUENT							4. UNITS		5. INTAKE (optional)		
	a. BELIEVED PRESENT	b. BELIEVED ABSENT	a. MAXIMUM DAILY VALUE		b. MAXIMUM 30 DAY VALUE (if available)		c. LONG TERM AVRG. VALUE (if available)		d. NO. OF ANALYSES	a. CONCENTRATION	b. MASS	a. LONG TERM AVERAGE VALUE		b. NO. OF ANALYSES
			(1) CONCENTRATION	(2) MASS	(1) CONCENTRATION	(2) MASS	(1) CONCENTRATION	(2) MASS				(1) CONCENTRATION	(2) MASS	
g. Nitrogen, Total Organic (as N)														
h. Oil and Grease														
i. Phosphorus (as P), Total (7723-14-0)														
j. Radioactivity														
(1) Alpha, Total														
(2) Beta, Total														
(3) Radium, Total														
(4) Radium 226, Total														
k. Sulfate (as SO_4) (14808-79-8)														
l. Sulfide (as S)														
m. Sulfite (as SO_3) (14265-45-3)														
n. Surfactants														
o. Aluminum, Total (7429-90-5)														
p. Barium, Total (7440-39-3)														
q. Boron, Total (7440-42-8)														
r. Cobalt, Total (7440-48-4)														
s. Iron, Total (7439-89-6)														
t. Magnesium, Total (7439-95-4)														
u. Molybdenum, Total (7439-98-7)														
v. Manganese, Total (7439-96-5)														
w. Tin, Total (7440-31-5)														
x. Titanium, Total (7440-32-6)														

 CONTINUE ON PAGE V-3

EPA I.D. NUMBER *(copy from Item 1 of Form 1)*	OUTFALL NUMBER

CONTINUED FROM PAGE 3 OF FORM 2-C

Form Approved OMB No. 158-R0173

PART C - If you are a primary industry and this outfall contains process wastewater, refer to Table 2c-2 in the instructions to determine which of the GC/MS fractions you must test for. Mark "X" in column 2-a for all such GC/MS fractions that apply to your industry and for ALL toxic metals, cyanides, and total phenols. If you are not required to mark column 2-a ***(secondary industries, non—process wastewater outfalls, and non—required GC/MS fractions)***, mark "X" in column 2-b for each pollutant you know or have reason to believe is present. Mark "X" in column 2-c for each pollutant you believe to be absent. If you mark either columns 2-a or 2-b for any pollutant, you must provide the results of at least one analysis for that pollutant. Note that there are seven pages to this part; please review each carefully. Complete one table ***(all seven pages)*** for each outfall. See instructions for additional details and requirements.

1. POLLUTANT AND CAS NUMBER *(if available)*	2. MARK 'X'			3. EFFLUENT							4. UNITS		5. INTAKE *(optional)*		
	a. TESTING REQUIRED	b. BELIEVED PRESENT	c. BELIEVED ABSENT	a. MAXIMUM DAILY VALUE		b. MAXIMUM 30 DAY VALUE *(if available)*		c. LONG TERM AVRG. VALUE *(if available)*		d. NO. OF ANALYSES	a. CONCENTRATION	b. MASS	a. LONG TERM AVERAGE VALUE		b. NO. OF ANALYSES
				(1) CONCENTRATION	(2) MASS	(1) CONCENTRATION	(2) MASS	(1) CONCENTRATION	(2) MASS				(1) CONCENTRATION	(2) MASS	
METALS, CYANIDE, AND TOTAL PHENOLS															
1M. Antimony, Total (7440-36-0)															
2M. Arsenic, Total (7440-38-2)															
3M. Beryllium, Total, 7440-41-7)															
4M. Cadmium, Total (7440-43-9)															
5M. Chromium, Total (7440-47-3)															
6M. Copper, Total (7550-50-8)															
7M. Lead, Total (7439-97-6)															
8M. Mercury, Total (7439-97-6)															
9M. Nickel, Total (7440-02-0)															
10M. Selenium, Total (7782-49-2)															
11M. Silver, Total (7440-22-4)															
12M. Thallium, Total (7440-28-0)															
13M. Zinc, Total (7440-66-6)															
14M. Cyanide, Total (57-12-5)															
15M. Phenols, Total															
DIOXIN															
2,3,7,8-Tetra-chlorodibenzo-P-Dioxin (1764-01-6)				DESCRIBE RESULTS											

EPA Form 3510-2C (6-80)

CONTINUE ON REVERSE

CONTINUED FROM THE FRONT

1. POLLUTANT AND CAS NUMBER (*if available*)	2. MARK 'X'			3. EFFLUENT							4. UNITS		5. INTAKE (*optional*)		
	a. TESTING REQUIRED	b. BELIEVED PRESENT	c. BELIEVED ABSENT	a. MAXIMUM DAILY VALUE		b. MAXIMUM 30 DAY VALUE (*if available*)		c. LONG TERM AVRG. VALUE (*if available*)		d. NO. OF ANALYSES	a. CONCENTRATION	b. MASS	a. LONG TERM AVERAGE VALUE		b. NO. OF ANALYSES
				(1) CONCENTRATION	(2) MASS	(1) CONCENTRATION	(2) MASS	(1) CONCENTRATION	(2) MASS				(1) CONCENTRATION	(2) MASS	
GC/MS FRACTION – VOLATILE COMPOUNDS															
1V. Acrolein (107-02-8)															
2V. Acrylonitrile (107-13-1)															
3V. Benzene (71-43-2)															
4V. Bis (*Chloromethyl*) Ether (542-88-1)															
5V. Bromoform (75-25-2)															
6V. Carbon Tetrachloride (56-23-5)															
7V. Chlorobenzene (108-90-7)															
8V. Chlorodibromomethane (124-48-1)															
9V. Chloroethane (75-00-3)															
10V. 2-Chloroethylvinyl Ether (110-75-8)															
11V. Chloroform (67-66-3)															
12V. Dichlorobromomethane (75-27-4)															
13V. Dichlorodifluoromethane (75-71-8)															
14V. 1,1-Dichloroethane (75-34-3)															
15V. 1,2-Dichloroethane (107-06-2)															
16V. 1,1-Dichloroethylene (75-35-4)															
17V. 1,2-Dichloropropane (78-87-5)															
18V. 1,2-Dichloropropylene (542-75-6)															
19V. Ethylbenzene (100-41-4)															
20V. Methyl Bromide (74-83-9)															
21V. Methyl Chloride (74-87-3)															

EPA Form 3510-2C (6-80)

CONTINUE ON PAGE V-5

Form Approved OMB No. 158-R0173

1. POLLUTANT AND CAS NUMBER *(if available)*	2. MARK 'X'			3. EFFLUENT							4. UNITS		5. INTAKE *(optional)*		
	a. TESTING REQUIRED	b. BELIEVED PRESENT	c. BELIEVED ABSENT	a. MAXIMUM DAILY VALUE		b. MAXIMUM 30 DAY VALUE *(if available)*		c. LONG TERM AVRG. VALUE *(if available)*		d. NO. OF ANALYSES	a. CONCENTRATION	b. MASS	a. LONG TERM AVERAGE VALUE		b. NO. OF ANALYSES
				(1) CONCENTRATION	(2) MASS	(1) CONCENTRATION	(2) MASS	(1) CONCENTRATION	(2) MASS				(1) CONCENTRATION	(2) MASS	
GC/MS FRACTION — VOLATILE COMPOUNDS *(continued)*															
22V. Methylene Chloride (75-09-2)															
23V. 1,1,2,2-Tetrachloroethane (79-34-5)															
24V. Tetrachloroethylene (127-18-4)															
25V. Toluene (108-88-3)															
26V. 1,2-Trans-Dichloroethylene (156-60-5)															
27V. 1,1,1-Trichloroethane (71-55-6)															
28V. 1,1,2-Trichloroethane (79-00-5)															
29V. Trichloroethylene (79-01-6)															
30V. Trichlorofluoromethane (75-69-4)															
31V. Vinyl Chloride (75-01-4)															
GC/MS FRACTION — ACID COMPOUNDS															
1A. 2-Chlorophenol (95-57-8)															
2A. 2,4-Dichlorophenol (120-83-2)															
3A. 2,4-Dimethylphenol (105-67-9)															
4A. 4,6-Dinitro-O-Cresol (534-52-1)															
5A. 2,4-Dinitrophenol (51-28-5)															
6A. 2-Nitrophenol (88-75-5)															
7A. 4-Nitrophenol (100-02-7)															
8A. P-Chloro-M-Cresol (59-50-7)															
9A. Pentachlorophenol (87-86-5)															
10A. Phenol (108-95-2)															
11A. 2,4,6-Trichlorophenol (88-06-2)															

CONTINUE ON REVERSE

CONTINUED FROM THE FRONT

1. POLLUTANT AND CAS NUMBER *(if available)*	2. MARK 'X'			3. EFFLUENT							4. UNITS		5. INTAKE *(optional)*		
	a. TESTING REQUIRED	b. BELIEVED PRESENT	c. BELIEVED ABSENT	a. MAXIMUM DAILY VALUE		b. MAXIMUM 30 DAY VALUE *(if available)*		c. LONG TERM AVRG. VALUE *(if available)*		d. NO. OF ANALYSES	a. CONCENTRATION	b. MASS	a. LONG TERM AVERAGE VALUE		b. NO. OF ANALYSES
				(1) CONCENTRATION	(2) MASS	(1) CONCENTRATION	(2) MASS	(1) CONCENTRATION	(2) MASS				(1) CONCENTRATION	(2) MASS	
GC/MS FRACTION – BASE/NEUTRAL COMPOUNDS															
1B. Acenaphthene (83-32-9)															
2B. Acenaphtylene (208-96-8)															
3B. Anthracene (120-12-7)															
4B. Benzidine (92-87-5)															
5B. Benzo *(a)* Anthracene (56-55-3)															
6B. Benzo *(a)* Pyrene (50-32-8)															
7B. 3,4-Benzo-fluoranthene (205-99-2)															
8B. Benzo *(ghi)* Perylene (191-24-2)															
9B. Benzo *(k)* Fluoranthene (207-08-9)															
10B. Bis *(2-Chloro-ethoxy)* Methane (111-91-1)															
11B. Bis *(2-Chloro-ethyl)* Ether (111-44-4)															
12B. Bis *(2-Chloro-isopropyl)* Ether (39638-32-9)															
13B. Bis *(2-Ethyl-hexyl)* Phthalate (117-81-7)															
14B. 4-Bromo-phenyl Phenyl Ether (101-55-3)															
15B. Butyl Benzyl Phthalate (85-68-7)															
16B. 2-Chloro-naphthalene (91-58-7)															
17B. 4-Chloro-phenyl Phenyl Ether (7005-72-3)															
18B. Chrysene (218-01-9)															
19B. Dibenzo *(a,h)* Anthracene (53-70-3)															
20B. 1,2-Dichloro-benzene (95-50-1)															
21B. 1,3-Dichloro-benzene (541-73-1)															

CONTINUE ON PAGE V-7

OUTFALL NUMBER

Form Approved OMB No. 158-R0173

CONTINUED FROM PAGE V-6

1. POLLUTANT AND CAS NUMBER (*if available*)	2. MARK 'X'			3. EFFLUENT								4. UNITS		5. INTAKE (*optional*)		
	a. TESTING REQUIRED	b. BELIEVED PRESENT	c. BELIEVED ABSENT	a. MAXIMUM DAILY VALUE		b. MAXIMUM 30 DAY VALUE (*if available*)		c. LONG TERM AVRG. VALUE (*if available*)		d. NO. OF ANALYSES	a. CONCENTRATION	b. MASS	a. LONG TERM AVERAGE VALUE		b. NO. OF ANALYSES
				(1) CONCENTRATION	(2) MASS	(1) CONCENTRATION	(2) MASS	(1) CONCENTRATION	(2) MASS				(1) CONCENTRATION	(2) MASS	
GC/MS FRACTION – BASE/NEUTRAL COMPOUNDS (*continued*)																
22B. 1,4-Dichlorobenzene (106-46-7)																
23B. 3,3'-Dichlorobenzidine (91-94-1)																
24B. Diethyl Phthalate (84-66-2)																
25B. Dimethyl Phthalate (131-11-3)																
26B. Di-N-Butyl Phthalate (84-74-2)																
27B. 2,4-Dinitrotoluene (121-14-2)																
28B. 2,6-Dinitrotoluene (606-20-2)																
29B. Di-N-Octyl Phthalate (117-84-0)																
30B. 1,2-Diphenylhydrazine (*as Azobenzene*) (122-66-7)																
31B. Fluoranthene (206-44-0)																
32B. Fluorene (86-73-7)																
33B. Hexachlorobenzene (118-71-1)																
34B. Hexachlorobutadiene (87-68-3)																
35B. Hexachlorocyclopentadiene (77-47-4)																
36B. Hexachloroethane (67-72-1)																
37B. Indeno (*1,2,3-cd*) Pyrene (193-39-5)																
38B. Isophorone (78-59-1)																
39B. Naphthalene (91-20-3)																
40B. Nitrobenzene (98-95-3)																
41B. N-Nitrosodimethylamine (62-75-9)																
42B. N-Nitrosodi-N-Propylamine (621-64-7)																

CONTINUE ON REVERSE

CONTINUED FROM THE FRONT

1. POLLUTANT AND CAS NUMBER (*if available*)	2. MARK 'X'			3. EFFLUENT							4. UNITS		5. INTAKE (*optional*)		
	a. TESTING REQUIRED	b. BELIEVED PRESENT	c. BELIEVED ABSENT	a. MAXIMUM DAILY VALUE		b. MAXIMUM 30 DAY VALUE (*if available*)		c. LONG TERM AVRG. VALUE (*if available*)		d. NO. OF ANALYSES	a. CONCENTRATION	b. MASS	a. LONG TERM AVERAGE VALUE		b. NO. OF ANALYSES
				(1) CONCENTRATION	(2) MASS	(1) CONCENTRATION	(2) MASS	(1) CONCENTRATION	(2) MASS				(1) CONCENTRATION	(2) MASS	
GC/MS FRACTION — BASE/NEUTRAL COMPOUNDS (*continued*)															
43B. N-Nitrosodiphenylamine (86-30-6)															
44B. Phenanthrene (85-01-8)															
45B. Pyrene (129-00-0)															
46B. 1,2,4 - Trichlorobenzene (120-82-1)															
GC/MS FRACTION — PESTICIDES															
1P. Aldrin (309-00-2)															
2P. α-BHC (319-84-6)															
3P. β-BHC (319-85-7)															
4P. γ-BHC (58-89-9)															
5P. δ-BHC (319-86-8)															
6P. Chlordane (57-74-9)															
7P. 4,4'-DDT (50-29-3)															
8P. 4,4'-DDE (72-55-9)															
9P. 4,4'-DDD (72-54-8)															
10P. Dieldrin (60-57-1)															
11P. α-Endosulfan (115-29-7)															
12P. β-Endosulfan (115-29-7)															
13P. Endosulfan Sulfate (1031-07-8)															
14P. Endrin (72-20-8)															
15P. Endrin Aldehyde (7421-93-4)															
16P. Heptachlor (76-44-8)															

CONTINUE ON PAGE V-9

EPA I.D. NUMBER (*copy from Item 1 of Form 1*)	OUTFALL NUMBER

CONTINUED FROM PAGE V-8

Form Approved OMB No. 158-R0173

1. POLLUTANT AND CAS NUMBER (*if available*)	2. MARK 'X'			3. EFFLUENT							4. UNITS		5. INTAKE (*optional*)		
	a. TESTING REQUIRED	b. BELIEVED PRESENT	c. BELIEVED ABSENT	a. MAXIMUM DAILY VALUE		b. MAXIMUM 30 DAY VALUE (*if available*)		c. LONG TERM AVRG. VALUE (*if available*)		d. NO. OF ANALYSES	a. CONCENTRATION	b. MASS	a. LONG TERM AVERAGE VALUE		b. NO. OF ANALYSES
				(1) CONCENTRATION	(2) MASS	(1) CONCENTRATION	(2) MASS	(1) CONCENTRATION	(2) MASS				(1) CONCENTRATION	(2) MASS	
GC/MS FRACTION – PESTICIDES (*continued*)															
17P. Heptachlor Epoxide (1024-57-3)															
18P. PCB-1242 (53469-21-9)															
19P. PCB-1254 (11097-69-1)															
20P. PCB-1221 (11104-28-2)															
21P. PCB-1232 (11141-16-5)															
22P. PCB-1248 (12672-29-6)															
23P. PCB-1260 (11096-82-5)															
24P. PCB-1016 (12674-11-2)															
25P. Toxaphene (8001-35-2)															

APPENDIX J
U.S. Army Corps of Engineers Sample Permit Form

APPLICATION FOR A DEPARTMENT OF THE ARMY PERMIT

For use of this form, see EP 1145-2-1

The Department of the Army permit program is authorized by Section 10 of the River and Harbor Act of 1899, Section 404 of P.L. 92-500 and Section 103 of P.L. 92-532. These laws require permits authorizing structures and work in or affecting navigable waters of the United States, the discharge of dredged or fill material into waters of the United States, and the transportation of dredged material for the purpose of dumping it into ocean waters. Information provided in ENG Form 4345 will be used in evaluating the application for a permit. Information in the application is made a matter of public record through issuance of a public notice. Disclosure of the information requested is voluntary; however, the data requested are necessary in order to communicate with the applicant and to evaluate the permit application. If necessary information is not provided, the permit application cannot be processed nor can a permit be issued.

One set of original drawings or good reproducible copies which show the location and character of the proposed activity must be attached to this application (see sample drawings and checklist) and be submitted to the District Engineer having jurisdiction over the location of the proposed activity. An application that is not completed in full will be returned.

1. Application number (To be assigned by Corps)

2. Date

Day Mo. Yr.

3. For Corps use only.

4. Name and address of applicant.

Telephone no. during business hours

A/C ( ) ________

A/C ( ) ________

5. Name, address and title of authorized agent.

Telephone no. during business hours

A/C ( ) ________

A/C ( ) ________

6. Describe in detail the proposed activity, its purpose and intended use (private, public, commercial or other) including description of the type of structures, if any to be erected on fills, or pile or float-supported platforms, the type, composition and quantity of materials to be discharged or dumped and means of conveyance, and the source of discharge or fill material. If additional space is needed, use Block 14.

7. Names, addresses and telephone numbers of adjoining property owners, lessees, etc., whose property also adjoins the waterway.

8. Location where proposed activity exists or will occur.

Address:

____

Street, road or other descriptive location

____

In or near city or town

____

County State Zip Code

Tax Assessors Description: (If known)

Map No. Subdiv. No. Lot No.

Sec. Twp. Rge.

9. Name of waterway at location of the activity.

10. Date activity is proposed to commence. ______________________________

Date activity is expected to be completed. ______________________________

11. Is any portion of the activity for which authorization is sought now complete? ☐ YES ☐ NO
If answer is "Yes" give reasons in the remark section. Month and year the activity was completed

______________________________. Indicate the existing work on the drawings.

12. List all approvals or certifications required by other federal, interstate, state or local agencies for any structures, construction, discharges, deposits or other activities described in this application.

Issuing Agency	Type Approval	Identification No.	Date of Application	Date of Approval

13. Has any agency denied approval for the activity described herein or for any activity directly related to the activity described herein?

☐ Yes ☐ No (If "Yes" explain in remarks)

14. Remarks (Checklist, Appendix H for additional information required for certain activities).

15. Application is hereby made for a permit or permits to authorize the activities described herein. I certify that I am familiar with the information contained in this application, and that to the best of my knowledge and belief such information is true, complete, and accurate. I further certify that I possess the authority to undertake the proposed activities.

____
Signature of Applicant or Authorized Agent

The application must be signed by the applicant; however, it may be signed by a duly authorized agent (named in Item 5) if this form is accompanied by a statement by the applicant designating the agent and agreeing to furnish upon request, supplemental information in support of the application.

18 U. S. C. Section 1001 provides that: Whoever, in any manner within the jurisdiction of any department or agency of The United States knowingly and willfully falsifies, conceals, or covers up by any trick, scheme, or device a material fact or makes any false, fictitious or fraudulent statements or representations or makes or uses any false writing or document knowing same to contain any false, fictitious or fraudulent statement or entry, shall be fined not more than $10,000 or imprisoned not more than five years, or both. Do not send a permit processing fee with this application. The appropriate fee will be assessed when a permit is issued.

APPENDIX K
Sample Development of Regional Impact Statement Forms

SOUTH FLORIDA DEVELOPMENT OF REGIONAL IMPACT
Application for Development Approval Under Section 380.06(6), Florida Statutes

The Development of Regional Impact (DRI) Application for Development Approval (ADA) is intended to provide a comprehensive view of a proposed development.

A. Applicants proposing residential Developments of Regional Impact must complete Part I and Part II of this Application. Applicants proposing nonresidential DRIs must complete Part I, Part II, and appropriate portions of Part III, as determined during the preapplication conference.

B. If a comprehensive DRI application is filed pursuant to Section 380.06, Florida Statutes, answer Part II separately for each type of DRI and for the total proposed development.

C. As provided for in Chapter 380, Florida Statutes, a Development of Regional Impact includes all development associated with it. Therefore, an applicant proposing a DRI that includes land uses not of DRI magnitude must include information regarding those uses in the appropriate locations of the Application.

D. All information supplied must be accurate, current, and complete.

E. Use the specified formats and units of measurement. If the specified format requires information by development phase, each phase must relate to those specified in Question 12-A, and final entries must correspond with Question 12-B.

F. Provide complete answers to the questions in the body of the Application. Reference to reports prepared for other purposes is not acceptable, although such reports may be attached as appendices.

G. Submit the Application in 8½″ × 11″ loose-leaf, 3-ring notebook form with a table of contents. This format makes it relatively easy to modify or add information to the Application. Maps contained in the Application should be 11″ × 14″.

H. Include a bibliography of information sources and the names, addresses, and telephone numbers of any consultants, agencies, or other persons who contributed to or completed sections of the Application.

I. Include all methodologies, models, assumptions, sources, and standards used in obtaining or developing the information in the Application.

J. "Region" is defined as Broward, Dade, and Monroe counties.

K. Copies of the completed Application must be submitted to: (1) the appropriate local government(s), (2) the South Florida Regional Planning Council, (3) the Division of Local Resource Management, Florida Department of Community Affairs, and (4) other State or Federal agencies participating in the review. Contact the appropriate local government and the Council for the number of copies required by each agency. An application cannot be considered formally submitted until each agency receives the appropriate number of copies.

PART I. APPLICATION INFORMATION

1. I, ______________________________, the undersigned owner (authorized representative) of ______________________________ (developer), hereby propose to undertake a Development of Regional Impact as defined in Section 380.06, Florida Statutes, and Chapter 27F-2.____, Florida Administrative Code. In support thereof I submit the following information con-

cerning ______________________________________ (name of development) which information is true and correct to the best of my knowledge.

______________ ____

Date Signature of Owner or Authorized Representative

2. **Applicant (name, address, phone).**
3. **Authorized Agent(s) (name, address, phone).**
4. **Names and addresses of all persons or entities having fee simple or lesser interest in the site. Lesser interest includes restrictive covenant, easement, or purchase option.**
5. **Attach a legal description of the development site. (Include section, township, and range or subdivision.)**
6. **Identify any adjacent property in which any person or entity listed in Question 4 above has an interest.**
7. **Specify the type of Development of Regional Impact and size (as defined in Chapter 27F-2, Florida Administrative Code). For residential DRIs, state the site area and number of dwelling units.**
8. **If you have received a DRI binding letter of interpretation or vested rights determination from the State pursuant to Section 380.06(4), Florida Statutes, attach a copy of the State response.**
9. **List the local government(s) with jurisdiction over the proposed development.**
10. **List agencies (local, state, and federal) from which approval and/or a permit must be obtained prior to initiation of development. Specify the permit(s) or approval(s) by agency. Provide copies of any submitted permit applications.**

PART II. GENERAL

11. **Maps A through K are part of the Application. Provide maps in the Application and selected presentation-scale maps. Determine the scale of presentation maps in consultation with the Regional Planning Council at the preapplication conference. Include a scale and north arrow on each map, and dates of preparation or revision. Questions 33, 35, 36, 38, 40, 41, and 43 also require specific map information.**

    Insert each map, at reduced scale, into the text where it is first referenced.

    A. A general location map (Map A).

    B. A recent aerial photo of site showing project boundaries (Map B). Specify date photo was taken on map.

    C. A topographic map with project boundaries identified (contour intervals, referenced to the 1929 NGVD, of from one to five feet should be determined in consultation with the Regional Planning Council and local government). Delineate 100-year flood prone areas (including hurricane flood zones and V-zones) and show major natural and man-made features.

    Provide a map or overlay showing topography upon project completion (Map(s) C).

D. A land use map showing existing uses on the site (cross referenced to Table 12.2) and within the primary impact area. Show any historical/archaeological sites. Provide, preferably as an overlay to this map, future land uses within the primary impact area (Map D).
E. A soils map of the site, preferably based on USDA Soil Conservation Service (SCS) published soil surveys (Map E).
F. A vegetation associations map showing the location and acreage of each association, based on the Level III vegetation types in *The Florida Land Use and Cover Classification System: A Technical Report*, which is available from the Regional Planning Council (Map F).
G. A master drainage plan for the site showing existing and proposed drainage areas, surface and subsurface retention areas, drainage structures, runoff routing scheme, drainage easements, canals, and other major drainage features. Use arrows to indicate direction of surface runoff and flow (Map G).
H. A master development plan for the site showing proposed land uses (cross referenced to Table 12.2), development phasing, major public facilities, utilities, easements, rights-of-way, roads, thoroughfares, and other significant elements. Include a map or overlay showing a schematic proposed landscape design and provide a list of the vegetation species proposed for use (Map(s) H).
I. A map showing existing and proposed public facilities (e.g., sewage, water supply, fire protection, public transit, solid waste disposal, hospitals, police, emergency medical facilities, etc.) that serve the site (Map I). Show existing and proposed pipeline and transmission line routes for water supply, sewage, electric power, and gas, as appropriate.
J. A map of the existing highway and transportation network within the primary impact area. The primary impact area includes the site and normally extends at least five miles beyond the development boundary; however, this area should be defined in consultation with the Regional Planning Council and clearly delineated on this map. Map J will be the base for the maps in Questions 31-A, 31-B, and 31-C (Map J).
K. A map of the primary market area for proposed commercial/retail development. The primary market area includes the site, divided into at least four geographic sectors, and normally extends at least ten miles beyond the development boundary; however, this area should be defined in consultation with the Regional Planning Council and local government staff and should be clearly delineated on this map. Developments similar to the proposed project should also be clearly indicated on this map (Map K).

12. General Project Description

A. Provide a brief summary of the major elements of the proposed development. Include all existing and proposed land uses ancillary to the project (e.g., a neighborhood shopping center in a residential DRI).
B. Complete Tables 12.1 and 12.2. (If the development has a proposed buildout of 10 years or less, show development in the first 5 years and subsequently. If the proposed buildout is greater than 10 years, show by 5-year increments.)
C. Describe the relationship of the project to existing zoning, comprehensive plans, and any special regulatory requirements (e.g., Biscayne Bay Aquatic Preserve Act).

TABLE 12.1: Phasing of Development

	Units[2]					
	Residential		Commercial	Other	Construction	
Phase[1]	(No.)	(Pop.)	(Sq. Ft.)	(Specify)	Beg.	End
1						
.						
.						
.						
N						
TOTAL						

[1]Expand table as necessary to accommodate additional units and phases.
[2]Use appropriate units for the type of development proposed.

TABLE 12.2: Existing and Proposed Land Uses[1]

	Land Use 1[3]: (specify)		Land Use 2[3]: (specify)		Total	
Phase[2]	Acres	% of Site	Acres	% of Site	Acres	% of Site
Existing						
Phase 1						
Phase .						
Phase .						
Phase N						
TOTAL						

[1]Expand table as necessary to accommodate additional land uses.
[2]Expand table as necessary to accommodate phases.
[3]Use Level III of the *Florida Land Use and Cover Classification System: A Technical Report*, available from the Council.

13. Environment and Natural Resources: Air

A. If an air quality permit has been completed, provide a copy of the permit application. If not, provide one-hour and eight-hour carbon monoxide concentrations projected by completing Table 13.1. Describe how the receptor stations used in generating concentrations yield worst-case conditions. Consult with the Council to determine if monitoring will be required to establish baseline data.

B. For any receptor location exceeding either the one-hour or eight-hour Florida Ambient Air Quality Standards, complete Table 13.2.

C. Specify what will be done to minimize emissions and mitigate adverse impacts. Specify the net change from present air quality.

14. Environment and Natural Resources: Land

A. Include each soil shown on Map E, in Table 14.1.

TABLE 13.1: Project Carbon Monoxide Emissions (Mg/M³)[1]

	One Hour Emissions Total Emissions Receptor Stations[2,3]					Eight Hour Emissions Total Emissions Receptor Stations[2,3]				
	1	2	3	4	N	1	2	3	4	N
Existing										
Phase 1										
Phase N[4]										

[1]Expand table as necessary to accommodate phases.
[2]Provide a location map to identify receptor stations and specify conditions that qualify.
[3]Consult with the Council staff to identify appropriate number and location of receptor stations.
[4]Emissions after buildout and occupancy.

TABLE 13.2: Percentage Contribution, by Source, for Receptor Stations Exceeding State Standards

	One-Hour Standard						Eight-Hour Standard					
	Receptor Station #1			Receptor Station #N[1]			Receptor Station #1			Receptor Station #N[1]		
	Area Sources	Line Sources	Project Source	Line Sources	Area Sources	Project Source	Area Sources	Line Sources	Project Source	Line Sources	Area Sources	Project Source
Existing												
Phase 1												
Phase N[2,3]												

[1]Expand table as necessary to accommodate stations.
[2]Assuming full occupancy at buildout.
[3]Expand table as necessary to accommodate phases.

TABLE 14.1: Soil Descriptions and Interpretations

Soil Name and Map Symbol[1]	Brief Soil Description	Depth to Bedrock	Seasonal High Water Table		Percolation Rate (in./hour)	Limitation for Low Buildings	Limitation for Pond Embankments
			Depth	Duration			
						*	*

[1]Place soil name in parentheses if information from the USDA Soil Conservation Service Soils Survey has not been field-checked.
*Appropriate responses include: slight, moderate, severe, and very severe, as defined by the Soil Conservation Service.

B. Where a soil presents a limitation to the type of structure(s) proposed in the development, state how the limitation will be overcome. Specify construction methods that would be used for building, road, and parking lot foundations and for lake or canal bank stabilization, as relevant.
C. Specify how wind and water soil erosion will be controlled during construction and after the development is completed.
D. Specify the source location(s), volume(s), type(s) of fill, and the volume and disposal location(s) for spoil.

15. Environment and Natural Resources: Water

A. Describe the existing ground and surface hydrologic conditions.
B. Specify probable project effects on water quality and quantity, and whether the proposed development site lies wholly or partially within the cone-of-infuence of any existing or proposed public potable water-supply well.
C. Complete Table 15.1.
D. Specify actions that will be taken to mitigate or minimize adverse water impacts.

TABLE 15.1: Existing Water Quality

Sample Location[1]	*Parameters*[2]
Groundwater	
1	
.	
.	
.	
N	
Surface Water	
1	
.	
.	
.	
N	

[1]Show the site(s) at which sampling/monitoring has been conducted.
[2]Consult with the Council for specific parameters to be used for this development.

16. Environment and Natural Resources: Wetlands

If there are wetlands on the site, specify:

A. Proposed alterations or disturbances to the wetlands,
B. Wetland areas that will be preserved in their natural or existing state and the methods that will accomplish this preservation, and
C. Actions that will be taken to mitigate or minimize impact on wetlands.

17. Environment and Natural Resources: Flood Prone Areas

If development is proposed within the 100-year flood prone area identified by FEMA, specify:

A. Applicable flood elevation for the site, and source (e.g., FEMA, County).
B. Finished first floor elevations for structures and for impervious surfaces (e.g., roads, parking lots), and
C. Construction techniques to be used to minimize flood hazard (e.g., silting, use of fill).

18. Environment and Natural Resources: Vegetation and Wildlife

A. Complete Table 18.1.

B. Complete Table 18.2 for wildlife species, including birds, reptiles and amphibians, fish, invertebrates, and mammals, that use the proposed development area.

C. Specify actions that will be undertaken to mitigate or minimize impact on vegetation and wildlife.

TABLE 18.1: Existing Vegetation Associations

		Major Species of the Vegetation Association		*On Rare or Endangered Species List*			
Vegetation Association[1]	*General Condition of the Vegetation Association*[2]	*Common Name*	*Scientific Name*	*List*[3]	*Category*[4]	*Acres on the Site*	*Acres to be Removed (−) or Added (+)*

[1]Use the *Florida Land Use and Cover Classification System.*
[2]For example, the general condition of the vegetation association might be described as "healthy and productive," or "heavily impacted by off-road vehicle use in the area," or "75% of plants are diseased."
[3]Appropriate lists to consult include those prepared by the State of Florida, the Florida Committee on Rare and Endangered Plants and Animals, the U.S. Department of the Interior, and Dade County.
[4]Categories include endangered, threatened, rare, species of special concern, and status undetermined.

TABLE 18.2: Wildlife

Wildlife Species					*On Rare or Endangered Species List*	
Common Name	*Scientific Name*	*Site Use*[2]	*Frequency of Site Use (# of months each year)*	*Season of Site Use*	*List*[3]	*Category*[4]

[1]Place an "N" in the column if the particular species uses the area for nesting and/or cover, an "F" if the habitat is used for feeding, and an "M" if the area is used during migration.
[2]Winter (W), Spring (Sp), Summer (S), Fall (F), or All Year (All).
[3]Appropriate lists to consult include those prepared by the State of Florida, the Florida Committee on Rare and Endangered Plants and Animals, the U.S. Department of the Interior, and Dade County.
[4]Categories include endangered, threatened, rare, species of special concern, and status undetermined.

19. Environment and Natural Resources: Historical and Archaeological Sites

A. Identify and provide a brief description of each site shown on Map D. Specify sites that have been evaluated or surveyed by the Florida Bureau of Historic Sites and Properties and the County archaeologist.

B. Include a letter from the Florida Division of Archives, History and Records Management, or from the County archaeologist indicating completeness of the list and whether a survey is desirable.

C. Specify measures that will be taken to protect and, where appropriate, provide survey, research, or public access to any site listed above.

20. Economy: Employment and Economic Characteristics

A. Complete Tables 20.1 and 20.2.

B. For non-residential and mixed-use developments, complete Table 20.3.

C. Complete Table 20.4 for each phase, estimating the total number of employees by three or four-digit SIC Code (col. 1, # Emp.) and the percentage of employees that will come from within the Region (col. 2, & Region). For mixed-use projects with a residential component, estimate the number of residents expected to be employed within the project (col. 3, # Res. Emp.).

TABLE 20.1: Construction Costs

	Phase 1		Phase N[1]		Total	
Cost Item	*$ Cost*	*% Spent in Region*[2]	*$ Cost*	*% Spent in Region*	*$ Cost*	*Spent in Region*
Land						
Labor[3]						
Materials						
Interest[4]						
Preliminary Planning						
Other (Legal Services, Administrative, Overhead)						
TOTAL						

[1]Expand table as necessary to accommodate phases.
[2]Regional encompasses Broward, Dade, and Monroe counties.
[3]Indicate average annual construction wage.
[4]Pre-completion interest rate.

TABLE 20.2: Construction Employment[1,2]

Phase[3]	*Total*
1	
.	
.	
.	
N	
TOTAL	

[1]Specify the average annual wage for construction employment.
[2]Specify employment in employee years (full-time equivalent jobs).
[3]Expand table as necessary to accommodate phases.

TABLE 20.3: Annual Operating Costs

	Phase 1		Phase N[1]		Total	
Cost Item	*$ Cost*	*% Spent in Region*[2]	*$ Cost*	*% Spent in Region*	*$ Cost*	*% Spent in Region*
Wages and Salaries						
Fringe Benefits						
Overhead[3]						
Advertising and Promotion						
Supplies and Equipment						
Other[4]						
TOTAL						

[1]Expand table as necessary to accommodate phases.
[2]Region encompasses Broward, Dade, and Monroe counties.
[3]Includes rents, taxes, property insurance, maintenance and repairs, utilities, communications, and other indirect costs.
[4]Specify the type of operating cost.

TABLE 20.4: Permanent Employment[1]

	$8,000			$8,000–14,999			$15,000–24,999			$25,000–49,999			Over $50,000		
SIC CODE (3-digit)	*# Emp.*	*% Region*	*# Res. Emp.*	*# Emp.*	*% Region*	*# Res. Emp.*	*# Emp.*	*% Region*	*# Res. Emp.*	*# Emp.*	*% Region*	*# Res. Emp.*	*# Emp.*	*% Region*	*# Res. Emp.*
Phase 1															
SIC Code 1															
.															
.															
.															
SIC Code N															
Subtotal															
Phase N[2]															
SIC Code 1															
.															
.															
.															
SIC Code N															
Subtotal															
Total															
SIC Code 1															
.															
.															
.															
SIC Code N															
TOTAL															

[1]Specify average annual wages.
[2]Expand table as necessary to accommodate phases.

D. If the number of employees will vary seasonally, show the average and seasonal peak employment in full-time equivalent (FTE) jobs at the project upon build-out. Repeat Table 20.5 for each project phase.
E. Complete Table 20.6.
F. If assistance will be sought from Federal, State, or other governmental funding programs, specify from what agency, under what program, and in what amount.
G. For public projects, specify the funding source(s) and amounts.
H. Complete Tables 20.7 and 20.8. Provide folio numbers for all land proposed for development (see Question 5).
I. Complete Figure 20.1.
J. Provide a copy of the market study prepared for the proposed development.
K. Complete Table 20.9.

TABLE 20.5: Permanent Seasonal Peak & Average Annual (FTE) Employment

SIC Code	*<$8,000*		*$8,000–14,999*		*$15,000–24,999*		*$25,000–49,999*		*Over $50,000*		*Total*	
(3-digit)	*Peak*	*Avg.*	*Peak*	*Avg.*	*Peak*	*Avg.*	*Peak*	*Avg.*	*Peak*	*Avg.*	*Peak*	*Avg.*
Phase 1												
.												
.												
.												
Phase N[1]												
TOTAL												

[1]Expand table as necessary to accommodate phases.

TABLE 20.6: Existing Economic Activity On-Site

SIC Code (3-digit)	*No. of Permanent Employees Working On-Site*	*Number of Seasonal Employees (if any) Working On-Site*	*No. of Acres*	*Dollar Value of Annual Production and/or Sale*
Existing				
Subtotal				
Phase 1				
.				
.				
.				
Phase N[1]				
Subtotal				
TOTAL				

[1]Expand table as necessary to accommodate phases.

TABLE 20.7: Local Ad Valorem Revenues

Phase	*Market Value ($)*	*Assessment Ratio*	*Non-Exempt Assessed Value ($)*	*Total Operating Millage Rate*[1]	*Ad Valorem Tax Yield ($)*
Pre-Development					
Land					
Buildings					
Phase 1					
.					
.					
.					
Phase N[2]					

[1]Show the individual operating millage rates for county, city, school board, and special assessment districts that apply to the project, as a footnote to this table.
[2]Expand table as necessary to accommodate phases.

TABLE 20.8: Local Non-Ad Valorem Revenues

Phase	*Tax and Permit Fee Payments*	*Assumptions of Calculation*[1]
Phase 1		
Building Permit Fees		
Water & Sewer Permit		
All other Permit & License Fees		
Utility Service Taxes		
Other (specify)		
Phase N[2]		

[1]Specify jurisdiction receiving revenues.
[2]Expand table as necessary to accommodate phases.

FIGURE 20.1: Market Area Data by Sector and by Phase[1]

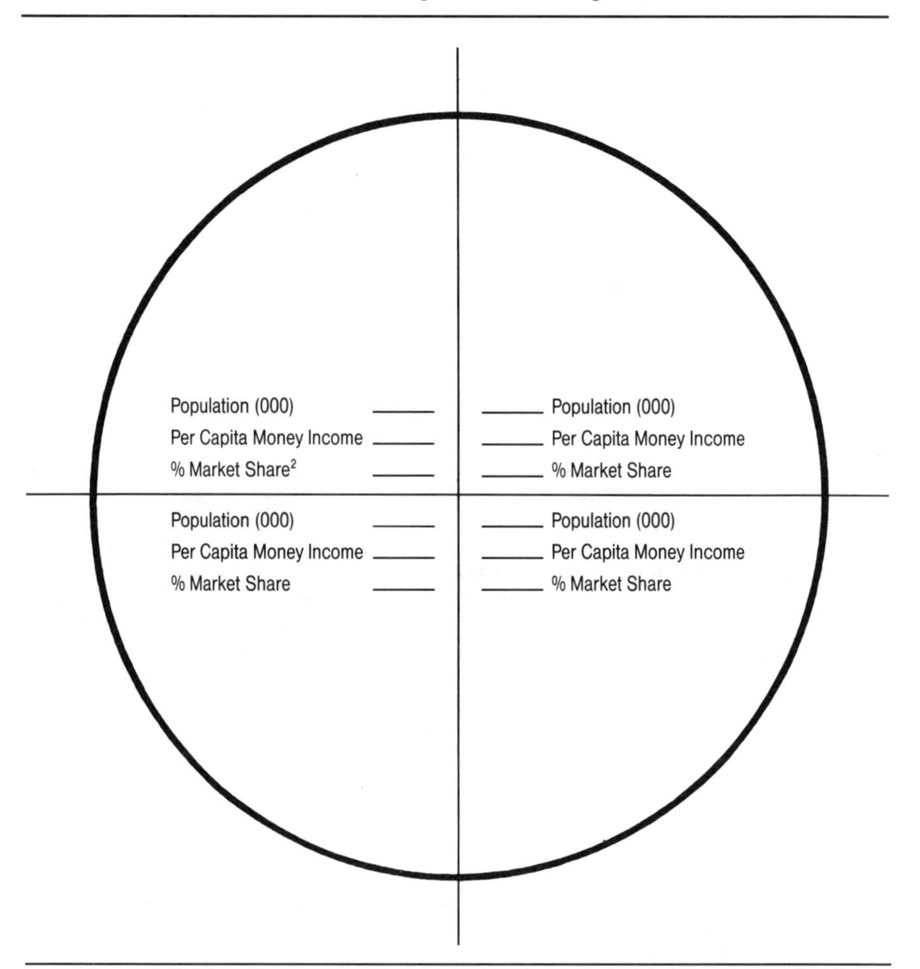

[1]Complete figure 20.1 (or provide a comparable graphic which delineates the market area boundaries and identified sectors within those boundaries) for the existing market area and for each project phase. The center of the figure corresponds to the site location, and the circle corresponds to the market area boundaries delineated in Map K.

[2]Anticipated sales of products in the proposed development as a percentage of the total sales of these products in the market area.

L. Complete Table 20.10 to show the estimated capital costs for additional public facilities needed to serve the proposed development.

M. Complete Table 20.11.

TABLE 20.9: Tenant Costs

			For Commercial Tenants				
Expected Tenants[1] (By Phase)[2]	*SIC Code (3-digit)*	*G.L.A.[3] in Sq. Ft.*	*Avg. Annual Sales/Sq. Ft.*	*Rate of Pctge. Rent[4]*	*Total Rent per Sq. Ft.[5]*	*Common Area Charges per Sq. Ft.[6]*	*Total Charges per Sq. Ft.[7]*
Phase 1							
1							
.							
.							
N							
Subtotal							
Phase N[2]							
Subtotal							
TOTAL							

[1]Include both private goods or services.
[2]Expand table as necessary to accommodate all phases.
[3]GROSS LEASABLE AREA (GLA) is the total floor area designed for tenant occupancy and exclusive use, including any basements, mezzanines, upper floors, etc., expressed in square feet and measured from the center line of joint partitions and from outside wall faces. Gross leasable area is all for which tenants pay rent.
[4]RATE OF PERCENTAGE RENT. For tenants under a lease provision which requires that they pay rent in an amount equal to a specified percentage of the tenants' sales. For tenants under a multiple rate schedule (for example, 6 percent of sales over $100,000, 5 percent of sales over $150,000, etc.) use only the initial rate (e.g., 6 percent).
[5]TOTAL RENT PER SQUARE FOOT. Total annual rent per square foot of GLA paid for by each individual tenant. Total rent is the sum of all forms of rent–guaranteed minimum rent, percentage rent, and combinations thereof–paid by each tenant.
[6]COMMON AREA CHARGES PER SQUARE FOOT OF GLA. Annual common area charges on the basis of square feet of GLA paid by each individual tenant classification and type. COMMON AREA is the total area within the development which is not designated for rental to tenants but is available to common use by all tenants or groups of tenants, their invitees, and adjacent stores. Parking and its appurtenances, malls, sidewalks, landscaped areas, public toilets, truck and service facilities, etc. are typically included in the common area.
[7]TOTAL CHARGES PER SQUARE FOOT. The total charges per square foot of GLA includes total rent, common area charges, and all other charges levied against the tenant, including taxes paid under escalator clauses, percentage returns on vending machines.

TABLE 20.10: Public Facility Capital Costs

Additional Public Facility or Trans. Improvement by Phase[1]	*Total Facility or Improvement Cost*	*Facility Construction Cost to be Borne by Developer*	*Market Value of Land Donated by Developer*	*Total Cost Borne by Developer*
Phase 1				
Subtotal				
Phase N[2]				
Subtotal				
TOTAL				

[1]Specify the type of improvement needed to serve the proposed development.
[2]Expand table if necessary to accommodate phases.

TABLE 20.11: Public Facility or Service Operating Costs

Public Facility or Service Operation by Phase[1]	*Annual Operating Cost*	*Cost Borne by Local Gov't.*	*Cost Borne by Developer*
Phase 1			
Subtotal			
Phase N[2]			
Subtotal			
TOTAL			

[1]Specify the type of public facility or service.
[2]Expand table as necessary to accommodate phases.

21. Wastewater Management

A. Complete Table 21.1.

B. If applicable, specify the characteristics of any effluents containing hazardous and toxic wastes and any special treatment needed to dispose of such effluents.

C. If on-site treatment and disposal is *not* intended, go to D. If on-site treatment and disposal is intended, specify:
  1) method and degree of treatment,
  2) the quality of treated effluent,
  3) treatment and hydraulic capacity,
  4) who will operate and maintain the treatment works,
  5) effluent disposal location,
  6) volume of sludge and methods for its treatment and disposal, and
  7) if spray irrigation is proposed, specify location and approximate area of spray fields, depth to water table, percolation rate, proposed rate of application, and back-up system capacity.

D. If septic tanks will *not* be used, go to E. If septic tanks will be used, specify location and number of units.

E. Attach a letter from the agency or firm providing off-site treatment specifying:
  1) projected excess capacity of the treatment and transmission facilities to which connection will be made at the end of each phase and upon completion of the project, and
  2) any commitments that have been made for projected excess capacity.

TABLE 21.1: Cumulative Average Daily Wastewater Flows[1]

	Residential				Industrial			Other (Specify)			Totals	
			Flows (MGD)			Flows (MGD)			Flows (MGD)			
Phase	Units	Pop.	Avg.	Peak	Sq. Ft.	Avg.	Peak	Units	Avg.	Peak	Avg.	Peak
1												
.												
.												
N[2]												
TOTAL												

[1]Specify source of generation rates.
[2]Expand table as necessary to accommodate phases.

22. Drainage

A. Complete Table 22.1.

B. Referencing Map G, provide drainage calculations, including the following minimum information:

1) drainage areas details,
2) major water control structures,
3) berms, vegetated swales, and surface retention areas,
4) location and description of internal canals and water bodies with typical sections, including depths and slopes,
5) location and description of proposed storm sewers, french drains, structural detention areas, retention areas, oil and grease separators, diversion structures, and other conveyance and storage facilities,
6) design storm (frequency, duration, total rainfall, and reason for using) and
7) a runoff routing scheme, including calculations with stage-storage and stage-discharge relationships.

C. Specify the volume and quality (nutrient and pollutant loadings) of runoff from the site in its existing condition and at the end of each phase of development. Specify provisions incorporated in the drainage system to minimize runoff from the site and to improve runoff water quality.

D. Specify who will operate and maintain the drainage system after development completion.

TABLE 22.1: Drainage Areas

Phase	Impervious Surfaces (Acres)	Surface Retention (Acres)[1]	Open Space (Acres)	Total Area (Acres)
1				
.				
.				
N[2]				
TOTAL				

[1]Category includes lakes, ponds, storage areas, etc.
[2]Expand table as necessary to accommodate phases.

23. Water Supply

A. Complete Table 23.1.

B. If on-site wells are planned, complete Table 23.2.

C. If an internal (well or lake pumpage) water supply system is proposed, specify who will operate and maintain it after development completion.

D. If off-site water supply is planned, attach a letter from the agency or firm providing that service outlining:

1) projected excess water supply capacity at the end of each phase and upon completion of the project, and
2) all commitments that have been made for projected excess capacity.

TABLE 23.1: Potable/Non-Potable Water Demand[1]

	Potable Water Demand (MGD)			*Non-Potable Water Demand (MGD)*			*Total Water Demand (MGD)*	
Phase	*Avg.*	*Peak*	*Source[2]*	*Avg.*	*Peak*	*Source*	*Avg.*	*Peak*
Existing								
Phase								
1								
.								
.								
N[3]								
TOTAL			N/A			N/A		

[1]Specify source of generation rates.
[2]Source refers to the source of water supply.
[3]Expand table as necessary to accommodate phases.

TABLE 23.2: On-Site Wells

			Pumping Rates					
Well Number (from Map H)	*Diameter (inches)*	*Depth (feet)*	*Avg. (MGD)*	*Max. (MGD)*	*Units Served[1]*	*Potable/ Non-Potable*	*Begin Operation (year)*	*Phase-Out (year)*

[1]Use units appropriate to the type of development, e.g., housing units, square footage, acres.

TABLE 24.1: Solid Waste Generation

	Domestic Solid Waste		*Industrial or Other Special Wastes (specify)*
Phase	*Cubic Yards/Day*	*Tons/Day*	*Tons/Day*
Existing			
Phase			
1			
.			
.			
N[1]			
TOTALS			

[1]Expand table as necessary to accommodate phases.

24. Solid Waste

A. Complete Table 24.1.

B. If on-site solid waste disposal is *not* proposed, go to C. If *on-site* solid waste disposal is proposed, specify:
   1) method of disposal and/or recycling,
   2) capacity and life of operation,
   3) disposal facility location and area,
   4) methods or techniques that will be used to prevent groundwater contamination, and
   5) facility operating entity.

C. If *off-site* solid waste disposal is proposed, attach a letter from the disposal facility operator outlining:
   1) projected excess disposal capacity at the end of each phase and at completion of the project, and
   2) any commitments that have been made for this projected capacity.

25. Energy

A. Complete Table 25.1.

B. If on-site power supply is *not* proposed, go to C. If on-site power generation is proposed, complete Table 25.2.

C. Attach letter(s) from off-site energy supplier(s) outlining:
   1) projected excess capacities of the electrical generation facility and transmission line to which connection will be made for each phase through completion of the project, and
   2) any commitments that have been made for this excess capacity.

D. Specify energy-conserving measures that have been incorporated into the site plan, building design, building operation, and equipment selection and operation.

TABLE 25.1: Energy/Fuel Sources[1]

Land Use Type by Phase	Energy Use	Energy/Fuel Source: Electricity	Utility Gas	Fuel	Oil	LP Gas	Solar	Other	Total
Phase 1									
Residential	Space Heating								
	Space Cooling								
	Water Heating								
	Cooking								
	Other (specify)								
Commercial	Space Heating								
	Space Cooling								
	Water Heating								
	Cooking								
	Other (specify)								
Industrial	Space Heating								
	Space Cooling								
	Water Heating								
	Cooking								
	Other (specify)								
Subtotal									
Phase N[2]									
.									
.									
.									
Subtotal									
TOTAL									

[1]Enter proportion of units (e.g., dwelling units, square feet, etc.) that will use energy/fuel source by energy use.
[2]Expand table as necessary to accommodate phases.

TABLE 25.2: Projected On-Site Power Generation

Facility by Phase	Fuel/Energy Source	End Use[1]	Average Daily Power Capacity	Description
Phase 1				
.				
.				
.				
Phase N[2]				

[1]Including emergency use.
[2]Expand table as necessary to accommodate phases.

26. Education

A. Complete Table 26.1.
B. Complete Table 26.2.
C. Attach a letter from the school board, accepting the projected school-age population, school requirements (including typical school capacities for elementary, middle, and high), and capital improvements needed to accommodate these students.

TABLE 26.1: Projected School-Age Children

		Elementary		*Middle*		*High*	
Housing Type by Phase	*No. of Dwelling Units*	*Generation Rate*[1]	*No. of Students*	*Generation Rate*	*No. of Students*	*Generation Rate*	*No. of Students*
Phase 1							
Single Family							
Townhouse							
Garden Apt.							
High Rise							
Mobile Home							
Subtotal							
Phase N[2]							
Single Family							
Townhouse							
Garden Apt.							
High Rise							
Mobile Home							
Subtotal							
TOTAL							

[1]Specify source of generation rate(s).
[2]Expand table as necessary to accommodate phases.

TABLE 26.2: Education Facility Needs

	Elementary		*Middle*		*High*	
Phase	*# Students Expected*	*# Schools Needed*	*# Students Expected*	*# Schools Needed*	*# Students Expected*	*# Schools Needed*
Phase 1						
.						
.						
.						
Phase N[1]						
TOTAL						

[1]Expand table as necessary to accommodate phases.

27. Recreation and Open Space

A. Complete Table 27.1.

TABLE 27.1: Recreation and Open Space

		Acres			Responsible Entity		
Facility Types by Phase	*Number of Facilities*	*Land*	*Water*	*Total*	*Ownership*	*Operation*	*Maintenance*
Phase 1							
Surface Water							
Right-of-way							
School Sites							
Elementary							
Middle							
High							
Parks							
Neighborhood							
Community							
Regional							
Private Recreation							
Preserve Area							
Landscaped Area							
Other (Describe)							
Subtotal							
Phase N[1]							
.							
.							
.							
Subtotal							
TOTAL							

[1]Expand table as necessary to accommodate phases.

28. Health Care

A. Attach a copy of your letter to the agency responsible for providing emergency medical services in the project area, notifying the agency of your proposed development.

B. Include the agency response, if any, and specify the response time to the proposed development.

29. Police

A. Attach a copy of your letter to the public agency or agencies responsible for providing police services in the project area, notifying the agency(ies) of your proposed development.

B. Specify response times and include the agency response(s), if any.

30. Fire

A. Attach a copy of your letter to the public agency or agencies responsible for providing fire protection to the project, notifying the agency(ies) of your proposed development.

B. Specify response times and include the agency response(s), if any.

31. Transportation

A. Existing Traffic

1) Complete Table 31.1.
Where traffic counts are not available from State and local agencies, consult with the Regional Planning Council on the need for supplementary counts to assess project impacts.

2) On Map J-1, show current directional peak-hour traffic volume to capacity ratios and levels of service for all roadway segments in Table 31.1. Adjacent to each, include arrows showing A.M. and P.M. peak-hour direction of flow.

B. Programmed and Planned Improvements

1) On Map J-2, highlight all roadway improvements or new facilities included in the adopted Transportation Improvement Program (TIP) or funded privately for completion prior to each phase of the project.

2) Complete Table 31.2.

TABLE 31.1: Existing Traffic

	Segment							
Roadway[1]	*From*	*To*	*Number of Lanes*[2]	*Direction*[3]	*Peak-Hour Volume*[4]	*Peak-Hour Capacity*[4]	*Peak-Hour V/C*[5]	*Peak-Hour LOS*[6]

[1]Reference all County, State, and Federal roadway designations.
[2]Use the following abbreviations:
divided roadway .... "D"
for freeway .... "F/W"
for one-way street ...... "O/W"
[3]Use the following abbreviations:
Northbound .... "N"
Southbound .... "S"
Eastbound .... "E"
Westbound .... "W"
[4]Consult Council staff on directional distribution ("d" factor) assumptions when directional counts are not available.
[5]Use unadjusted peak-hour service volumes for LOS "C" shown in "Guidelines and Standards for preparing a DRI Application for Development Approval," available from the Council.
[6]Use the following V/C ratios to determine level of service:

V/C	*Level of Service*
.70	A
.71– .90	B
.91–1.10	C
1.11–1.30	D
1.31–1.60	E
1.61+	F

3) Attach response letters from the appropriate agencies stating the current status of any improvements identified above as "programmed."

C. Traffic Projections

1) Background Traffic

Provide peak-hour traffic projections for background traffic at all count locations identified in Table 31.1 for each phase of the development, by completing Table 31.3. Use the capacity of each publicly programmed or privately funded roadway improvement and project background traffic coincident with each phase of the project. Do not increase capacities for improvements identified in the response to Q. 31-E as "recommended."

a) Describe projection method(s), source(s), and assumptions.

b) On Map J-3, show projected directional peak-hour traffic volume to capacity ratios and levels of service for each roadway segment in Table 31.3. Adjacent to each, provide arrows showing A.M. and P.M. peak-hour direction of flow.

2) Other Committed Development Traffic

Show projected traffic to be generated by other committed development within the impact area, coincident with the phasing of the proposed development, for the count locations identified in Table 31.1. Attach letters from

TABLE 31.2: Programmed[1] and Planned[2] Improvements

	Programmed Improvements				*Planned Improvements*			
Phase	*Location*	*Type of Improvement*	*Cost*[3]	*Year of Construction*	*Location*	*Type of Improvement*	*Cost*[3]	*Year of Construction*
1								
.								
.								
.								
N[4]								

[1]Included in adopted County Transportation Improvement Program or known to be funded privately.
[2]Included in Broward County *Financially-Feasible Plan* or Dade County *Long Range Transportation Element.*
[3]Use current-year dollars.
[4]Expand table as necessary to accommodate phases.

TABLE 31.3: Future Background Traffic

	Segment							
Roadway	*From*	*To*	*Number of Lanes*	*Direction*	*Peak-Hour Volume*	*Peak-Hour Capacity*	*Peak-Hour V/C*	*Peak-Hour LOS*
Phase								
1								
.								
.								
.								
N[1]								

[1]Expand table as necessary to accommodate phases.

the local government(s) of jurisdiction and the Regional Planning Council that specify the other committed developments.

a) Specify projection method(s), source(s), assumptions, and provide calculations. Complete Table 31.4, listing all developments, their location, scale, phasing, A.M. and P.M. peak-hour trip generation rates, and A.M. and P.M. peak-hour trip generation. Use of any rate other than standard ITE trip rates requires a concurring letter from the Council.

b) Provide two maps of the traffic impact area for each committed development (J-5 series) that show the location of the development and the number and percentage distribution of A.M. and P.M. peak-hour trips.

c) Complete Table 31.5.

d) On Map J-5, show projected directional peak-hour traffic volume to capacity ratios and levels of service for each roadway segments in Table 31.5. Adjacent to each, provide arrows showing A.M. and P.M. peak-hour direction of flow.

3) Project Traffic

a) Provide phase-specific, project-generated peak-hour traffic projections for the same count locations in Table 31.1.

b) Specify all methods, assumptions, and standards used, including trip generation rates, modal split, peak hour factors, vehicle occupancy rates, etc. Provide all calculations.

c) Complete Table 31.6.

TABLE 31.4: Committed Developments

					Generation Rates		Vehicle Trips	
Development Name	*Location*	*Land Use*	*Scale*	*Phasing*	*AM Peak*	*PM Peak*	*AM Peak*	*PM Peak*

TABLE 31.5: Future Background and Committed Development Traffic

	Segment							
Roadway	*From*	*To*	*Number of Lanes*	*Direction*	*Peak-Hour Volume*	*Peak-Hour Capacity*	*Peak-Hour V/C*	*Peak-Hour LOS*
Phase								
1								
.								
.								
.								
N[1]								

[1]Expand table as necessary to accommodate phases.

d) Provide two maps of the traffic impact area showing the number and percentage of external project traffic assigned to the roadway system for both the A.M. and P.M. peak hours.
e) Complete Table 31.7.
f) On Map J-7, show projected directional peak-hour volume to capacity ratios and levels of service for each roadway segment in Table 31.7 and, for each, include arrows showing A.M. and P.M. peak-hour direction of flow.

D. Critical Intersections

1. For the roadway segments identified in Table 31.7 as operating below average daily LOS "C," identify the intersections that are critical to traffic flow in con-

TABLE 31.6: Trip Generation

	AM Peak-Hour Trips				*PM Peak-Hour Trips*			
Land Use	*Transit*	*Internal*[1]	*External*	*Total*	*Transit*	*Internal*	*External*	*Total*
Phase 1								
1								
.								
.								
N								
Subtotal								
Phase N[2]								
1								
.								
.								
N								
Subtotal								
TOTAL								

[1]Attach a letter from the Council verifying any internal orientation factor applied. Categorize trips into the following trip purposes by land use when internal trips are shown: 1) Work, 2) Shopping, 3) Recreation, 4) Education, and 5) Other (specify).
[2]Expand table as necessary to accommodate phases.

TABLE 31.7: Future Background, Committed Development, and Project Traffic

	Segment							
Roadway	*From*	*To*	*Number of Lanes*	*Direction*	*Peak-Hour Volume*	*Peak-Hour Capacity*	*Peak-Hour V/C*	*Peak-Hour LOS*
Phase								
1								
.								
.								
.								
N[1]								

[1]Expand table as necessary to accommodate phases.

sultation with the Council and provide detailed peak-hour capacity analyses (using the TRB Circular 212 methodology) for each. Provide capacity analysis worksheets for all intersections. Both A.M. and P.M. peak-hour analyses are needed for all expressway intersections, one-way streets, or intersections where A.M. volumes are found to be higher than P.M. volumes for the following scenarios:

- existing traffic (without recommended improvements),
- background plus committed development traffic (without recommended improvements),
- total traffic (without recommended improvements), and
- total traffic (with recommended improvements).

2. Complete Table 31.8.
3. Provide a large-scale map of the impact area showing projected peak-hour volumes for all movements in all critical intersections identified in Table 31.8 for the following scenarios:
   - existing traffic
   - background traffic
   - committed development traffic
   - project traffic
4. Traffic Composition at Critical Intersections.
   Complete Table 31.9 for all Critical Intersections.

TABLE 31.8: Total Traffic Peak-Hour Analysis

Phase	*Critical Intersection*	*Time Period*[1]	*Level of Service*
1			
.			
.			
.			
N[2]			

[1]AM or PM.
[2]Expand table as necessary to accommodate phases.

TABLE 31.9: Critical Intersections

			Traffic Component (%)			
Phase	*Critical Intersection*	*Time Period*[1]	*Existing*	*Background Growth*	*Committed Development*	*Project*
1						
.						
.						
.						
N[2]						

[1]AM or PM.
[2]Expand table as necessary to accommodate phases.

E. Recommended Improvements
Show existing geometrics and provide a conceptual design and cost estimate, including any necessary right-of-way acquisition, for each improvement and/or modification required to bring intersections and roadways projected to operate below peak-hour level of service "C" up to this operational standard ("D" in downtown Miami or Fort Lauderdale). Complete Table 31.10 for all recommended improvements.

F. Parking
1. Describe the parking to be provided, including the type of facility (e.g., open-air, enclosed garage, etc.) the number of spaces, and the local requirements. If applicable, calculate the number of parking spaces required using standards outlined in *Shared Parking,* ULI 1983.

G. Mass Transit
1. Describe provisions that will be made for access other than by private automobile. If special pedestrian or bicycle facilities are planned, use Map J (or Map H if appropriate) as a base to show the location of such facilities and describe their characteristics (including safety provisions at roadway crossings, lighting, user amenities, etc.).
2. Specify the type and frequency of any current public transit service. Using Map J (or Map H if appropriate) show relevant routes and stops.
3. If transit service (e.g., commuter or shopper bus or tram) is expected within this project, specify the type and frequency of service, route locations, fares, capital and operating costs, and methods of financing.

32. Housing

A. Complete Table 32.1.
B. Complete Table 32.2.

TABLE 31.10: Recommended Improvements

Phase[1]	Recommended Improvement	Right-of-Way Cost[2]	Construction Cost[2]	% of Traffic In Movement Requiring Improvement[3]		
				Background	Committed Development	Project
1						
.						
.						
.						
N						

[1]Expand table as necessary to accommodate phases.
[2]Use current-year dollars.
[3]Use data for buildout year of project and adjust percentages to total 100 percent.

TABLE 32.1: Owner Housing Units

Market Value $[1]	Housing Type	Net Density (units/acre)	Average Square Feet	1 Bedroom[2]			2 Bedrooms			3 Bedrooms			4 Bedrooms			Total		
				Total[3]	Permanent Employees	Seasonal	Total	Permanent Employees	Seasonal	Total	Permanent Employees	Seasonal	Total	Permanent Employees	Seasonal	Total	Permanent Employees	Seasonal
Phase 1																		
Less than 25,000																		
25,000 – 49,999																		
50,000 – 79,999																		
80,000 – 124,999																		
125,000 – 199,999																		
200,000 or more																		
Subtotal																		
Phase N[4]																		
.																		
.																		
.																		
Subtotal																		
TOTAL																		

[1]For each market value, specify owner housing types (e.g., single family attached and detached, mobile home, etc.).
[2]For each bedroom category proposed, estimate total units, number of units to be provided for permanent employees of the development, and seasonal (second home) units.
[3]Total category may not equal the sum of "Perm. Emp." and "Seasonal."
[4]Expand table as necessary to accommodate phases.

TABLE 32.2: Rental Housing Units

Market Rental $	Housing Type[1]	Net Density (units/acre)	Average Square Feet	Efficiency[2]			1 Bedroom			2 Bedrooms			3 Bedrooms			Total		
				Total[3]	Permanent Employees	Seasonal	Total	Permanent Employees	Seasonal	Total	Permanent Employees	Seasonal	Total	Permanent Employees	Seasonal	Total	Permanent Employees	Seasonal
Phase 1																		
Less than $200																		
$200 – $299																		
$300 – $399																		
$400 – $499																		
$500 – $750																		
$750 or more																		
Subtotal																		
.																		
.																		
.																		
Phase N[4]																		
.																		
.																		
.																		
Subtotal																		
TOTAL																		

[1]For each rent range, specify rental housing types (e.g., garden apartments, duplexes, etc.).
[2]For each bedroom category proposed, estimate total units, number of units to be provided for permanent employees of the development, and seasonal (second home) units.
[3]Total category may not equal the sum of "Perm. Emp." and "Seasonal."
[4]Expand table as necessary to accommodate phases.

PART III. SPECIFIC DRI INFORMATION

33. Airports

A. Provide a copy of any proposed or approved Airport Layout Plan.

B. If authorization has been requested under the Federal Airport and Airway Development Act of 1970, Title 49, United States Code, Section 1701 et. seq., attach a copy of the application and FAA action, if any.

C. For the proposed development, specify existing and projected:
1) types of aircraft using the facility,
2) annual enplaned passengers, and
3) types of cargo and annual tonnage.

D. Complete Figure 33.1-33.3, showing either aircraft landings and takeoffs separately or total operations.

E. Using Map D as a base, show flight patterns over existing and future land uses. Provide, as an overlay to this map, current and projected LDN (Day-Night Average Sound Level) contours using the following ranges of noise exposure: LDN 65-74 and LDN 75+. Specify steps that will be taken to mitigate noise impacts exceeding 65+ LDN in the surrounding community.

FIGURE 33.1: Diurnal/Nocturnal Distribution of Aircraft Operations[1]

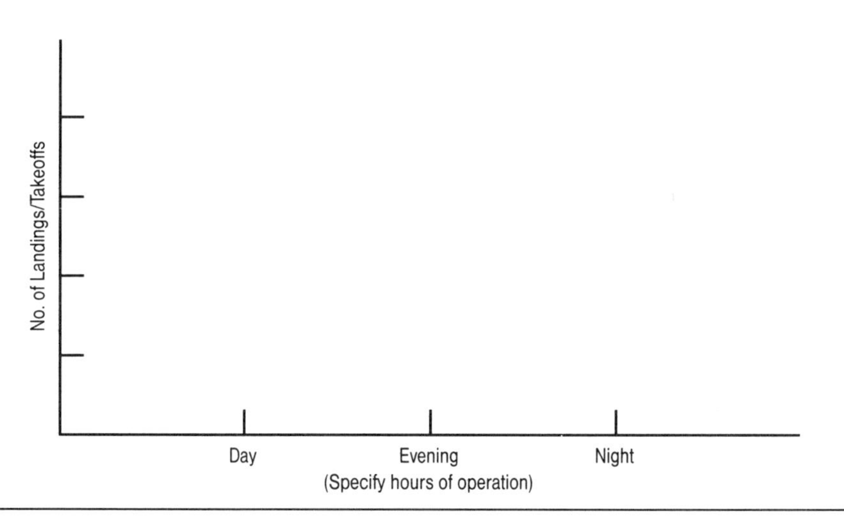

[1]Complete for all project phases.

FIGURE 33.2: Daily Distribution of Aircraft Operations[1,2]

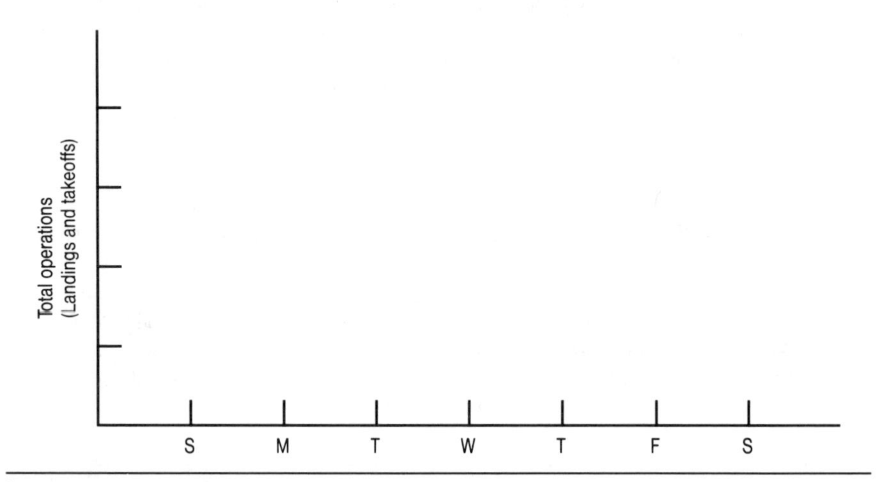

[1]Complete for all project phases.
[2]Plot separate curves for peak, seasonal, and off-peak operations.

FIGURE 33.3.: Monthly Distribution of Aircraft Operations[1]

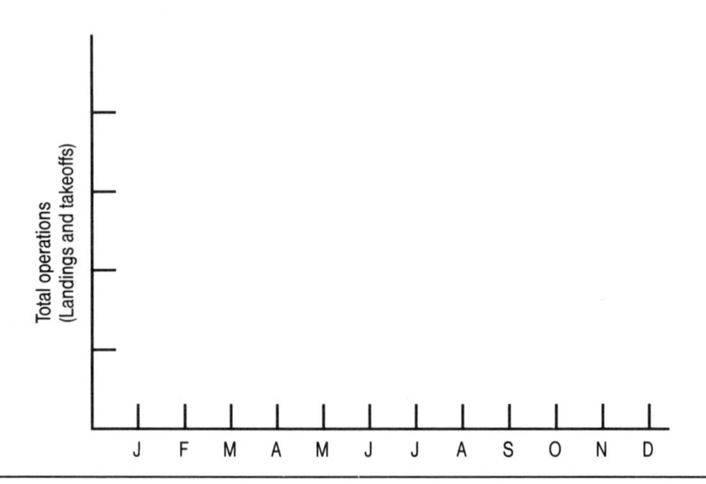

[1]Complete for all project phases.

34. Attractions and Recreation Facilities

A. Complete Figures 34.1-34.3.

B. Complete Table 34.4.

FIGURE 34.1: Monthly Distribution of Attendance[1]

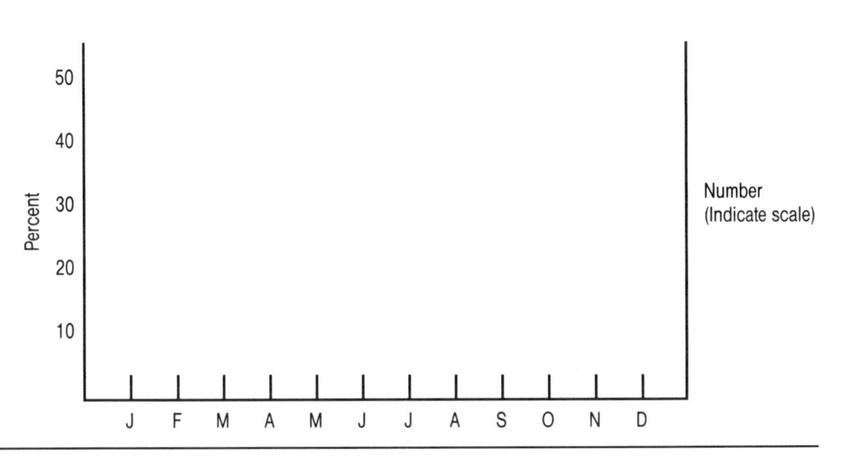

[1]Complete for all project phases.

FIGURE 34.2: Daily Distribution of Attendance[1,2]

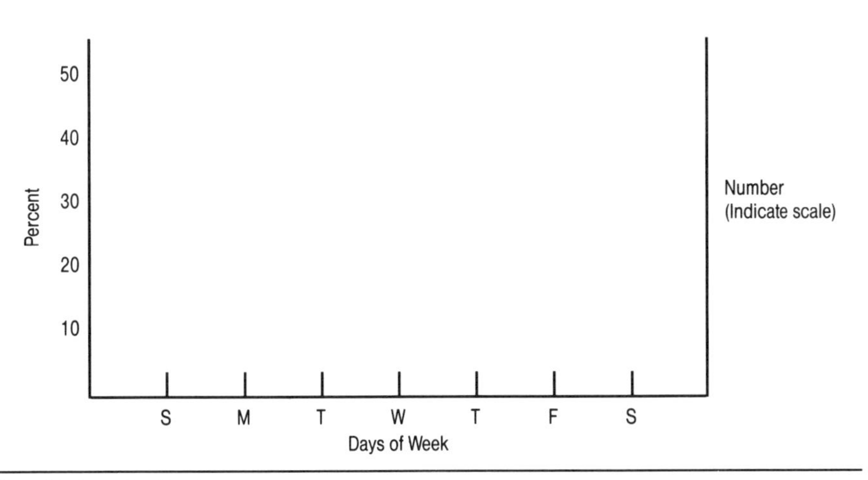

[1]Complete figure for all project phases.

[2]Plot separate curves for peak and off-peak season.

FIGURE 34.3: Peak Day Hourly Distribution of Attendance by Arrival and Departure[1,2]

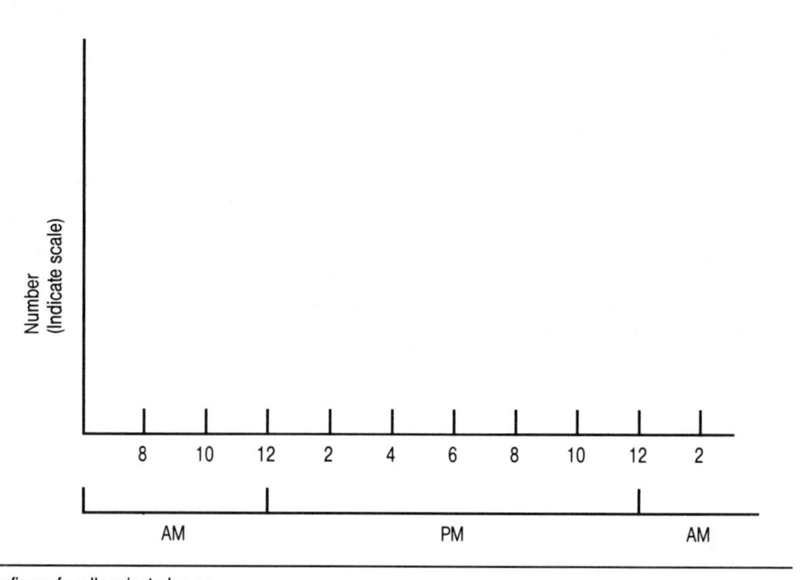

[1]Complete figure for all project phases.
[2]Plot separate curves for arrivals and for departures by peak and off-peak season (4 graphs).

TABLE 34.4: Projected Daily Energy Use

Energy-Consuming Facility	*Area Sq. Ft.*	*Functional/Safety Lighting KWH*	*Decorative Lighting KWH*	*Air Conditioning KWH*	*Cooking KWH*	*Water Heating KWH*	*Space Heating KWH*	*Misc. Electrical KWH*	*Cooking (Gas) Cu. Ft.*	*Water Heating (Gas) Cu. Ft.*	*Space Heating (Gas) Cu. Ft.*	*Miscellaneous (Gas) Cu. Ft.*	*Other (Specify)*
Phase 1													
.													
.													
.													
Subtotal													
Phase N[1]													
.													
.													
.													
Subtotal													
TOTAL													

[1]Expand table as necessary to accommodate phases.

35. Hospitals

A. If an application for a certificate of need under Chapter 381.494, F.S., has been prepared, provide a copy of the completed application along with comments by the Department of Health and Rehabilitative Service, if any.

B. If the proposed facility is to be a part of a medical complex, specify the related facilities to be provided.

C. For the proposed development, specify:

1) design capacity by development phase,
2) types of medical services to be provided, i.e., outpatient, emergency, etc., and
3) projected utilization rate by development phase.

D. Using Map I as a base, locate the hospital and its service area in relation to existing medical facilities.

36. Industrial Plants and Industrial Parks

A. Complete Table 36.1.

B. Specify the other products and/or services that are needed to support the anticipated industrial or commercial activity.

TABLE 36.1: Projected Occupants

SIC Code (3-digit)	*Number of Firms*	*Number of Employees*	*Gross Sales[1] ($/Yr.)*	*Square Footage*
Phase 1				
xxx[2]				
.				
.				
.				
Subtotal				
Phase N[3]				
.				
.				
.				
Subtotal				
TOTAL				

[1]Firms producing intermediate goods may provide "value added" figures in place of gross sales.
[2]Use "XXX-EX" to indicate an expansion of a firm already existing within the Region, "XXX-EXO" to show an expansion of firm from outside the Region, and "XXX-N" to indicate a newly formed firm.
[3]Expand table as necessary to accommodate phases.

TABLE 37.1: Projected Extraction

Phase	*Extracted Materials*		*Overburden*	*Spoil*	
(years)	*Type*	*Amount*	*(Cu. Yds.)*	*(Cu. Yds.)*	*Shipment Mode*
Existing					
Phase 1					
.					
.					
.					
Subtotal					
Phase N[1]					
.					
.					
.					
Subtotal					
TOTAL					

[1]Expand table as necessary to accommodate phases.

37. Mining Operations

A. Complete Table 37.1.

B. On Map H, identify the proposed area to be mined each year and total area to be altered by mining, roads, spoil, deposit, processing, and other mining operations.

C. Provide a Reuse Plan (with accompanying reuse plan map), including the annual reclamation schedule, the portion (in acres) of total mined and disturbed area that will be reused, and proposed reuse activities showing the Florida Land Use and Cover Classification System codes referenced in the response to Question 12.

38. Office Parks

A. Complete Table 38.1.

TABLE 38.1: Projected Occupants

SIC Code (3-digit)	Number of Firms	Number of Employees	Square Footage
Phase 1			
xxx[1]			
.			
.			
.			
Subtotal			
Phase N[2]			
.			
.			
.			
Subtotal			
TOTAL			

[1]Use "XXX-EX" to indicate an expansion of a firm already existing within the Region, "XXX-EXO" to show an expansion of firm from outside the Region, and "XXX-N" to indicate a newly-formed firm.
[2]Expand table as necessary to accommodate phases.

39. Petroleum Storage Facilities

A. Identify the types and amounts of petroleum to be stored.
B. On Map H, show the tanks and tank capacities (in barrels), the distances between the tanks, diked areas (in square feet), and dike height(s).
C. Describe measures to be used to insure the impermeability of the site.
D. Referring to Maps G and H, identify measures to minimize vapor emissions and petroleum spillages resulting from normal operations, accidental tank spills, loading and unloading procedures, or other project-related activities.
E. Specify what measures and facilities are available to handle oil spill clean-up.

40. Ports

Commercial Ports

A. On Tables 40.1 and 40.2, project the capacity of the proposed port for cargo by type and number of passengers by development phase.

TABLE 40.1: Commodity Statistics (in tons)

	Development Phase							
	Current Year 19--		Phase 1 19--		Phase N[1] 19--		Total	
Cargo Type	Import	Export	Import	Export	Import	Export	Import	Export
Container								
Trailer								
Breakbulk								
Liquid Bulk								
TOTAL								

[1]Expand table as necessary to accommodate phases.

TABLE 40.2: Passengers

	Development Phase			
Passengers	*Current Year*	*Phase 1 19--*	*Phase N[1] 19--*	*Total*
Departing				
Arriving				
Intransit				
TOTAL				

[1]Expand table as necessary to accommodate phases.

B. If the operation of the port will include the handling and/or storage of petroleum products, answer question 39.
C. On Map H, identify the existing and proposed harbor depth, depth of the main channel and turning basins, the number of berths or slips, and the linear feet of berthing space.

Marinas

D. Provide a site plan, with appropriate cross sections, at an approximate scale of $1'' = 100'$. Details shown should include: existing shoreline(s), mean high and low waterlines, water depths, structure dimensions, dredge and fill locations and volumes, number of slips or "tie-ups," distance to navigation channel(s), location of sewage pumpout and fuel facilities.
E. Describe existing hydrodynamic conditions, the effects of project on water circulation, and any actions that will be undertaken to mitigate or minimize adverse impacts.
F. Specify the type, acreage, location, and general condition of seagrasses in the area.
G. Specify actions that will be taken to mitigate or minimize adverse impacts.

41. Schools

A. Complete Table 41.1.
B. Complete Tables 41.2 and 41.3.
C. List the degree programs to be offered.
D. Identify the amount and source of all government and private revenues.

TABLE 41.1: Enrollment

Phase	*Number of Students*[1]
Existing	
Phase 1	
.	
.	
.	
Phase N[2]	
TOTAL	

[1]Show total number, not full-time equivalent students.
[2]Expand table as necessary to accommodate phases.

TABLE 41.2: Student Attendance by Day of Week

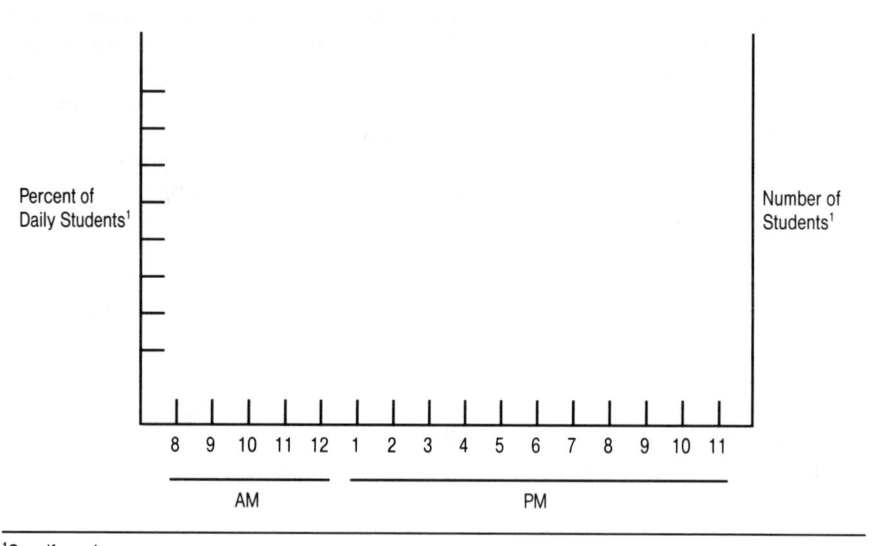

[1]Specify scale.

TABLE 41.3: Student Attendance by Hour of Day[1,2]

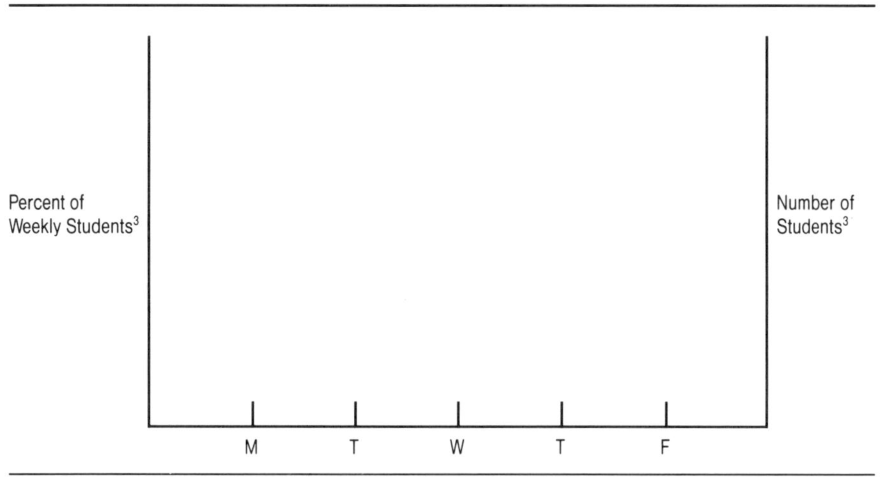

[1]Monday and Thursday.
[2]Plot separate curves for arrivals and departures.
[3]Specify scale.

42. Shopping Centers

A. Complete Table 42.1.

TABLE 42.1: Projected Tenants[1]

SIC Code (3-digit)	*Number of Firms*	*Total Sq. Ft.*	*Number of Employees*	*Gross Sales ($/Yr.)*
Phase 1				
xxx[1]				
.				
.				
.				
Subtotal				
Phase N[2]				
.				
.				
.				
Subtotal				
TOTAL				

[1]Use "XXX-EX" to indicate an expansion of a firm already existing in the Region, "XXX-EXO" to show an expansion of a firm from outside the Region, and "XXX-N" to indicate a newly formed firm.
[2]Expand as necessary to accommodate phases.

Index

Page numbers in *italic* indicate figures.

NOTES

NOTES

NOTES